Cambridge IGCSE®

Core Mathematics

Fourth edition

Ric Pimentel
Terry Wall

The Publishers would like to thank the following for permission to reproduce copyright material.

Photo credits
pp.2–3 © Aleksandra Antic/Shutterstock; **p.3** © Inv. Id P. Lund. inv. 35, reproduced with the kind permission of Lund University Library, Sweden; **pp.106–107** © katjen/Shutterstock; **p.107** © Eduard Kim/Shutterstock; **pp.164–5** © Halfpoint/Shutterstock; **p.165** © Georgios Kollidas – Fotolia; **pp.184–5** © ESB Professional/Shutterstock; **p.185** © Alexey Pavluts – Fotolia; **pp.232–3** © WitR/Shutterstock; **p.233** © uwimages – Fotolia; **pp.274–5** © 3Dsculptor/Shutterstock; **p.275** © Dinodia Photos/Alamy Stock Photo; **pp.298–9** © Anton Petrus/Shutterstock; **p.299** © Matěj Bat'ha via Wikipedia Commons (https://creativecommons.org/licenses/by-sa/2.5/deed.en); **pp.320–1** © Harvepino/Shutterstock; **p.321** © Bernard 63 – Fotolia; **pp.340–1** © Shutterstock; **p.341** © Jason Butcher/Getty Images.

All exam-style questions and sample answers in this title were written by the authors.

Orders: please contact Bookpoint Ltd, 130 Park Drive, Milton Park, Abingdon, Oxon OX14 4SE. Telephone: (44) 01235 827720. Fax: (44) 01235 400401. Email education@bookpoint.co.uk Lines are open from 9 a.m. to 5 p.m., Monday to Saturday, with a 24-hour message answering service. You can also order through our website: www.hoddereducation.com

First edition published 1997

Second edition published 2006

Third edition published 2013

This fourth edition published 2018 by

Hodder Education,
An Hachette UK Company
Carmelite House
50 Victoria Embankment
London EC4Y 0DZ

www.hoddereducation.com

Impression number 10 9 8 7 6 5 4 3 2 1

Year 2022 2021 2020 2019 2018

Cover photo © Maxal Tamor/Shutterstock

Illustrations by © Pantek Media and Integra Software Services

Typeset in Times Ten LT Std Roman 10/12 by Integra Software Servises Pvt. Ltd., Pondicherry, India

Printed in Slovenia

A catalogue record for this title is available from the British Library.

ISBN: 978 1 5104 2166 0

Contents

Introduction

This book has been written for all students of Cambridge IGCSE® and IGCSE (9–1) Mathematics syllabuses (0580/0980). It carefully and precisely follows the syllabus from Cambridge Assessment International Education. It provides the detail and guidance that are needed to support you throughout the course and help you to prepare for your examinations.

How to use this book

To make your study of mathematics as rewarding and successful as possible, this Cambridge endorsed textbook offers the following important features:

Learning objectives

- Each topic starts with an outline of the subject material and syllabus objectives to be covered.

Organisation

- Topics follow the order of the syllabus and are divided into chapters. Within each chapter there is a blend of teaching, worked examples and exercises to help you build confidence and develop the skills and knowledge you need. At the end of each chapter there are comprehensive student assessments. You will also find short sets of informal, digital questions linked to the **Student eTextbook**, which offer practice in topic areas that students often find difficult.

ICT, mathematical modelling and problem solving

- The syllabus specifically refers to 'Applying mathematical techniques to solve problems', and this is fully integrated into the exercises and assessments in the book. There are also sections called 'Mathematical investigations and ICT', which include problem-solving questions and ICT activities (although the latter are not part of the examination). On the **Student eTextbook** there is a selection of **videos** which offer support in problem-solving strategies and encourage reflective practice.

Callouts

These commentaries provide additional explanations and encourage full understanding of mathematical principles.

Worked examples

The worked examples cover important techniques and question styles. They are designed to reinforce the explanations, and give you step-by-step help for solving problems.

Exercises

These appear throughout the text, and allow you to apply what you have learned. There are plenty of routine questions covering important examination techniques.

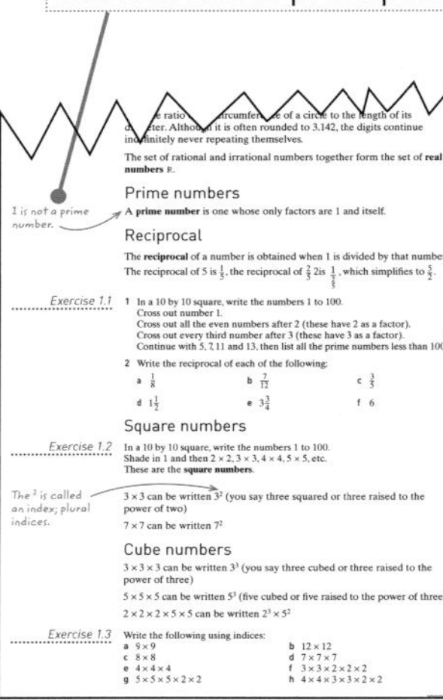

The set of rational and irrational numbers together form the set of **real numbers** ℝ.

Prime numbers

A **prime number** is one whose only factors are 1 and itself.

Reciprocal

The **reciprocal** of a number is obtained when 1 is divided by that number.

Square numbers

Cube numbers

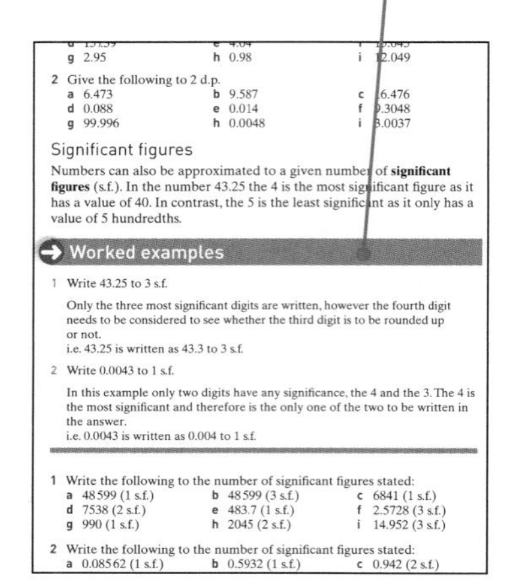

g 2.95 h 0.98 i 12.049

2 Give the following to 2 d.p.

a 6.473 b 9.587 c 16.476
d 0.088 e 0.014 f 9.3048
g 99.996 h 0.0048 i 3.0037

Significant figures

Numbers can also be approximated to a given number of **significant figures** (s.f.). In the number 43.25 the 4 is the most significant figure as it has a value of 40. In contrast, the 5 is the least significant as it only has a value of 5 hundredths.

Worked examples

1 Write 43.25 to 3 s.f.

Only the three most significant digits are written, however the fourth digit needs to be considered to see whether the third digit is to be rounded up or not.

i.e. 43.25 is written as 43.3 to 3 s.f.

2 Write 0.0043 to 1 s.f.

In this example only two digits have any significance, the 4 and the 3. The 4 is the most significant and therefore is the only one of the two to be written in the answer.

i.e. 0.0043 is written as 0.004 to 1 s.f.

1 Write the following to the number of significant figures stated:

a 48 599 (1 s.f.) b 48 599 (3 s.f.) c 6841 (1 s.f.)
d 7538 (2 s.f.) e 483.7 (1 s.f.) f 2.5728 (3 s.f.)
g 990 (1 s.f.) h 2045 (2 s.f.) i 14.952 (3 s.f.)

2 Write the following to the number of significant figures stated:

a 0.085 62 (1 s.f.) b 0.5932 (1 s.f.) c 0.942 (2 s.f.)

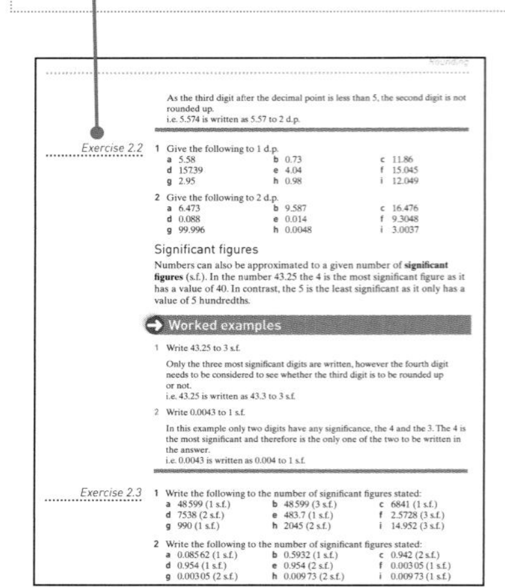

Mathematical investigations and ICT

More real world problem solving activities are provided at the end of each section to put what you've learned into practice.

Student assessments

End-of-chapter questions to test your understanding of the key topics and help to prepare you for your exam.

TOPIC 1

Mathematical investigations and ICT

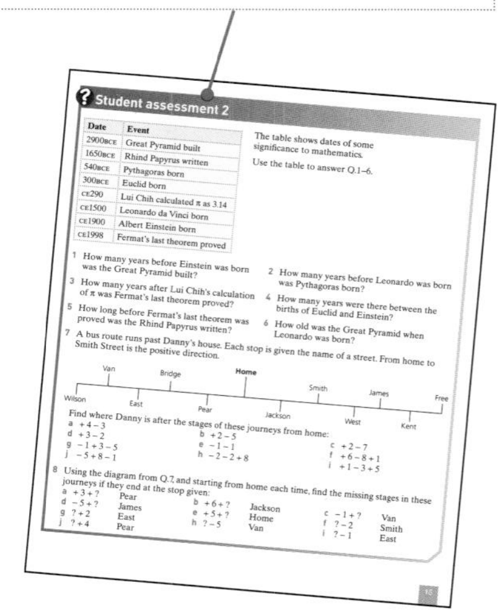

Student assessment 2

Assessment

For Cambridge IGCSE Core Mathematics there are two examination papers, Paper 1 and Paper 3. You may use a scientific calculator for both papers.

	Length	Type of questions
Paper 1	1 hour	Short-answer questions
Paper 3	2 hours	Structured questions

Examination techniques

Make sure you check the instructions on the question paper, the length of the paper and the number of questions you have to answer. In the case of Cambridge IGCSE® Mathematics examinations you will have to answer every question as there will be no choice.

Allocate your time sensibly between each question. Every year, good students let themselves down by spending too long on some questions and too little time (or no time at all) on others.

Make sure you show your working to show how you've reached your answer.

Command words

The command words that may appear in your question papers are listed below. The command word will relate to the context of the question.

Command word	What it means
Calculate	work out from given facts, figures or information, generally using a calculator
Construct*	make an accurate drawing
Describe	state the points of a topic / give characteristics and main features
Determine	establish with certainty
Explain	set out purposes or reasons / make the relationships between things evident / provide why and/or how and support with relevant evidence
Give	produce an answer from a given source or recall/memory
Plot	mark point(s) on a graph
Show (that)	provide structured evidence that leads to a given result
Sketch	make a simple freehand drawing showing the key features
Work out	calculate from given facts, figures or information with or without the use of a calculator
Write	give an answer in a specific form
Write down	give an answer without significant working

*Note: 'construct' is also used in the context of equations or expressions. When you construct an equation, you build it using information that you have been given or you have worked out. For example, you might construct an equation in the process of solving a word problem.

From the authors

Mathematics comes from the Greek word meaning *knowledge* or *learning*. Galileo Galilei (1564–1642) wrote 'the universe cannot be read until we learn the language in which it is written. It is written in mathematical language.' Mathematics is used in science, engineering, medicine, art, finance, etc., but mathematicians have always studied the subject for pleasure. They look for patterns in nature, for fun, as a game or a puzzle.

A mathematician may find that his or her puzzle solving helps to solve 'real life' problems. But trigonometry was developed without a 'real life' application in mind, before it was then applied to navigation and many other things. The algebra of curves was not 'invented' to send a rocket to Jupiter.

The study of mathematics is across all lands and cultures. A mathematician in Africa may be working with another in Japan to extend work done by a Brazilian in the USA.

People in all cultures have tried to understand the world around them, and mathematics has been a common way of furthering that understanding, even in cultures which have left no written records.

Each topic in this textbook has an introduction that tries to show how, over thousands of years, mathematical ideas have been passed from one culture to another. So, when you are studying from this textbook, remember that you are following in the footsteps of earlier mathematicians who were excited by the discoveries they had made. These discoveries changed our world.

You may find some of the questions in this book difficult. It is easy when this happens to ask the teacher for help. Remember though that mathematics is intended to stretch the mind. If you are trying to get physically fit, you do not stop as soon as things get hard. It is the same with mental fitness. Think logically. Try harder. In the end you are responsible for your own learning. Teachers and textbooks can only guide you. Be confident that you can solve that difficult problem.

Ric Pimentel and Terry Wall

TOPIC 1

Number

Contents

Course

C1.1
Identify and use natural numbers, integers (positive, negative and zero), prime numbers, square and cube numbers, common factors and common multiples, rational and irrational numbers (e.g. π, $\sqrt{2}$), real numbers, reciprocals.

C1.2
Understand notation of Venn diagrams.
Definition of sets
e.g. $A = \{x: x \text{ is a natural number}\}$
$B = \{a, b, c, \ldots\}$

C1.3
Calculate squares, square roots, cubes and cube roots and other powers and roots of numbers.

C1.4
Use directed numbers in practical situations.

C1.5
Use the language and notation of simple vulgar and decimal fractions and percentages in appropriate contexts.
Recognise equivalence and convert between these forms.

C1.6
Order quantities by magnitude and demonstrate familiarity with the symbols $=, \neq, >, <, \geqslant, \leqslant$.

C1.7
Understand the meaning of indices (fractional, negative and zero) and use the rules of indices.
Use the standard form $A \times 10^n$ where n is a positive or negative integer, and $1 \leqslant A < 10$.

C1.8
Use the four rules for calculations with whole numbers, decimals and fractions (including mixed numbers and improper fractions), including correct ordering of operations and use of brackets.

C1.9
Make estimates of numbers, quantities and lengths, give approximations to specified numbers of significant figures and decimal places and round off answers to reasonable accuracy in the context of a given problem.

C1.10
Give appropriate upper and lower bounds for data given to a specified accuracy.

C1.11
Demonstrate an understanding of ratio and proportion.

Calculate average speed.

Use common measures of rate.

C1.12
Calculate a given percentage of a quantity.

Express one quantity as a percentage of another.

Calculate percentage increase or decrease.

C1.13
Use a calculator efficiently.

Apply appropriate checks of accuracy.

C1.14
Calculate times in terms of the 24-hour and 12-hour clock.

Read clocks, dials and timetables.

C1.15
Calculate using money and convert from one currency to another.

C1.16
Use given data to solve problems on personal and household finance involving earnings, simple interest and compound interest.

Extract data from tables and charts.

C1.17
Extended curriculum only.

The development of number

In Africa, bones have been discovered with marks cut into them that are probably tally marks. These tally marks may have been used for counting time, such as numbers of days or cycles of the moon, or for keeping records of numbers of animals. A tallying system has no place value, which makes it hard to show large numbers.

The earliest system like ours (known as base 10) dates to 3100BCE in Egypt. Many ancient texts, for example texts from Babylonia (modern Iraq) and Egypt, used zero. Egyptians used the word *nfr* to show a zero balance in accounting. Indian texts used a Sanskrit word, *shunya*, to refer to the idea of the number zero. By the 4th century BCE, the people of south-central Mexico began to use a true zero. It was represented by a shell picture and became a part of Mayan numerals. By CE130, Ptolemy was using a symbol, a small circle, for zero. This Greek zero was the first use of the zero we use today.

Fragment of a Greek papyrus, showing an early version of the zero sign

The idea of negative numbers was recognised as early as 100BCE in the Chinese text *Jiuzhang Suanshu* (*Nine Chapters on the Mathematical Art*). This is the earliest known mention of negative numbers in the East. In the 3rd century BCE in Greece, Diophantus had an equation whose solution was negative. He said that the equation gave an absurd result.

European mathematicians did not use negative numbers until the 17th century, although Fibonacci allowed negative solutions in financial problems where they could be debts or losses.

1 Number and language

Natural numbers

A child learns to count 'one, two, three, four, …' These are sometimes called the counting numbers or whole numbers.

The child will say 'I am three', or 'I live at number 73'.

If we include the number zero, then we have the set of numbers called the **natural numbers**.

The set of natural numbers $\mathbb{N} = \{0, 1, 2, 3, 4, \ldots\}$.

Integers

On a cold day, the temperature may drop to 4 °C at 10 p.m. If the temperature drops by a further 6 °C, then the temperature is 'below zero'; it is −2 °C.

If you are overdrawn at the bank by \$200, this might be shown as −\$200.

The set of **integers** $\mathbb{Z} = \{\ldots, -3, -2, -1, 0, 1, 2, 3, \ldots\}$.

$\mathbb{Z}$ is therefore an extension of $\mathbb{N}$. Every natural number is an integer.

Rational numbers

A child may say 'I am three'; she may also say 'I am three and a half', or even 'three and a quarter'. $3\frac{1}{2}$ and $3\frac{1}{4}$ are **rational numbers**. All rational numbers can be written as a fraction whose denominator is not zero. All **terminating decimals** and **recurring decimals** are rational numbers as they can also be written as fractions, e.g.

$$0.2 = \frac{1}{5} \quad 0.3 = \frac{3}{10} \quad 7 = \frac{7}{1} \quad 1.53 = \frac{153}{100} \quad 0.\dot{2} = \frac{2}{9}$$

The set of rational numbers $\mathbb{Q}$ is an extension of the set of integers.

Irrational numbers

Numbers which cannot be expressed as a fraction are not rational numbers; they are **irrational numbers**.

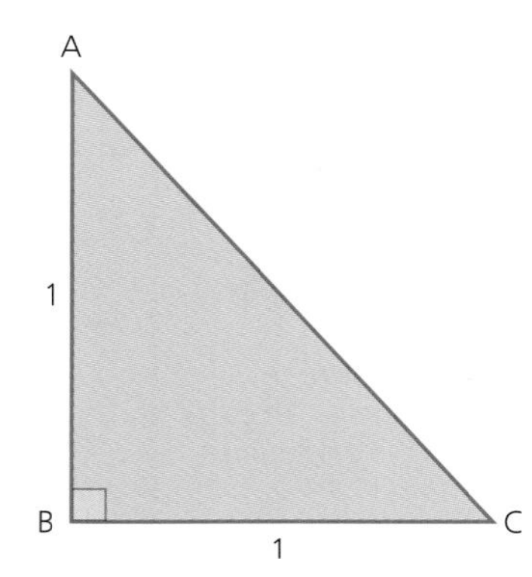

Using Pythagoras' rule in the diagram to the left, the length of the hypotenuse AC is found as:

$$AC^2 = 1^2 + 1^2$$
$$AC^2 = 2$$
$$AC = \sqrt{2}$$

$\sqrt{2} = 1.41421356\ldots$. The digits in this number do not recur or repeat. This is a property of all irrational numbers. Another example of an irrational number you will come across is π (pi).

π is the ratio of the circumference of a circle to the length of its diameter. Although it is often rounded to 3.142, the digits continue indefinitely never repeating themselves.

The set of rational and irrational numbers together form the set of **real numbers** $\mathbb{R}$.

Prime numbers

1 is not a prime number.

A **prime number** is one whose only factors are 1 and itself.

Reciprocal

The **reciprocal** of a number is obtained when 1 is divided by that number. The reciprocal of 5 is $\frac{1}{5}$, the reciprocal of $\frac{2}{5}$ is $\frac{1}{\frac{2}{5}}$, which simplifies to $\frac{5}{2}$.

Exercise 1.1

1 In a 10 by 10 square, write the numbers 1 to 100.
Cross out number 1.
Cross out all the even numbers after 2 (these have 2 as a factor).
Cross out every third number after 3 (these have 3 as a factor).
Continue with 5, 7, 11 and 13, then list all the prime numbers less than 100.

2 Write the reciprocal of each of the following:

a $\frac{1}{8}$ b $\frac{7}{12}$ c $\frac{3}{5}$

d $1\frac{1}{2}$ e $3\frac{3}{4}$ f 6

Square numbers

Exercise 1.2

In a 10 by 10 square, write the numbers 1 to 100.
Shade in 1 and then 2×2, 3×3, 4×4, 5×5, etc.
These are the **square numbers**.

The ^{2} is called an index; plural indices.

3×3 can be written 3^2 (you say three squared or three raised to the power of two)

7×7 can be written 7^2

Cube numbers

$3 \times 3 \times 3$ can be written 3^3 (you say three cubed or three raised to the power of three)

$5 \times 5 \times 5$ can be written 5^3 (five cubed or five raised to the power of three)

$2 \times 2 \times 2 \times 5 \times 5$ can be written $2^3 \times 5^2$

Exercise 1.3

Write the following using indices:

a 9×9 b 12×12
c 8×8 d $7 \times 7 \times 7$
e $4 \times 4 \times 4$ f $3 \times 3 \times 2 \times 2 \times 2$
g $5 \times 5 \times 5 \times 2 \times 2$ h $4 \times 4 \times 3 \times 3 \times 2 \times 2$

Factors

The **factors** of 12 are all the numbers which will divide exactly into 12,

i.e. 1, 2, 3, 4, 6 and 12.

Exercise 1.4

List all the factors of the following numbers:

a 6	**b** 9	**c** 7	**d** 15	**e** 24
f 36	**g** 35	**h** 25	**i** 42	**j** 100

Prime factors

The factors of 12 are 1, 2, 3, 4, 6 and 12.

Of these, 2 and 3 are prime numbers, so 2 and 3 are the **prime factors** of 12.

Exercise 1.5

List the prime factors of the following numbers:

a 15	**b** 18	**c** 24	**d** 16	**e** 20
f 13	**g** 33	**h** 35	**i** 70	**j** 56

An easy way to find prime factors is to divide by the prime numbers in order, smallest first.

Worked examples

1 Find the prime factors of 18 and express it as a product of prime numbers:

	18
2	9
3	3
3	1

$18 = 2 \times 3 \times 3$ or 2×3^2

2 Find the prime factors of 24 and express it as a product of prime numbers:

	24
2	12
2	6
2	3
3	1

$24 = 2 \times 2 \times 2 \times 3$ or $2^3 \times 3$

3 Find the prime factors of 75 and express it as a product of prime numbers:

	75
3	25
5	5
5	1

$75 = 3 \times 5 \times 5$ or 3×5^2

Exercise 1.6

Find the prime factors of the following numbers and express them as a product of prime numbers:

a 12	**b** 32	**c** 36	**d** 40	**e** 44
f 56	**g** 45	**h** 39	**i** 231	**j** 63

Highest common factor

The factors of 12 are 1, 2, 3, 4, 6, 12.

The factors of 18 are 1, 2, 3, 6, 9, 18.

So the **highest common factor** (HCF) can be seen by inspection to be 6.

Exercise 1.7

Find the HCF of the following numbers:

a 8, 12	**b** 10, 25	**c** 12, 18, 24
d 15, 21, 27	**e** 36, 63, 108	**f** 22, 110
g 32, 56, 72	**h** 39, 52	**i** 34, 51, 68
j 60, 144		

Multiples

Multiples of 5 are 5, 10, 15, 20, etc.

The **lowest common multiple** (LCM) of 2 and 3 is 6, since 6 is the smallest number divisible by 2 and 3.

The LCM of 3 and 5 is 15. The LCM of 6 and 10 is 30.

Exercise 1.8

1 Find the LCM of the following numbers:

a 3, 5	**b** 4, 6	**c** 2, 7	**d** 4, 7
e 4, 8	**f** 2, 3, 5	**g** 2, 3, 4	**h** 3, 4, 6
i 3, 4, 5	**j** 3, 5, 12		

2 Find the LCM of the following numbers:

a 6, 14	**b** 4, 15	**c** 2, 7, 10	**d** 3, 9, 10
e 6, 8, 20	**f** 3, 5, 7	**g** 4, 5, 10	**h** 3, 7, 11
i 6, 10, 16	**j** 25, 40, 100		

Rational and irrational numbers

Earlier in this chapter you learnt about rational and irrational numbers.

A **rational number** is any number which can be expressed as a fraction. Examples of some rational numbers and how they can be expressed as a fraction are:

$$0.2 = \frac{1}{5} \quad 0.3 = \frac{3}{10} \quad 7 = \frac{7}{1} \quad 1.53 = \frac{153}{100} \quad 0.\dot{2} = \frac{2}{9}$$

An **irrational number** cannot be expressed as a fraction. Examples of irrational numbers include:

$$\sqrt{2},\ \sqrt{5},\ 6 - \sqrt{3},\ \pi$$

In summary

Rational numbers include:

- whole numbers
- fractions
- recurring decimals
- terminating decimals.

Irrational numbers include:

- the square root of any number other than square numbers
- a decimal which neither repeats nor terminates (e.g. π).

Exercise 1.9

1 For each of the numbers shown below state whether it is rational or irrational:

a 1.3 **b** $0.\dot{6}$ **c** $\sqrt{3}$

d $-2\frac{3}{5}$ **e** $\sqrt{25}$ **f** $\sqrt[3]{8}$

g $\sqrt{7}$ **h** 0.625 **i** $0.\dot{1}\dot{1}$

2 For each of the numbers shown below state whether it is rational or irrational:

a $\sqrt{4} \times \sqrt{3}$ **b** $\sqrt{2} + \sqrt{3}$ **c** $\sqrt{2} \times \sqrt{3}$

d $\frac{\sqrt{8}}{\sqrt{2}}$ **e** $\frac{2\sqrt{5}}{\sqrt{20}}$ **f** $4 + (\sqrt{9} - 4)$

3 Look at these shapes and decide if the measurements required are rational or irrational. Give reasons for your answer.

a

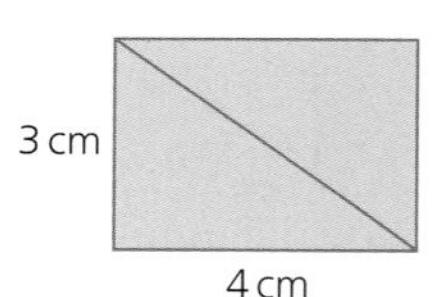

Length of diagonal

b

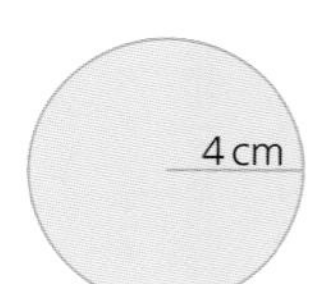

Circumference of circle

c

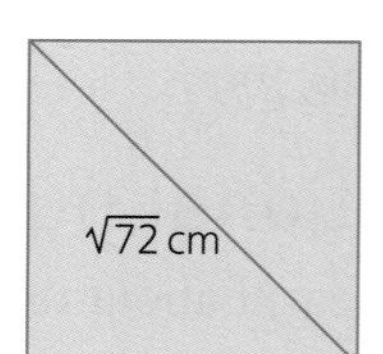

Side length of square

d

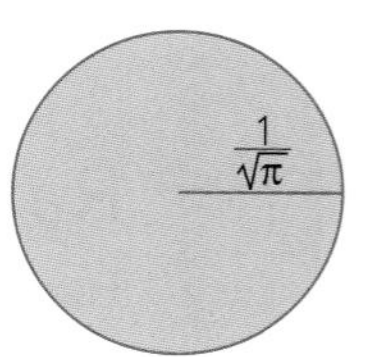

Area of circle

Calculating squares

This is a square of side 1 cm.

This is a square of side 2 cm.
It has four squares of side 1 cm in it.

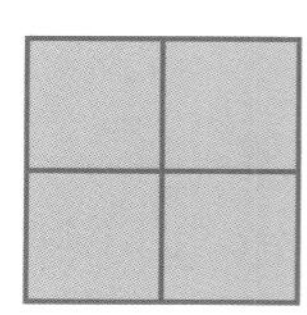

Exercise 1.10

Calculate how many squares of side 1 cm there would be in squares of side:

a 3 cm **b** 5 cm **c** 8 cm **d** 10 cm
e 11 cm **f** 12 cm **g** 7 cm **h** 13 cm
i 15 cm **j** 20 cm

In index notation, the square numbers are $1^2, 2^2, 3^2, 4^2$, etc. 4^2 is read as '4 squared'.

Worked example

This square is of side 1.1 units.

Its area is 1.1×1.1 units^2.

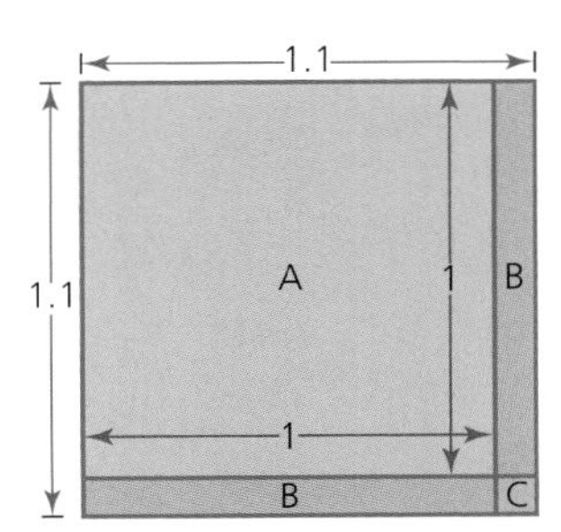

$A = 1 \times 1 = 1$

$B = 1 \times 0.1 = 0.1$

$B = 1 \times 0.1 = 0.1$

$C = 0.1 \times 0.1 = 0.01$

Total = 1.21 units^2

Exercise 1.11

1 Draw diagrams and use them to find the area of squares of side:

a 2.1 units **b** 3.1 units **c** 1.2 units
d 2.2 units **e** 2.5 units **f** 1.4 units

2 Use long multiplication to work out the area of squares of side:

a 2.4 **b** 3.3 **c** 2.8 **d** 6.2
e 4.6 **f** 7.3 **g** 0.3 **h** 0.8
i 0.1 **j** 0.9

3 Check your answers to Q.1 and 2 by using the x^2 key on a calculator.

Using a graph

Exercise 1.12

1 Copy and complete the table for the equation $y = x^2$.

x	0	1	2	3	4	5	6	7	8
y				9				49	

Plot the graph of $y = x^2$. Use your graph to find the value of the following:

a 2.5^2 **b** 3.5^2 **c** 4.5^2 **d** 5.5^2
e 7.2^2 **f** 6.4^2 **g** 0.8^2 **h** 0.2^2
i 5.3^2 **j** 6.3^2

2 Check your answers to Q.1 by using the x^2 key on a calculator.

Square roots

The orange square (overleaf) contains 16 squares. It has sides of length 4 units.

So the square root of 16 is 4.

This can be written as $\sqrt{16} = 4$.

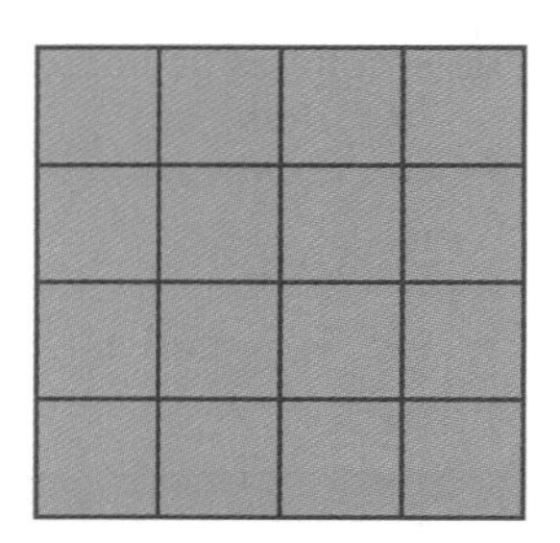

Note that $4 \times 4 = 16$ so 4 is the square root of 16.

However, -4×-4 is also 16 so -4 is also the square root of 16.

By convention, $\sqrt{16}$ means 'the positive square root of 16' so $\sqrt{16} = 4$ but the square root of 16 is ± 4, i.e. $+4$ or -4.

Note that -16 has no square root since any integer squared is positive.

Exercise 1.13

1 Find the following:

a $\sqrt{25}$ b $\sqrt{9}$ c $\sqrt{49}$ d $\sqrt{100}$
e $\sqrt{121}$ f $\sqrt{169}$ g $\sqrt{0.01}$ h $\sqrt{0.04}$
i $\sqrt{0.09}$ j $\sqrt{0.25}$

2 Use the $\sqrt{\ }$ key on your calculator to check your answers to Q.1.

3 Calculate the following:

a $\sqrt{\frac{1}{9}}$ b $\sqrt{\frac{1}{16}}$ c $\sqrt{\frac{1}{25}}$ d $\sqrt{\frac{1}{49}}$
e $\sqrt{\frac{1}{100}}$ f $\sqrt{\frac{4}{9}}$ g $\sqrt{\frac{9}{100}}$ h $\sqrt{\frac{49}{81}}$
i $\sqrt{2\frac{7}{9}}$ j $\sqrt{6\frac{1}{4}}$

Using a graph

Exercise 1.14

1 Copy and complete the table below for the equation $y = \sqrt{x}$.

x	0	1	4	9	16	25	36	49	64	81	100
y											

Plot the graph of $y = \sqrt{x}$. Use your graph to find the approximate values of the following:

a $\sqrt{70}$ b $\sqrt{40}$ c $\sqrt{50}$ d $\sqrt{90}$
e $\sqrt{35}$ f $\sqrt{45}$ g $\sqrt{55}$ h $\sqrt{60}$
i $\sqrt{2}$ j $\sqrt{3}$ k $\sqrt{20}$ l $\sqrt{30}$
m $\sqrt{12}$ n $\sqrt{75}$ o $\sqrt{115}$

2 Check your answers to Q.1 above by using the $\sqrt{\ }$ key on a calculator.

Cubes of numbers

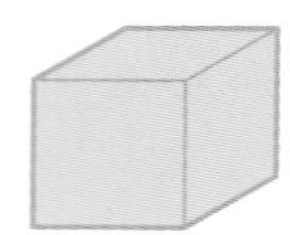

The small cube has sides of 1 unit and occupies 1 cubic unit of space.

The large cube has sides of 2 units and occupies 8 cubic units of space.

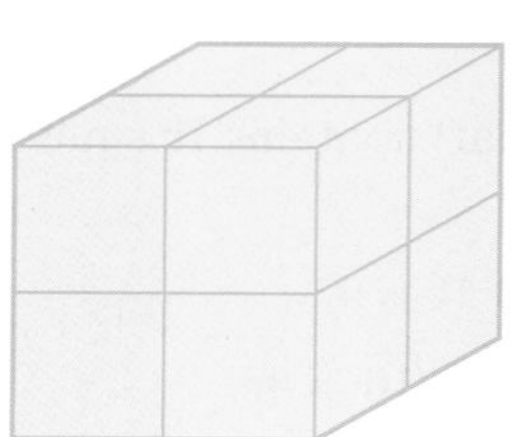

That is, $2 \times 2 \times 2$.

Exercise 1.15

How many cubic units would be occupied by cubes of side:

a 3 units **b** 5 units **c** 10 units
d 4 units **e** 9 units **f** 100 units?

In index notation, the **cube numbers** are $1^3, 2^3, 3^3, 4^3$, etc. 4^3 is read as '4 cubed'.

Some calculators have an x^3 key. On others, to find a cube you multiply the number by itself three times.

Exercise 1.16

1 Copy and complete the table below:

Number	1	2	3	4	5	6	7	8	9	10
Cube			27							

2 Use a calculator to find the following:

a 11^3 **b** 0.5^3 **c** 1.5^3 **d** 2.5^3
e 20^3 **f** 30^3 **g** $3^3 + 2^3$ **h** $(3 + 2)^3$
i $7^3 + 3^3$ **j** $(7 + 3)^3$

Cube roots

$\sqrt[3]{}$ is read as 'the cube root of ...'.

$\sqrt[3]{64}$ is 4, since $4 \times 4 \times 4 = 64$.

Note that $\sqrt[3]{64}$ is not -4

since $-4 \times -4 \times -4 = -64$

but $\sqrt[3]{-64}$ is -4.

Exercise 1.17

Find the following cube roots:

a $\sqrt[3]{8}$ **b** $\sqrt[3]{125}$ **c** $\sqrt[3]{27}$ **d** $\sqrt[3]{0.001}$
e $\sqrt[3]{0.027}$ **f** $\sqrt[3]{216}$ **g** $\sqrt[3]{1000}$ **h** $\sqrt[3]{1\,000\,000}$
i $\sqrt[3]{-8}$ **j** $\sqrt[3]{-27}$ **k** $\sqrt[3]{-1000}$ **l** $\sqrt[3]{-1}$

Further powers and roots

We have seen that the square of a number is the same as raising that number to the power of 2. For example, the square of 5 is written as 5^2 and means 5×5. Similarly, the cube of a number is the same as raising that number to the power of 3. For example, the cube of 5 is written as 5^3 and means $5 \times 5 \times 5$.

Numbers can be raised by other powers too. Therefore, 5 raised to the power of 6 can be written as 5^6 and means $5 \times 5 \times 5 \times 5 \times 5 \times 5$.

You will find a button on your calculator to help you to do this. On most calculators, it will look like y^x.

We have also seen that the square root of a number can be written using the $\sqrt{}$ symbol. Therefore, the square root of 16 is written as $\sqrt{16}$ and is ± 4, because both $4 \times 4 = 16$ and $-4 \times -4 = 16$.

The **cube root** of a number can be written using the $\sqrt[3]{}$ symbol. Therefore, the cube root of 125 is written as $\sqrt[3]{125}$ and is 5 because $5 \times 5 \times 5 = 125$.

Numbers can be rooted by other values as well. The fourth root of a number can be written using the symbol $\sqrt[4]{}$. Therefore, the fourth root of 625 can be expressed as $\sqrt[4]{625}$ and is ± 5 because both $5 \times 5 \times 5 \times 5 = 625$ and $(-5) \times (-5) \times (-5) \times (-5) = 625$.

You will find a button on your calculator to help you to calculate with roots too. On most calculators, it will look like $\sqrt[x]{y}$.

Exercise 1.18

Work out:

- **a** 6^4
- **b** $3^5 + 2^4$
- **c** $(3^4)^2$
- **d** $0.1^6 \div 0.01^4$
- **e** $\sqrt[4]{2401}$
- **f** $\sqrt[8]{256}$
- **g** $(\sqrt[5]{243})^3$
- **h** $(\sqrt[9]{36})^9$
- **i** $2^7 \times \sqrt{\frac{1}{4}}$
- **j** $\sqrt[6]{\frac{1}{64}} \times 2^7$
- **k** $\sqrt[4]{5^4}$
- **l** $(\sqrt[10]{59\,049})^2$

Directed numbers

Worked example

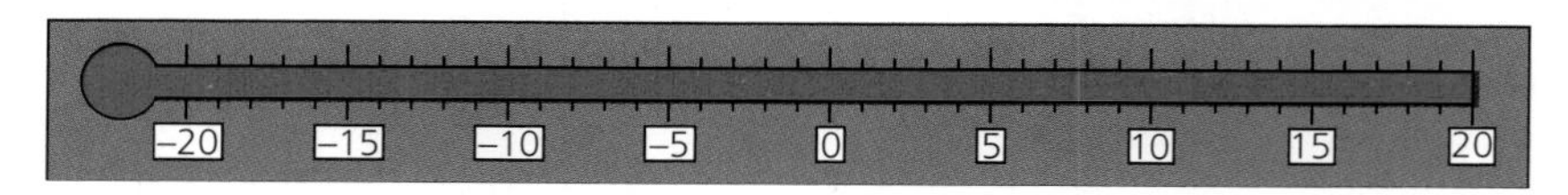

The diagram shows the scale of a thermometer. The temperature at 0400 was −3 °C. By 0900 it had risen by 8 °C. What was the temperature at 0900?

$(-3)° + (8)° = (5)°$

Exercise 1.19

1 Find the new temperature if:

- **a** The temperature was −5° C, and rises 9° C.
- **b** The temperature was −12 °C, and rises 8 °C.
- **c** The temperature was +14 °C, and falls 8 °C.
- **d** The temperature was −3 °C, and falls 4 °C.
- **e** The temperature was −7 °C, and falls 11 °C.
- **f** The temperature was 2 °C, it falls 8 °C, then rises 6 °C.
- **g** The temperature was 5 °C, it falls 8 °C, then falls a further 6 °C.
- **h** The temperature was −2 °C, it falls 6 °C, then rises 10 °C.
- **i** The temperature was 20 °C, it falls 18 °C, then falls a further 8 °C.
- **j** The temperature was 5 °C below zero and falls 8 °C.

2 Mark lives in Canada. Every morning before school he reads a thermometer to find the temperature in the garden. The thermometer below shows the results for 5 days in winter.

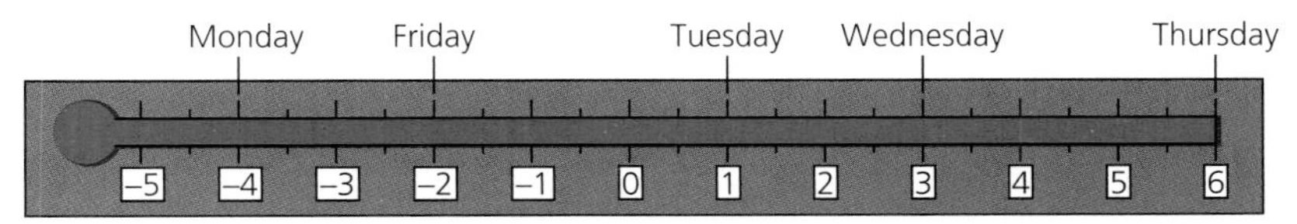

Find the change in temperature between:

a Monday and Friday
b Monday and Thursday
c Tuesday and Friday
d Thursday and Friday
e Monday and Tuesday.

3 The highest temperature ever recorded was in Libya. It was 58 °C. The lowest temperature ever recorded was –88 °C in Antarctica. What is the temperature difference?

4 Julius Caesar was born in 100BCE and was 56 years old when he died. In what year did he die?

5 Marcus Flavius was born in 20BCE and died in CE42. How old was he when he died?

6 Rome was founded in 753BCE. The last Roman city, Constantinople, fell in CE1453. How long did the Roman Empire last?

7 My bank account shows a credit balance of $105. Describe my balance as a positive or negative number after each of these transactions is made in sequence:

a rent $140
b car insurance $283
c 1 week's salary $230
d food bill $72
e credit transfer $250

8 A lift in the Empire State Building in New York has stopped somewhere close to the halfway point. Call this 'floor zero'. Show on a number line the floors it stops at as it makes the following sequence of journeys:

a up 75 floors
b down 155 floors
c up 110 floors
d down 60 floors
e down 35 floors
f up 100 floors

9 A hang-glider is launched from a mountainside. It climbs 650 m and then starts its descent. It falls 1220 m before landing.

a How far below its launch point was the hang-glider when it landed?
b If the launch point was at 1650 m above sea level, at what height above sea level did it land?

10 The average noon temperature in Sydney in January is +32 °C. The average midnight temperature in Boston in January is –12 °C. What is the temperature difference between the two cities?

11 The temperature in Madrid on New Year's Day is –2 °C. The temperature in Moscow on the same day is –14 °C. What is the temperature difference between the two cities?

Exercise 1.19 (cont)

12 The temperature inside a freezer is –8 °C. To defrost it, the temperature is allowed to rise by 12 °C. What will the temperature be after this rise?

13 A plane flying at 8500 m drops a sonar device onto the ocean floor. If the sonar falls a total of 10 200 m, how deep is the ocean at this point?

14 The roof of an apartment block is 130 m above ground level. The car park beneath the apartment is 35 m below ground level. How high is the roof above the floor of the car park?

15 A submarine is at a depth of 165 m. If the ocean floor is 860 m from the surface, how far is the submarine from the ocean floor?

Student assessment 1

1 List the prime factors of the following numbers:
a 28　b 38

2 Find the lowest common multiple of the following numbers:
a 6, 10　b 7, 14, 28

3 The diagram shows a square with a side length of $\sqrt{6}$ cm.

√6 cm
√6 cm

Explain, giving reasons, whether the following are rational or irrational:
a The perimeter of the square.　b The area of the square.

4 Find the value of:
a 9^2　b 15^2　c $(0.2)^2$　d $(0.7)^2$

5 Draw a square of side 2.5 units. Use it to find $(2.5)^2$.

6 Calculate:
a $(3.5)^2$　b $(4.1)^2$　c $(0.15)^2$

7 Copy and complete the table for $y = \sqrt{x}$.

x	0	1	4	9	16	25	36	49
y								

Plot the graph of $y = \sqrt{x}$. Use your graph to find:
a $\sqrt{7}$　b $\sqrt{30}$　c $\sqrt{45}$

8 Without using a calculator, find:
a $\sqrt{225}$　b $\sqrt{0.01}$　c $\sqrt{0.81}$
d $\sqrt{\frac{9}{25}}$　e $\sqrt{5\frac{4}{9}}$　f $\sqrt{2\frac{23}{49}}$

9 Without using a calculator, find:
a 4^3　b $(0.1)^3$　c $\left(\frac{2}{3}\right)^3$

10 Without using a calculator, find:

a $\sqrt[3]{27}$ b $\sqrt[3]{1000000}$ c $\sqrt[3]{\frac{64}{125}}$

11 Using a calculator if necessary work out:

a $3^5 \div 3^7$ b $5^4 \times \sqrt[4]{625}$ c $\sqrt[7]{2187} \div 3^3$

Student assessment 2

Date	Event
2900BCE	Great Pyramid built
1650BCE	Rhind Papyrus written
540BCE	Pythagoras born
300BCE	Euclid born
CE290	Lui Chih calculated π as 3.14
CE1500	Leonardo da Vinci born
CE1900	Albert Einstein born
CE1998	Fermat's last theorem proved

The table shows dates of some significance to mathematics.

Use the table to answer Q.1–6.

1 How many years before Einstein was born was the Great Pyramid built?

2 How many years before Leonardo was born was Pythagoras born?

3 How many years after Lui Chih's calculation of π was Fermat's last theorem proved?

4 How many years were there between the births of Euclid and Einstein?

5 How long before Fermat's last theorem was proved was the Rhind Papyrus written?

6 How old was the Great Pyramid when Leonardo was born?

7 A bus route runs past Danny's house. Each stop is given the name of a street. From home to Smith Street is the positive direction.

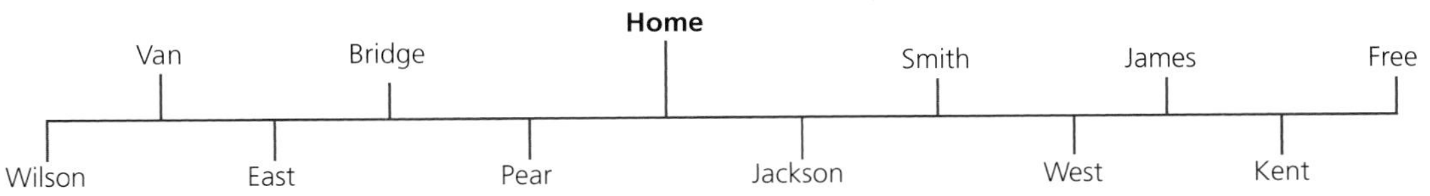

Find where Danny is after the stages of these journeys from home:

a $+4-3$ b $+2-5$ c $+2-7$
d $+3-2$ e $-1-1$ f $+6-8+1$
g $-1+3-5$ h $-2-2+8$ i $+1-3+5$
j $-5+8-1$

8 Using the diagram from Q.7, and starting from home each time, find the missing stages in these journeys if they end at the stop given:

a $+3+?$ Pear b $+6+?$ Jackson c $-1+?$ Van
d $-5+?$ James e $+5+?$ Home f $?-2$ Smith
g $?+2$ East h $?-5$ Van i $?-1$ East
j $?+4$ Pear

2 Accuracy

Approximation

In many instances exact numbers are not necessary or even desirable. In those circumstances approximations are given. The approximations can take several forms. Common types of approximation are dealt with in this chapter.

Rounding

If 28 617 people attend a gymnastics competition, this figure can be reported to various levels of accuracy.

To the nearest 10 000 this figure would be rounded up to 30 000.

To the nearest 1000 the figure would be rounded up to 29 000.

To the nearest 100 the figure would be rounded down to 28 600.

In this type of situation it is unlikely that the exact number would be reported.

Exercise 2.1

1 Round these numbers to the nearest 1000:

a 68 786	**b** 74 245	**c** 89 000
d 4020	**e** 99 500	**f** 999 999

2 Round these numbers to the nearest 100:

a 78 540	**b** 6858	**c** 14 099
d 8084	**e** 950	**f** 2984

3 Round these numbers to the nearest 10:

a 485	**b** 692	**c** 8847
d 83	**e** 4	**f** 997

Decimal places

A number can also be approximated to a given number of **decimal places** (d.p.). This refers to the number of digits written after a decimal point.

Worked examples

1 Write 7.864 to 1 d.p.

The answer needs to be written with one digit after the decimal point. However, to do this, the second digit after the decimal point needs to be considered. If it is 5 or more then the first digit is rounded up.
i.e. 7.864 is written as 7.9 to 1 d.p.

2 Write 5.574 to 2 d.p.

The answer here is to be given with two digits after the decimal point. In this case the third digit after the decimal point needs to be considered.

As the third digit after the decimal point is less than 5, the second digit is not rounded up.
i.e. 5.574 is written as 5.57 to 2 d.p.

Exercise 2.2

1 Give the following to 1 d.p.

a 5.58	b 0.73	c 11.86
d 157.39	e 4.04	f 15.045
g 2.95	h 0.98	i 12.049

2 Give the following to 2 d.p.

a 6.473	b 9.587	c 16.476
d 0.088	e 0.014	f 9.3048
g 99.996	h 0.0048	i 3.0037

Significant figures

Numbers can also be approximated to a given number of **significant figures** (s.f.). In the number 43.25 the 4 is the most significant figure as it has a value of 40. In contrast, the 5 is the least significant as it only has a value of 5 hundredths.

Worked examples

1 Write 43.25 to 3 s.f.

Only the three most significant digits are written, however the fourth digit needs to be considered to see whether the third digit is to be rounded up or not.
i.e. 43.25 is written as 43.3 to 3 s.f.

2 Write 0.0043 to 1 s.f.

In this example only two digits have any significance, the 4 and the 3. The 4 is the most significant and therefore is the only one of the two to be written in the answer.
i.e. 0.0043 is written as 0.004 to 1 s.f.

Exercise 2.3

1 Write the following to the number of significant figures stated:

a 48 599 (1 s.f.)	b 48 599 (3 s.f.)	c 6841 (1 s.f.)
d 7538 (2 s.f.)	e 483.7 (1 s.f.)	f 2.5728 (3 s.f.)
g 990 (1 s.f.)	h 2045 (2 s.f.)	i 14.952 (3 s.f.)

2 Write the following to the number of significant figures stated:

a 0.085 62 (1 s.f.)	b 0.5932 (1 s.f.)	c 0.942 (2 s.f.)
d 0.954 (1 s.f.)	e 0.954 (2 s.f.)	f 0.003 05 (1 s.f.)
g 0.003 05 (2 s.f.)	h 0.009 73 (2 s.f.)	i 0.009 73 (1 s.f.)

Appropriate accuracy

In many instances, calculations carried out using a calculator produce answers which are not whole numbers. A calculator will give the answer to as many decimal places as will fit on its screen. In most cases this degree of accuracy is neither desirable nor necessary. Unless specifically asked for, answers should not be given to more than two decimal places. Indeed, one decimal place is usually sufficient.

Worked example

Calculate $4.64 \div 2.3$, giving your answer to an appropriate degree of accuracy.

The calculator will give the answer to $4.64 \div 2.3$ as 2.017 391 3. However, the answer given to 1 d.p. is sufficient.

Therefore $4.64 \div 2.3 = 2.0$ (1 d.p.).

Estimating answers to calculations

Even though many calculations can be done quickly and effectively on a calculator, often an **estimate** for an answer can be a useful check. This is done by rounding each of the numbers so that the calculation becomes relatively straightforward.

Worked examples

1 Estimate the answer to 57×246.

Here are two possibilities:

a $60 \times 200 = 12\,000$

b $50 \times 250 = 12\,500$.

2 Estimate the answer to $6386 \div 27$.

$6000 \div 30 = 200$.

3 Estimate the answer to $\sqrt[3]{120} \times 48$.

As $\sqrt[3]{125} = 5$, $\sqrt[3]{120} \approx 5$

Therefore $\sqrt[3]{120} \times 48 \approx 5 \times 50$
≈ 250

≈ is used to state that the actual answer is approximately equal to the answer shown.

4 Estimate the answer to $\frac{2^5 \times \sqrt[4]{600}}{8}$

An approximate answer can be calculated using the knowledge that $2^5 = 32$ and $\sqrt[4]{625} = 5$

Therefore $\frac{2^5 \times \sqrt[4]{600}}{8} \approx \frac{30 \times 5}{8} \approx \frac{150}{8}$
≈ 20

Exercise 2.4

1 Without using a calculator, estimate the answers to:

a 62×19 **b** 270×12 **c** 55×60
d 4950×28 **e** 0.8×0.95 **f** 0.184×475

2 Without using a calculator, estimate the answers to:

a $3946 \div 18$ **b** $8287 \div 42$ **c** $906 \div 27$
d $5520 \div 13$ **e** $48 \div 0.12$ **f** $610 \div 0.22$

3 Without using a calculator, estimate the answers to:

a $78.45 + 51.02$ **b** $168.3 - 87.09$ **c** 2.93×3.14

d $84.2 \div 19.5$ **e** $\frac{4.3 \times 752}{15.6}$ **f** $\frac{(9.8)^3}{(2.2)^2}$

g $\frac{\sqrt[3]{78} \times 6}{5^3}$ **h** $\frac{38 \times 6^3}{\sqrt[4]{9900}}$ **i** $\sqrt[4]{25} \times \sqrt[4]{25}$

4 Using estimation, identify which of the following are definitely incorrect. Explain your reasoning clearly.

a $95 \times 212 = 20\,140$ **b** $44 \times 17 = 748$

c $689 \times 413 = 28\,457$ **d** $142\,656 \div 8 = 17\,832$

e $77.9 \times 22.6 = 2512.54$ **f** $\frac{8.42 \times 46}{0.2} = 19\,366$

5 Estimate the shaded area of the following shapes. Do *not* work out an exact answer.

a

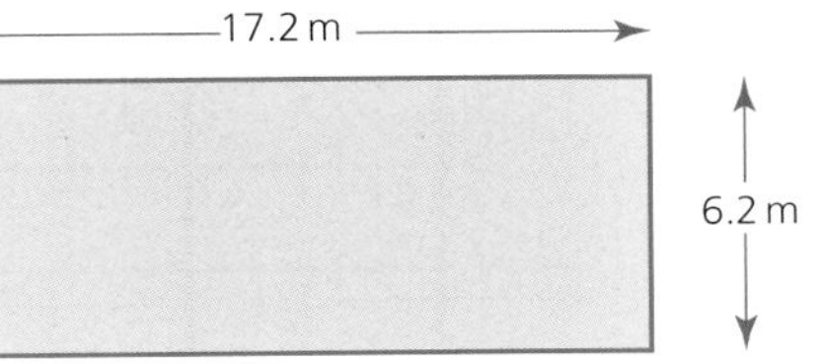

b

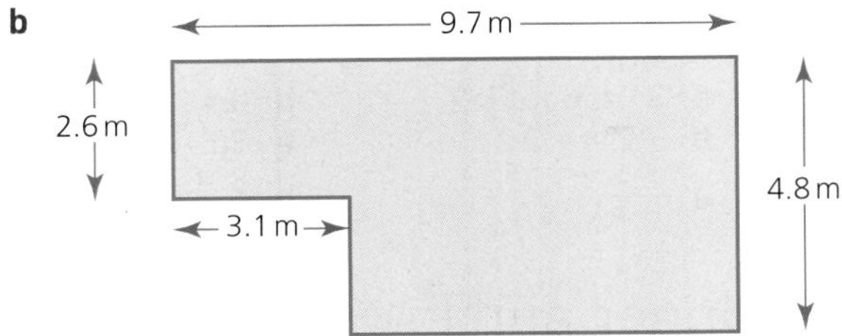

c

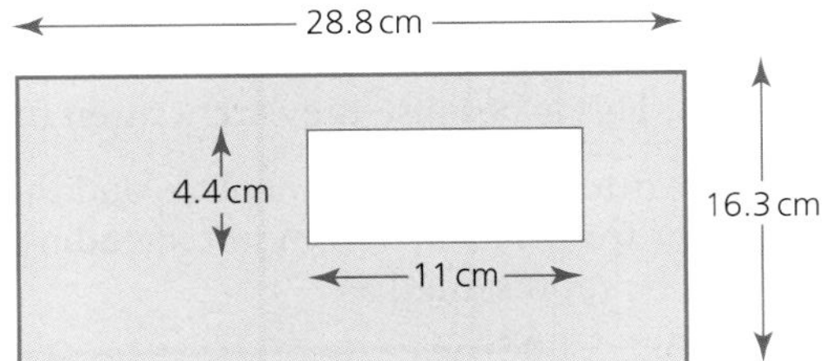

Exercise 2.4 (cont)

6 Estimate the volume of each solid. Do *not* work out an exact answer.

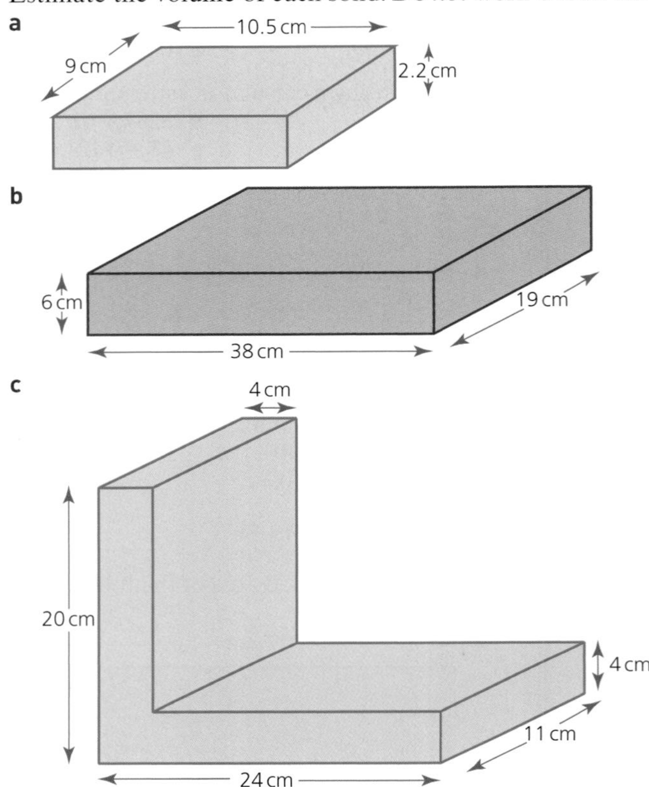

7 Calculate the following, giving your answer to an appropriate degree of accuracy:

a 23.456×17.89
b 0.4×12.62
c 18×9.24
d $76.24 \div 3.2$
e 7.6^2
f 16.42^3
g $\frac{2.3 \times 3.37}{4}$
h $\frac{8.31}{2.02}$
i $9.2 \div 4^2$

Upper and lower bounds

Numbers can be written to different degrees of accuracy. For example, 4.5, 4.50 and 4.500, although appearing to represent the same number, do not. This is because they are written to different degrees of accuracy.

4.5 is written to one decimal place and therefore could represent any number from 4.45 up to but not including 4.55. On a number line this would be represented as:

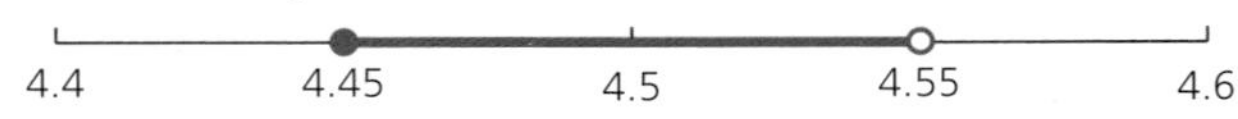

As an inequality where x represents the number, 4.5 would be expressed as

$$4.45 \leqslant x < 4.55$$

4.45 is known as the **lower bound** of 4.5, whilst 4.55 is known as the **upper bound**.

4.50 on the other hand is written to two decimal places and only numbers from 4.495 up to but not including 4.505 would be rounded to 4.50. This therefore represents a much smaller range of numbers than those which would be rounded to 4.5. Similarly the range of numbers being rounded to 4.500 would be even smaller.

Worked example

A girl's height is given as 162 cm to the nearest centimetre.

a Work out the lower and upper bounds within which her height can lie.
Lower bound = 161.5 cm
Upper bound = 162.5 cm

b Represent this range of numbers on a number line.

c If the girl's height is h cm, express this range as an inequality.

$161.5 \leq h < 162.5$

Exercise 2.5

1 Each of the following numbers is expressed to the nearest whole number.
   i Give the upper and lower bounds of each.
   ii Using x as the number, express the range in which the number lies as an inequality.
   a 8 **b** 71 **c** 146
   d 200 **e** 1

2 Each of the following numbers is correct to one decimal place.
   i Give the upper and lower bounds of each.
   ii Using x as the number, express the range in which the number lies as an inequality.
   a 2.5 **b** 14.1 **c** 2.0
   d 20.0 **e** 0.5

3 Each of the following numbers is correct to two significant figures.
   i Give the upper and lower bounds of each.
   ii Using x as the number, express the range in which the number lies as an inequality.
   a 5.4 **b** 0.75 **c** 550
   d 6000 **e** 0.012 **f** 10000

4 The mass of a sack of vegetables is given as 7.8 kg.
   a Illustrate the lower and upper bounds of the mass on a number line.
   b Using M kg for the mass, express the range of values in which M must lie as an inequality.

Exercise 2.5 (cont)

5 At a school sports day, the winning time for the 100 m race was given as 12.1 s.
   a Illustrate the lower and upper bounds of the time on a number line.
   b Using T seconds for the time, express the range of values in which T must lie as an inequality.

6 The capacity of a swimming pool is given as 740 m^3 correct to two significant figures.
   a Calculate the lower and upper bounds of the pool's capacity.
   b Using x cubic metres for the capacity, express the range of values in which x must lie as an inequality.

7 A farmer measures the dimensions of his rectangular field to the nearest 10 m. The length is recorded as 570 m and the width is recorded as 340 m.
   a Calculate the lower and upper bounds of the length.
   b Using W metres for the width, express the range of values in which W must lie as an inequality.

Student assessment 1

1 Round the following numbers to the degree of accuracy shown in brackets:
   a 2841 (nearest 100)
   b 7286 (nearest 10)
   c 48 756 (nearest 1000)
   d 951 (nearest 100)

2 Round the following numbers to the number of decimal places shown in brackets:
   a 3.84 (1 d.p.)
   b 6.792 (1 d.p.)
   c 0.8526 (2 d.p.)
   d 1.5849 (2 d.p.)
   e 9.954 (1 d.p.)
   f 0.0077 (3 d.p.)

3 Round the following numbers to the number of significant figures shown in brackets:
   a 3.84 (1 s.f.)
   b 6.792 (2 s.f.)
   c 0.7765 (1 s.f.)
   d 9.624 (1 s.f.)
   e 834.97 (2 s.f.)
   f 0.004 51 (1 s.f.)

4 A cuboid's dimensions are given as 12.32 cm by 1.8 cm by 4.16 cm. Calculate its volume, giving your answer to an appropriate degree of accuracy.

5 Estimate the answers to the following. Do *not* work out an exact answer.
   a $\frac{5.3 \times 11.2}{2.1}$
   b $\frac{(9.8)^2}{(4.7)^2}$
   c $\frac{18.8 \times (7.1)^2}{(3.1)^2 \times (4.9)^2}$

Student assessment 2

1 The following numbers are expressed to the nearest whole number. Illustrate on a number line the range in which each must lie.
 a 7 b 40 c 300

2 The following numbers are expressed correct to two significant figures. Representing each number by the letter x, express the range in which each must lie using an inequality.
 a 210 b 64 c 300

3 A school measures the dimensions of its rectangular playing field to the nearest metre. The length was recorded as 350 m and the width as 200 m. Express the ranges in which the length and width lie using inequalities.

4 A boy's mass was measured to the nearest 0.1 kg. If his mass was recorded as 58.9 kg, illustrate on a number line the range within which it must lie.

5 An electronic clock is accurate to $\frac{1}{1000}$ of a second. The duration of a flash from a camera is timed at 0.004 second. Express the upper and lower bounds of the duration of the flash using inequalities.

6 The following numbers are rounded to the degree of accuracy shown in brackets. Express the lower and upper bounds of these numbers as an inequality.
 a $x = 4.83$ (2 d.p.) b $y = 5.05$ (2 d.p.) c $z = 10.0$ (1 d.p.)

3 Calculations and order

Ordering

The following symbols have a specific meaning in mathematics:

$=$ is equal to

$\neq$ is not equal to

$>$ is greater than

$\geqslant$ is greater than or equal to

$<$ is less than

$\leqslant$ is less than or equal to

$x \geqslant 3$ states that x is greater than or equal to 3, i.e. x can be 3, 4, 4.2, 5, 5.6, etc.

$3 \leqslant x$ states that 3 is less than or equal to x, i.e. x can be 3, 4, 4.2, 5, 5.6, etc.

Therefore:

$5 > x$ can be rewritten as $x < 5$, i.e. x can be 4, 3.2, 3, 2.8, 2, 1, etc.

$-7 \leqslant x$ can be rewritten as $x \geqslant -7$, i.e. x can be $-7, -6, -5$, etc.

These inequalities can also be represented on a number line:

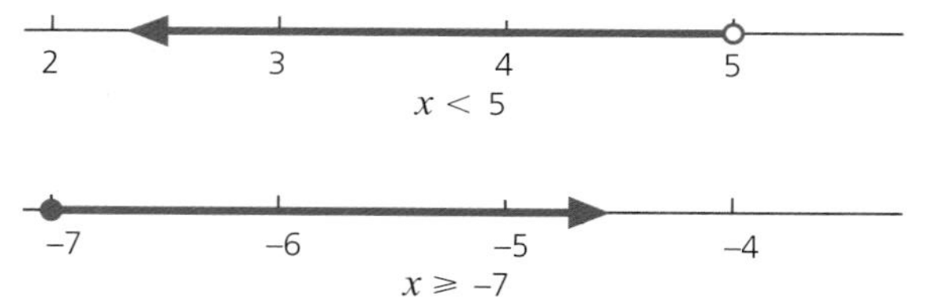

Note that ○⟶ implies that the number is not included in the solution whilst ●⟶ implies that the number is included in the solution.

Worked examples

1 Write $a > 3$ in words.

a is greater than 3.

2 Write 'x is greater than or equal to 8' using appropriate symbols.

$x \geqslant 8$

3 Write 'V is greater than 5, but less than or equal to 12' using the appropriate symbols.

$5 < V \leqslant 12$

4 Integers, fractions, decimals and percentages

Fractions

A single unit can be broken into equal parts called **fractions**, e.g. $\frac{1}{2}, \frac{1}{3}, \frac{1}{6}$. If, for example, the unit is broken into ten equal parts and three parts are then taken, the fraction is written as $\frac{3}{10}$. That is, three parts out of ten parts.

In the fraction $\frac{3}{10}$:

The three is called the **numerator**.

The ten is called the **denominator**.

A **proper fraction** has its numerator less than its denominator, e.g. $\frac{3}{4}$.

An **improper fraction** has its numerator more than its denominator, e.g. $\frac{9}{2}$. Another name for an improper fraction is a **vulgar fraction**.

A **mixed number** is made up of a whole number and a proper fraction, e.g. $4\frac{1}{5}$.

Exercise 4.1

1 Copy these fractions and indicate which number is the numerator and which is the denominator.

a $\frac{2}{3}$ **b** $\frac{15}{22}$ **c** $\frac{4}{3}$ **d** $\frac{5}{2}$

2 Separate the following into three sets: 'proper fractions', 'improper fractions' and 'mixed numbers'.

a $\frac{2}{3}$ **b** $\frac{15}{22}$ **c** $\frac{4}{3}$ **d** $\frac{5}{2}$

e $1\frac{1}{2}$ **f** $2\frac{3}{4}$ **g** $\frac{7}{4}$ **h** $\frac{7}{11}$

i $7\frac{1}{4}$ **j** $\frac{5}{6}$ **k** $\frac{6}{5}$ **l** $1\frac{1}{5}$

m $\frac{1}{10}$ **n** $2\frac{7}{8}$ **o** $\frac{5}{3}$

A fraction of an amount

Worked examples

1 Find $\frac{1}{5}$ of 35.

This means 'divide 35 into five equal parts'.

$\frac{1}{5}$ of 35 is = 7.

2 Find $\frac{3}{5}$ of 35.

Since $\frac{1}{5}$ of 35 is 7, $\frac{3}{5}$ of 35 is 7×3.

That is, 21.

Exercise 4.2

1 Evaluate:

a $\frac{1}{5}$ of 40 b $\frac{3}{5}$ of 40 c $\frac{1}{9}$ of 36 d $\frac{5}{9}$ of 36

e $\frac{1}{8}$ of 72 f $\frac{7}{8}$ of 72 g $\frac{1}{4}$ of 60 h $\frac{5}{12}$ of 60

i $\frac{1}{4}$ of 8 j $\frac{3}{4}$ of 8

2 Evaluate:

a $\frac{3}{4}$ of 12 b $\frac{4}{5}$ of 20 c $\frac{4}{9}$ of 45 d $\frac{5}{8}$ of 64

e $\frac{3}{11}$ of 66 f $\frac{9}{10}$ of 80 g $\frac{5}{7}$ of 42 h $\frac{8}{9}$ of 54

i $\frac{7}{8}$ of 240 j $\frac{4}{5}$ of 65

Changing a mixed number to a vulgar fraction

Worked examples

1 Change $2\frac{3}{4}$ to a vulgar fraction.

$$1 = \frac{4}{4}$$

$$2 = \frac{8}{4}$$

$$2\frac{3}{4} = \frac{8}{4} + \frac{3}{4} = \frac{11}{4}$$

2 Change $3\frac{5}{8}$ to a vulgar fraction.

$$3\frac{5}{8} = \frac{24}{8} + \frac{5}{8} = \frac{24+5}{8} = \frac{29}{8}$$

Exercise 4.3 Change the following mixed numbers to vulgar fractions:

a $4\frac{2}{3}$ b $3\frac{3}{5}$ c $5\frac{7}{8}$ d $2\frac{5}{6}$

e $8\frac{1}{2}$ f $9\frac{5}{7}$ g $6\frac{4}{9}$ h $4\frac{3}{5}$

i $5\frac{4}{11}$ j $7\frac{6}{7}$ k $4\frac{3}{10}$ l $11\frac{3}{13}$

Changing a vulgar fraction to a mixed number

Worked example

Change $\frac{27}{4}$ to a mixed number.

$$\begin{aligned}\frac{27}{4} &= \frac{24+3}{4} \\ &= \frac{24}{4} + \frac{3}{4} \\ &= 6\frac{3}{4}\end{aligned}$$

Exercise 4.4

Change the following vulgar fractions to mixed numbers:

a $\frac{29}{4}$ **b** $\frac{33}{5}$ **c** $\frac{41}{6}$ **d** $\frac{53}{8}$

e $\frac{49}{9}$ **f** $\frac{17}{12}$ **g** $\frac{66}{7}$ **h** $\frac{33}{10}$

i $\frac{19}{2}$ **j** $\frac{73}{12}$

Decimals

H	T	U	·	$\frac{1}{10}$	$\frac{1}{100}$	$\frac{1}{1000}$
		3	.	2	7	
		0	.	0	3	8

3.27 is 3 units, 2 tenths and 7 hundredths

i.e. $3.27 = 3 + \frac{2}{10} + \frac{7}{100}$

0.038 is 3 hundredths and 8 thousandths

i.e. $0.038 = \frac{3}{100} + \frac{8}{1000}$

Note that 2 tenths and 7 hundredths is equivalent to 27 hundredths

i.e. $\frac{2}{10} + \frac{7}{100} = \frac{27}{100}$

and that 3 hundredths and 8 thousandths is equivalent to 38 thousandths

i.e. $\frac{3}{100} + \frac{8}{1000} = \frac{38}{1000}$

A **decimal fraction** is a fraction between 0 and 1 in which the denominator is a power of 10 and the numerator is an integer.

$\frac{3}{10}, \frac{23}{100}, \frac{17}{1000}$ are all examples of decimal fractions.

$\frac{1}{20}$ is not a decimal fraction because the denominator is not a power of 10.

$\frac{11}{10}$ is not a decimal fraction because its value is greater than 1.

Exercise 4.5

1 Make a table similar to the one you have just seen. List the digits in the following numbers in their correct position:

a 6.023 **b** 5.94 **c** 18.3
d 0.071 **e** 2.001 **f** 3.56

2 Write these fractions as decimals:

a $4\frac{5}{10}$ **b** $6\frac{3}{10}$ **c** $17\frac{8}{10}$ **d** $3\frac{7}{100}$

e $9\frac{27}{100}$ **f** $11\frac{36}{100}$ **g** $4\frac{6}{1000}$ **h** $5\frac{27}{1000}$

i $4\frac{356}{1000}$ **j** $9\frac{204}{1000}$

3 Evaluate the following without using a calculator:

a $2.7 + 0.35 + 16.09$ **b** $1.44 + 0.072 + 82.3$
c $23.8 - 17.2$ **d** $16.9 - 5.74$
e $121.3 - 85.49$ **f** $6.03 + 0.5 - 1.21$
g $72.5 - 9.08 + 3.72$ **h** $100 - 32.74 - 61.2$
i $16.0 - 9.24 - 5.36$ **j** $1.1 - 0.92 - 0.005$

Percentages

A fraction whose denominator is 100 can be expressed as a percentage.

$\frac{29}{100}$ can be written as 29% $\frac{45}{100}$ can be written as 45%

Exercise 4.6

Write the fractions as percentages:

a $\frac{39}{100}$ **b** $\frac{42}{100}$ **c** $\frac{63}{100}$ **d** $\frac{5}{100}$

Changing a fraction to a percentage

By using equivalent fractions to change the denominator to 100, other fractions can be written as percentages.

Worked example

Change $\frac{3}{5}$ to a percentage.

$\frac{3}{5} = \frac{3}{5} \times \frac{20}{20} = \frac{60}{100}$

$\frac{60}{100}$ can be written as 60%

Exercise 4.7

1 Express each of the following as a fraction with denominator 100, then write them as percentages:

a $\frac{29}{50}$　b $\frac{17}{25}$　c $\frac{11}{20}$　d $\frac{3}{10}$

e $\frac{23}{25}$　f $\frac{19}{50}$　g $\frac{3}{4}$　h $\frac{2}{5}$

2 Copy and complete the table of equivalents:

Fraction	Decimal	Percentage
$\frac{1}{10}$		
	0.2	
		30%
$\frac{4}{10}$		
	0.5	
		60%
	0.7	
$\frac{4}{5}$		
	0.9	
$\frac{1}{4}$		
		75%

The four rules

Calculations with whole numbers

Addition, subtraction, multiplication and division are mathematical operations.

Long multiplication

When carrying out long multiplication, it is important to remember place value.

Worked example

$184 \times 37 =$

$$
\begin{array}{lrl}
184 \times 37 & 184 & \\
 & \times\ 37 & \\
\hline
 & 1288 & (184 \times 7) \\
 & 5520 & (184 \times 30) \\
\hline
 & 6808 & (184 \times 37)
\end{array}
$$

Short division

Worked example

453 ÷ 6 =

453 ÷ 6

```
   7 5 r3
6|4 5 ³3
```

It is usual, however, to give the final answer in decimal form rather than with a remainder. The division should therefore be continued:

453 ÷ 6

```
   7 5 . 5
6|4 5 ³3 . ³0
```

Long division

Worked example

Calculate 7184 ÷ 23 to one decimal place (1 d.p.).

```
     3 1 2 . 3 4
23 | 7 1 8 4 . 0 0
     6 9
       2 8
       2 3
         5 4
         4 6
           8 0
           6 9
           1 1 0
             9 2
             1 8
```

Therefore 7184 ÷ 23 = 312.3 to 1 d.p.

Mixed operations

When a calculation involves a mixture of operations, the order of the operations is important. Multiplications and divisions are done first, whilst additions and subtractions are done afterwards. To override this, brackets need to be used.

Worked examples

1 $3 + 7 \times 2 - 4$
$= 3 + 14 - 4$
$= 13$

2 $(3 + 7) \times 2 - 4$
$= 10 \times 2 - 4$
$= 20 - 4$
$= 16$

3 $3 + 7 \times (2 - 4)$
$= 3 + 7 \times (-2)$
$= 3 - 14$
$= -11$

4 $(3 + 7) \times (2 - 4)$
$= 10 \times (-2)$
$= -20$

Exercise 4.8

1 Evaluate the answer to:
a $3 + 5 \times 2 - 4$ **b** $6 + 4 \times 7 - 12$ **c** $3 \times 2 + 4 \times 6$
d $4 \times 5 - 3 \times 6$ **e** $8 \div 2 + 18 \div 6$ **f** $12 \div 8 + 6 \div 4$

2 Copy these equations and put brackets in the correct places to make them correct:
a $6 \times 4 + 6 \div 3 = 20$ **b** $6 \times 4 + 6 \div 3 = 36$ **c** $8 + 2 \times 4 - 2 = 12$
d $8 + 2 \times 4 - 2 = 20$ **e** $9 - 3 \times 7 + 2 = 44$ **f** $9 - 3 \times 7 + 2 = 54$

3 Without using a calculator, work out the solutions to these multiplications:
a 63×24 **b** 531×64 **c** 785×38
d 164×253 **e** 144×144 **f** 170×240

4 Work out the remainders in these divisions:
a $33 \div 7$ **b** $68 \div 5$ **c** $72 \div 7$
d $430 \div 9$ **e** $156 \div 5$ **f** $687 \div 10$

5 **a** The sum of two numbers is 16, their product is 63. What are the two numbers?
b When a number is divided by 7 the result is 14 remainder 2. What is the number?
c The difference between two numbers is 5, their product is 176. What are the numbers?
d How many 9s can be added to 40 before the total exceeds 100?
e A length of rail track is 9 m long. How many complete lengths will be needed to lay 1 km of track?
f How many 35 cent stamps can be bought for 10 dollars?

6 Work out the following long divisions to 1 d.p.
a $7892 \div 7$ **b** $45\,623 \div 6$ **c** $9452 \div 8$
d $4564 \div 4$ **e** $7892 \div 15$ **f** $79\,876 \div 24$

Fractions

Equivalent fractions

$\frac{1}{2}$

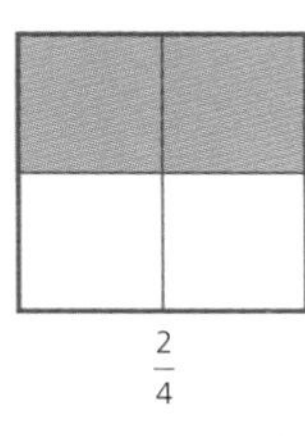

$\frac{2}{4}$

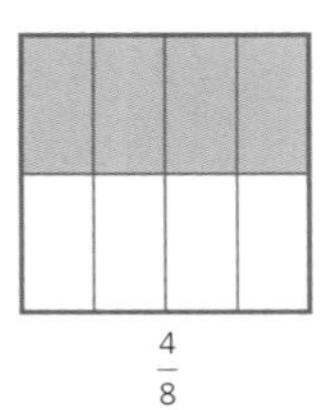

$\frac{4}{8}$

It should be apparent that $\frac{1}{2}$, $\frac{2}{4}$ and $\frac{4}{8}$ are **equivalent fractions**.

Similarly, $\frac{1}{3}$, $\frac{2}{6}$, $\frac{3}{9}$ and $\frac{4}{12}$ are equivalent, as are $\frac{1}{5}$, $\frac{10}{50}$ and $\frac{20}{100}$.

Equivalent fractions are mathematically the same as each other. In the diagrams above $\frac{1}{2}$ is mathematically the same as $\frac{4}{8}$. However $\frac{1}{2}$ is a simplified form of $\frac{4}{8}$.

When carrying out calculations involving fractions it is usual to give your answer in its **simplest form**. Another way of saying 'simplest form' is '**lowest terms**'.

Worked examples

1 Write $\frac{4}{22}$ in its simplest form.

Divide both the numerator and the denominator by their highest common factor.

The highest common factor of both 4 and 22 is 2.

Dividing both 4 and 22 by 2 gives $\frac{2}{11}$.

Therefore $\frac{2}{11}$ is $\frac{4}{22}$ written in its simplest form.

2 Write $\frac{12}{40}$ in its lowest terms.

Divide both the numerator and the denominator by their highest common factor.

The highest common factor of both 12 and 40 is 4.

Dividing both 12 and 40 by 4 gives $\frac{3}{10}$.

Therefore $\frac{3}{10}$ is $\frac{12}{40}$ written in its lowest terms.

Exercise 4.9

1 Copy the following sets of equivalent fractions and fill in the blanks:

a $\frac{2}{5} = \frac{4}{} = \frac{}{20} = \frac{}{50} = \frac{16}{}$

b $\frac{3}{8} = \frac{6}{} = \frac{}{24} = \frac{15}{} = \frac{}{72}$

c $\frac{}{7} = \frac{8}{14} = \frac{12}{} = \frac{}{56} = \frac{36}{}$

d $\frac{5}{} = \frac{}{27} = \frac{20}{36} = \frac{}{90} = \frac{55}{}$

2 Express the fractions in their lowest terms:

a $\frac{5}{10}$ b $\frac{7}{21}$ c $\frac{8}{12}$

d $\frac{16}{36}$ e $\frac{75}{100}$ f $\frac{81}{90}$

3 Write these improper fractions as mixed numbers, e.g. $\frac{15}{4} = 3\frac{3}{4}$

a $\frac{17}{4}$ b $\frac{23}{5}$ c $\frac{8}{3}$

d $\frac{19}{3}$ e $\frac{12}{3}$ f $\frac{43}{12}$

4 Write these mixed numbers as improper fractions, e.g. $3\frac{4}{5} = \frac{19}{5}$

a $6\frac{1}{2}$ b $7\frac{1}{4}$ c $3\frac{3}{8}$

d $11\frac{1}{9}$ e $6\frac{4}{5}$ f $8\frac{9}{11}$

Addition and subtraction of fractions

For fractions to be either added or subtracted, the denominators need to be the same.

Worked examples

1 $\frac{3}{11} + \frac{5}{11} = \frac{8}{11}$

2 $\frac{7}{8} + \frac{5}{8} = \frac{12}{8} = 1\frac{1}{2}$

3 $\frac{1}{2} + \frac{1}{3} = \frac{3}{6} + \frac{2}{6} = \frac{5}{6}$

4 $\frac{4}{5} - \frac{1}{3} = \frac{12}{15} - \frac{5}{15} = \frac{7}{15}$

When dealing with calculations involving mixed numbers, it is sometimes easier to change them to improper fractions first.

Worked examples

1 $5\frac{3}{4} - 2\frac{5}{8}$

$= \frac{23}{4} - \frac{21}{8}$

$= \frac{46}{8} - \frac{21}{8}$

$= \frac{25}{8} = 3\frac{1}{8}$

2 $1\frac{4}{7} + 3\frac{3}{4}$

$= \frac{11}{7} + \frac{15}{4}$

$= \frac{44}{28} + \frac{105}{28}$

$= \frac{149}{28} = 5\frac{9}{28}$

Exercise 4.10

Evaluate each of the following and write the answer as a fraction in its simplest form:

1 a $\frac{3}{5}+\frac{4}{5}$ b $\frac{3}{11}+\frac{7}{11}$ c $\frac{2}{3}+\frac{1}{4}$
d $\frac{3}{5}+\frac{4}{9}$ e $\frac{8}{13}+\frac{2}{5}$ f $\frac{1}{2}+\frac{2}{3}+\frac{3}{4}$

2 a $\frac{1}{8}+\frac{3}{8}+\frac{5}{8}$ b $\frac{3}{7}+\frac{5}{7}+\frac{4}{7}$ c $\frac{1}{3}+\frac{1}{2}+\frac{1}{4}$
d $\frac{1}{5}+\frac{1}{3}+\frac{1}{4}$ e $\frac{3}{8}+\frac{3}{5}+\frac{3}{4}$ f $\frac{3}{13}+\frac{1}{4}+\frac{1}{2}$

3 a $\frac{3}{7}-\frac{2}{7}$ b $\frac{4}{5}-\frac{7}{10}$ c $\frac{8}{9}-\frac{1}{3}$
d $\frac{7}{12}-\frac{1}{2}$ e $\frac{5}{8}-\frac{2}{5}$ f $\frac{3}{4}-\frac{2}{5}+\frac{7}{10}$

4 a $\frac{3}{4}+\frac{1}{5}-\frac{2}{3}$ b $\frac{3}{8}+\frac{7}{11}-\frac{1}{2}$ c $\frac{4}{5}-\frac{3}{10}+\frac{7}{20}$
d $\frac{9}{13}+\frac{1}{3}-\frac{4}{5}$ e $\frac{9}{10}-\frac{1}{5}-\frac{1}{4}$ f $\frac{8}{9}-\frac{1}{3}-\frac{1}{2}$

5 a $2\frac{1}{2}+3\frac{3}{4}$ b $3\frac{3}{5}+1\frac{7}{10}$ c $6\frac{1}{2}-3\frac{2}{5}$
d $8\frac{5}{8}-2\frac{1}{2}$ e $5\frac{7}{8}-4\frac{3}{4}$ f $3\frac{1}{4}-2\frac{5}{9}$

6 a $2\frac{1}{2}+1\frac{1}{4}+1\frac{3}{8}$ b $2\frac{4}{5}+3\frac{1}{8}+1\frac{3}{10}$ c $4\frac{1}{2}-1\frac{1}{4}-3\frac{5}{8}$
d $6\frac{1}{2}-2\frac{3}{4}-3\frac{2}{5}$ e $2\frac{4}{7}-3\frac{1}{4}-1\frac{3}{5}$ f $4\frac{7}{20}-5\frac{1}{5}+2\frac{2}{5}$

Multiplication and division of fractions

Worked examples

1 $\frac{3}{4}\times\frac{2}{3}$

$=\frac{6}{12}$

$=\frac{1}{2}$

2 $3\frac{1}{2}\times4\frac{4}{7}$

$=\frac{7}{2}\times\frac{32}{7}$

$=\frac{224}{14}$

$=16$

The reciprocal of a number is obtained when 1 is divided by that number. The reciprocal of 5 is $\frac{1}{5}$, the reciprocal of $\frac{2}{5}$ is $\frac{1}{\frac{2}{5}}=\frac{5}{2}$, etc.

Worked examples

Dividing fractions is the same as multiplying by the reciprocal.

1 $\frac{3}{8} \div \frac{3}{4}$

$= \frac{3}{8} \times \frac{4}{3}$

$= \frac{12}{24}$

$= \frac{1}{2}$

2 $5\frac{1}{2} \div 3\frac{2}{3}$

$= \frac{11}{2} \div \frac{11}{3}$

$= \frac{11}{2} \times \frac{3}{11}$

$= \frac{3}{2}$

Exercise 4.11

1 Write the reciprocal of:

a $\frac{3}{4}$ **b** $\frac{5}{9}$ **c** 7

d $\frac{1}{9}$ **e** $2\frac{3}{4}$ **f** $4\frac{5}{8}$

2 Write the reciprocal of:

a $\frac{1}{8}$ **b** $\frac{7}{12}$ **c** $\frac{3}{5}$

d $1\frac{1}{2}$ **e** $3\frac{3}{8}$ **f** 6

3 Evaluate:

a $\frac{3}{8} \times \frac{4}{9}$ **b** $\frac{2}{3} \times \frac{9}{10}$ **c** $\frac{5}{7} \times \frac{4}{15}$

d $\frac{3}{4}$ of $\frac{8}{9}$ **e** $\frac{5}{6}$ of $\frac{3}{10}$ **f** $\frac{7}{8}$ of $\frac{2}{5}$

4 Evaluate:

a $\frac{5}{8} \div \frac{3}{4}$ **b** $\frac{5}{6} \div \frac{1}{3}$ **c** $\frac{4}{5} \div \frac{7}{10}$

d $1\frac{5}{8} \div \frac{2}{5}$ **e** $\frac{3}{7} \div 2\frac{1}{7}$ **f** $1\frac{1}{4} \div 1\frac{7}{8}$

5 Evaluate:

a $\frac{3}{4} \times \frac{4}{5}$ **b** $\frac{7}{8} \times \frac{2}{3}$ **c** $\frac{3}{4} \times \frac{4}{7} \times \frac{3}{10}$

d $\frac{4}{5} \div \frac{2}{3} \times \frac{7}{10}$ **e** $\frac{1}{2}$ of $\frac{3}{4}$ **f** $4\frac{1}{5} \div 3\frac{1}{9}$

6 Evaluate:

a $\left(\frac{3}{8} \times \frac{4}{5}\right) + \left(\frac{1}{2} \text{ of } \frac{3}{5}\right)$ **b** $\left(1\frac{1}{2} \times 3\frac{3}{4}\right) - \left(2\frac{3}{5} \div 1\frac{1}{2}\right)$

c $\left(\frac{3}{5} \text{ of } \frac{4}{9}\right) + \left(\frac{4}{9} \text{ of } \frac{3}{5}\right)$ **d** $\left(1\frac{1}{3} \times 2\frac{5}{8}\right)^2$

Changing a fraction to a decimal

To change a fraction to a decimal, divide the numerator by the denominator.

Worked examples

1 Change $\frac{5}{8}$ to a decimal.

$$8\overline{)5.0^{2}0^{4}0}\quad = 0.625$$

2 Change $2\frac{3}{5}$ to a decimal.

This can be represented as $2 + \frac{3}{5}$

$$5\overline{)3.0}\quad = 0.6$$

Therefore $2\frac{3}{5} = 2 + 0.6 = 2.6$

Exercise 4.12

1 Change the fractions to decimals:

a $\frac{3}{4}$ **b** $\frac{4}{5}$ **c** $\frac{9}{20}$ **d** $\frac{17}{50}$

e $\frac{1}{3}$ **f** $\frac{3}{8}$ **g** $\frac{7}{16}$ **h** $\frac{2}{9}$

2 Change the mixed numbers to decimals:

a $2\frac{3}{4}$ **b** $3\frac{3}{5}$ **c** $4\frac{7}{20}$ **d** $6\frac{11}{50}$

e $5\frac{2}{3}$ **f** $6\frac{7}{8}$ **g** $5\frac{9}{16}$ **h** $4\frac{2}{8}$

Changing a decimal to a fraction

Changing a decimal to a fraction is done by knowing the 'value' of each of the digits in any decimal.

Worked examples

1 Change 0.45 from a decimal to a fraction.

units	.	tenths	hundredths
0	.	4	5

0.45 is therefore equivalent to 4 tenths and 5 hundredths, which in turn is the same as 45 hundredths.

Therefore $0.45 = \frac{45}{100} = \frac{9}{20}$

2 Change 2.325 from a decimal to a fraction.

units	.	tenths	hundredths	thousandths
2	.	3	2	5

Therefore $2.325 = 2\frac{325}{1000} = 2\frac{13}{40}$

Exercise 4.13

1 Change the decimals to fractions:

a 0.5 b 0.7 c 0.6
d 0.75 e 0.825 f 0.05
g 0.050 h 0.402 i 0.0002

2 Change the decimals to mixed numbers:

a 2.4 b 6.5 c 8.2
d 3.75 e 10.55 f 9.204
g 15.455 h 30.001 i 1.0205

Student assessment 1

1 Copy the numbers. Circle improper fractions and underline mixed numbers:

a $\frac{3}{11}$ b $5\frac{3}{4}$ c $\frac{27}{8}$ d $\frac{3}{7}$

2 Evaluate:

a $\frac{1}{3}$ of 63 b $\frac{3}{8}$ of 72 c $\frac{2}{5}$ of 55 d $\frac{3}{13}$ of 169

3 Change the mixed numbers to vulgar fractions:

a $2\frac{3}{5}$ b $3\frac{4}{9}$ c $5\frac{5}{8}$

4 Change the improper fractions to mixed numbers:

a $\frac{33}{5}$ b $\frac{47}{9}$ c $\frac{67}{11}$

5 Copy the set of equivalent fractions and fill in the missing numerators:

$$\frac{2}{3} = \frac{}{6} = \frac{}{12} = \frac{}{18} = \frac{}{27} = \frac{}{30}$$

6 Write the fractions as decimals:

a $\frac{35}{100}$ b $\frac{275}{1000}$ c $\frac{675}{100}$ d $\frac{35}{1000}$

7 Write the following as percentages:

a $\frac{3}{5}$ b $\frac{49}{100}$ c $\frac{1}{4}$ d $\frac{9}{10}$

e $1\frac{1}{2}$ f $3\frac{27}{100}$ g $\frac{5}{100}$ h $\frac{7}{20}$

i 0.77 j 0.03 k 2.9 l 4

Student assessment 2

1 Evaluate:

a $6 \times 4 - 3 \times 8$

b $15 \div 3 + 2 \times 7$

2 The product of two numbers is 72, and their sum is 18. What are the two numbers?

3 How many days are there in 42 weeks?

4 Work out 368×49.

5 Work out $7835 \div 23$ giving your answer to 1 d.p.

6 Copy these equivalent fractions and fill in the blanks:

$$\frac{24}{36} = \frac{}{12} = \frac{4}{} = \frac{}{30} = \frac{60}{}$$

7 Evaluate:

a $2\frac{1}{2} - \frac{4}{5}$

b $3\frac{1}{2} \times \frac{4}{7}$

8 Change the fractions to decimals:

a $\frac{7}{8}$

b $1\frac{2}{5}$

9 Change the decimals to fractions. Give each fraction in its simplest form.

a 6.5

b 0.04

c 3.65

d 3.008

10 Write the reciprocals of the following numbers:

a $\frac{5}{9}$

b $3\frac{2}{5}$

c 0.1

5

Further percentages

You should already be familiar with the percentage equivalents of simple fractions and decimals as outlined in the table:

Fraction	Decimal	Percentage
$\frac{1}{2}$	0.5	50%
$\frac{1}{4}$	0.25	25%
$\frac{3}{4}$	0.75	75%
$\frac{1}{8}$	0.125	12.5%
$\frac{3}{8}$	0.375	37.5%
$\frac{5}{8}$	0.625	62.5%
$\frac{7}{8}$	0.875	87.5%
$\frac{1}{10}$	0.1	10%
$\frac{2}{10}$ or $\frac{1}{5}$	0.2	20%
$\frac{3}{10}$	0.3	30%
$\frac{4}{10}$ or $\frac{2}{5}$	0.4	40%
$\frac{6}{10}$ or $\frac{3}{5}$	0.6	60%
$\frac{7}{10}$	0.7	70%
$\frac{8}{10}$ or $\frac{4}{5}$	0.8	80%
$\frac{9}{10}$	0.9	90%

Simple percentages

Worked examples

1 Of 100 sheep in a field, 88 are ewes.

a What percentage of the sheep are ewes?
88 out of 100 are ewes = 88%

b What percentage are not ewes?
12 out of 100 are not ewes = 12%

2 A gymnast scored marks out of 10 from five judges.

They were: 8.0, 8.2, 7.9, 8.3 and 7.6.

Express these marks as percentages.

$\frac{8.0}{10} = \frac{80}{100} = 80\%$ $\frac{8.2}{10} = \frac{82}{100} = 82\%$ $\frac{7.9}{10} = \frac{79}{100} = 79\%$

$\frac{8.3}{10} = \frac{83}{100} = 83\%$ $\frac{7.6}{10} = \frac{76}{100} = 76\%$

3 Convert the percentages into fractions and decimals:

a 27%

$\frac{27}{100} = 0.27$

b 5%

$\frac{5}{100} = \frac{1}{20} = 0.05$

Exercise 5.1

1 In a survey of 100 cars, 47 were white, 23 were blue and 30 were red. Express each of these numbers as a percentage of the total.

2 $\frac{7}{10}$ of the surface of the Earth is water. Express this as a percentage.

3 There are 200 birds in a flock. 120 of them are female. What percentage of the flock are:
a female? **b** male?

4 Write these percentages as fractions of 100:
a 73% **b** 28% **c** 10% **d** 25%

5 Write these fractions as percentages:
a $\frac{27}{100}$ **b** $\frac{3}{10}$ **c** $\frac{7}{50}$ **d** $\frac{1}{4}$

6 Convert the percentages to decimals:
a 39% **b** 47% **c** 83%
d 7% **e** 2% **f** 20%

7 Convert the decimals to percentages:
a 0.31 **b** 0.67 **c** 0.09
d 0.05 **e** 0.2 **f** 0.75

Calculating a percentage of a quantity

Worked examples

1 Find 25% of 300 m.
25% can be written as 0.25.
$0.25 \times 300\,\text{m} = 75\,\text{m}$.

2 Find 35% of 280 m.
35% can be written as 0.35.
$0.35 \times 280\,\text{m} = 98\,\text{m}$.

Exercise 5.2

1 Write the percentage equivalent of these fractions:

a $\frac{1}{4}$ b $\frac{2}{3}$ c $\frac{5}{8}$

d $1\frac{4}{5}$ e $4\frac{9}{10}$ f $3\frac{7}{8}$

2 Write the decimal equivalent of the following:

a $\frac{3}{4}$ b 80% c $\frac{1}{5}$

d 7% e $1\frac{7}{8}$ f $\frac{1}{6}$

3 Evaluate:

a 25% of 80 b 80% of 125 c 62.5% of 80

d 30% of 120 e 90% of 5 f 25% of 30

4 Evaluate:

a 17% of 50 b 50% of 17 c 65% of 80

d 80% of 65 e 7% of 250 f 250% of 7

5 In a class of 30 students, 20% have black hair, 10% have blonde hair and 70% have brown hair. Calculate the number of students with:

a black hair b blonde hair c brown hair.

6 A survey conducted among 120 schoolchildren looked at which type of meat they preferred. 55% said they preferred chicken, 20% said they preferred lamb, 15% preferred goat and 10% were vegetarian. Calculate the number of children in each category.

7 A survey was carried out in a school to see what nationality its students were. Of the 220 students in the school, 65% were Australian, 20% were Pakistani, 5% were Greek and 10% belonged to other nationalities. Calculate the number of students of each nationality.

8 A shopkeeper keeps a record of the numbers of items he sells in one day. Of the 150 items he sold, 46% were newspapers, 24% were pens, 12% were books whilst the remaining 18% were other items. Calculate the number of each item he sold.

Expressing one quantity as a percentage of another

To express one quantity as a percentage of another, write the first quantity as a fraction of the second and then multiply by 100.

Worked example

In an examination, a girl obtains 69 marks out of 75. Express this result as a percentage.

$\frac{69}{75} \times 100\% = 92\%$

Exercise 5.3

1 For each of the following express the first quantity as a percentage of the second.

a 24 out of 50	b 46 out of 125
c 7 out of 20	d 45 out of 90
e 9 out of 20	f 16 out of 40
g 13 out of 39	h 20 out of 35

2 A hockey team plays 42 matches. It wins 21, draws 14 and loses the rest. Express each of these results as a percentage of the total number of games played.

3 Four candidates stood in an election:

A received 24500 votes	B received 18200 votes
C received 16300 votes	D received 12000 votes

Express each of these as a percentage of the total votes cast.

4 A car manufacturer produces 155000 cars a year. The cars are available in six different colours. The numbers sold of each colour were:

Red	55000	Silver	10200
Blue	48000	Green	9300
White	27500	Black	5000

Express each of these as a percentage of the total number of cars produced. Give your answers to 1 d.p.

Percentage increases and decreases

Worked examples

1 A doctor has a salary of \$18000 per month. If her salary increases by 8%, calculate:

a the amount extra she receives per month
Increase = 8% of \$18000 = $0.08 \times$ \$18000 = \$1440

b her new monthly salary.
New salary = old salary + increase = \$18000 + \$1440 per month
= \$19 440 per month

2 A garage increases the price of a truck by 12%. If the original price was \$14500, calculate its new price.

The original price represents 100%, therefore the increase can be represented as 112%.

New price = 112% of \$14500
= $1.12 \times$ \$14500
= \$16240

3 A shop is having a sale. It sells a set of tools costing \$130 at a 15% discount. Calculate the sale price of the tools.

The old price represents 100%, therefore the new price can be represented as (100 – 15)% = 85%.

85% of \$130 = $0.85 \times$ \$130
= \$110.50

Exercise 5.4

1 Increase the following by the given percentage:
- a 150 by 25%
- b 230 by 40%
- c 7000 by 2%
- d 70 by 250%
- e 80 by 12.5%
- f 75 by 62%

2 Decrease the following by the given percentage:
- a 120 by 25%
- b 40 by 5%
- c 90 by 90%
- d 1000 by 10%
- e 80 by 37.5%
- f 75 by 42%

3 In the following questions the first number is increased to become the second number. Calculate the percentage increase in each case.
- a $50 \rightarrow 60$
- b $75 \rightarrow 135$
- c $40 \rightarrow 84$
- d $30 \rightarrow 31.5$
- e $18 \rightarrow 33.3$
- f $4 \rightarrow 13$

4 In the following questions the first number is decreased to become the second number. Calculate the percentage decrease in each case.
- a $50 \rightarrow 25$
- b $80 \rightarrow 56$
- c $150 \rightarrow 142.5$
- d $3 \rightarrow 0$
- e $550 \rightarrow 352$
- f $20 \rightarrow 19$

5 A farmer increases the yield on his farm by 15%. If his previous yield was 6500 tonnes, what is his present yield?

6 The cost of a computer in a computer store is discounted by 12.5% in a sale. If the computer was priced at $7800, what is its price in the sale?

7 A winter coat is priced at $100. In the sale its price is discounted by 25%.
- a Calculate the sale price of the coat.
- b After the sale its price is increased by 25% again. Calculate the coat's price after the sale.

8 A farmer takes 250 chickens to be sold at a market. In the first hour he sells 8% of his chickens. In the second hour he sells 10% of those that were left.
- a How many chickens has he sold in total?
- b What percentage of the original number did he manage to sell in the two hours?

9 The number of fish on a fish farm increases by approximately 10% each month. If there were originally 350 fish, calculate to the nearest 100 how many fish there would be after 12 months.

Student assessment 1

1 Copy the table and fill in the missing values:

Fraction	Decimal	Percentage
$\frac{3}{4}$		
	0.8	
$\frac{5}{8}$		
	1.5	

2 Find 40% of 1600 m.

3 A shop increases the price of a television set by 8%. If the original price was $320, what is the new price?

4 A car loses 55% of its value after four years. If it cost $22 500 when new, what is its value after the four years?

5 Express the first quantity as a percentage of the second.
- a 40 cm, 2 m
- b 25 mins, 1 hour
- c 450 g, 2 kg
- d 3 m, 3.5 m
- e 70 kg, 1 tonne
- f 75 cl, 2.5 litres

6 A house is bought for 75 000 rand, then resold for 87 000 rand. Calculate the percentage profit.

7 A pair of shoes is priced at \$45. During a sale the price is reduced by 20%.
  a Calculate the sale price of the shoes.
  b What is the percentage increase in the price if after the sale it is once again restored to \$45?

8 The population of a town increases by 5% each year. If in 2007 the population was 86 000, in which year is the population expected to exceed 100 000 for the first time?

Student assessment 2

1 Copy the table and fill in the missing values:

Fraction	Decimal	Percentage
	0.25	
$\frac{3}{5}$		
		$62\frac{1}{2}\%$
$2\frac{1}{4}$		

2 Find 30% of 2500 m.

3 In a sale, a shop reduces its prices by 12.5%. What is the sale price of a desk previously costing 2400 Hong Kong dollars?

4 In the last six years the value of a house has increased by 35%. If it cost \$72 000 six years ago, what is its value now?

5 Express the first quantity as a percentage of the second.
  a 35 mins, 2 hours
  b 650 g, 3 kg
  c 5 m, 4 m
  d 15 s, 3 mins
  e 600 kg, 3 tonnes
  f 35 cl, 3.5 litres

6 Shares in a company are bought for \$600. After a year, the same shares are sold for \$550. Calculate the percentage depreciation.

7 In a sale the price of a jacket originally costing \$850 is reduced by \$200. Any item not sold by the last day of the sale is reduced by a further 50%. If the jacket is sold on the last day of the sale:
  a calculate the price it is finally sold for
  b calculate the overall percentage reduction in price.

8 Each day the population of a type of insect increases by approximately 10%. How many days will it take for the population to double?

9 A piece of wood is cut in the ratio 2 : 3. What fraction of the length is the longer piece?

10 If the original piece of wood in Q.9 is 80 cm long, how long is the shorter piece?

11 A gas pipe is 7 km long. A valve is positioned in such a way that it divides the length of the pipe in the ratio 4 : 3. Calculate the distance of the valve from each end of the pipe.

12 The sizes of the angles of a quadrilateral are in the ratio 1 : 2 : 3 : 3. Calculate the size of each angle.

13 The angles of a triangle are in the ratio 3 : 5 : 4. Calculate the size of each angle.

14 A millionaire leaves 1.4 million dollars in his will to be shared between his three children in the ratio of their ages. If they are 24, 28 and 32 years old, calculate to the nearest dollar the amount they will each receive.

15 A small company makes a profit of $8000. This is divided between the directors in the ratio of their initial investments. If Alex put $20 000 into the firm, Maria $35 000 and Ahmet $25 000, calculate the amount of the profit they will each receive.

Inverse proportion

Sometimes an increase in one quantity causes a decrease in another quantity. For example, if fruit is to be picked by hand, the more people there are picking the fruit, the less time it will take. The time taken is said to be **inversely proportional** to the number of people picking the fruit.

Worked examples

1 If 8 people can pick the apples from the trees in 6 days, how long will it take 12 people?

8 people take 6 days.
1 person will take 6×8 days.
Therefore 12 people will take $\frac{6 \times 8}{12}$ days, i.e. 4 days.

2 A cyclist averages a speed of 27 km/h for 4 hours. At what average speed would she need to cycle to cover the same distance in 3 hours?

Completing it in 1 hour would require cycling at 27×4 km/h.

Completing it in 3 hours requires cycling at $\frac{27 \times 4}{3}$ km/h; that is 36 km/h.

Exercise 6.5

1 A teacher shares sweets among 8 students so that they get 6 each. How many sweets would they each have got if there had been 12 students?

2 The table represents the relationship between the speed and the time taken for a train to travel between two stations.

Speed (km/h)	60			120	90	50	10
Time (h)	2	3	4				

Copy and complete the table.

3 Six people can dig a trench in 8 hours.
  a How long would it take:
    i 4 people ii 12 people iii 1 person?
  b How many people would it take to dig the trench in:
    i 3 hours ii 16 hours iii 1 hour?

4 Chairs in a hall are arranged in 35 rows of 18.
  a How many rows would there be with 21 chairs to a row?
  b How many chairs would each row have if there were 15 rows?

5 A train travelling at 100 km/h takes 4 hours for a journey. How long would it take a train travelling at 60 km/h?

6 A worker in a sugar factory packs 24 cardboard boxes with 15 bags of sugar in each. If he had boxes which held 18 bags of sugar each, how many fewer boxes would be needed?

7 A swimming pool is filled in 30 hours by two identical pumps. How much quicker would it be filled if five similar pumps were used instead?

Compound measures

A **compound measure** is one made up of two or more other measures. The most common ones are speed, density and population density. **Speed** is a compound measure as it is measured using distance and time.

$$\text{Speed} = \frac{\text{Distance}}{\text{Time}}$$

Units of speed include metres per second (m/s) or kilometres per hour (km/h).

The relationship between speed, distance and time is often presented as:

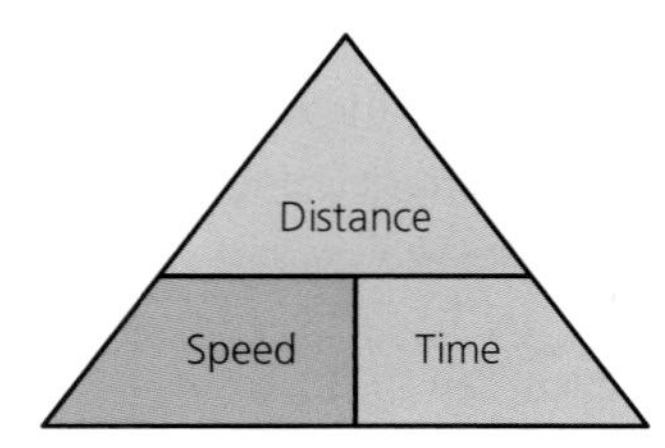

i.e. $\text{Speed} = \frac{\text{Distance}}{\text{Time}}$

$\text{Distance} = \text{Speed} \times \text{Time}$

$\text{Time} = \frac{\text{Distance}}{\text{Speed}}$

Similarly, **Average Speed** $= \frac{\text{Total Distance}}{\text{Total Time}}$

Density, which is a measure of the mass of a substance per unit of its volume, is calculated using the formula:

$$\text{Density} = \frac{\text{Mass}}{\text{Volume}}$$

Units of density include kilograms per cubic metre (kg/m^3) or grams per millilitre (g/ml).

The relationship between density, mass and volume, like speed, can also be presented in a helpful diagram as shown:

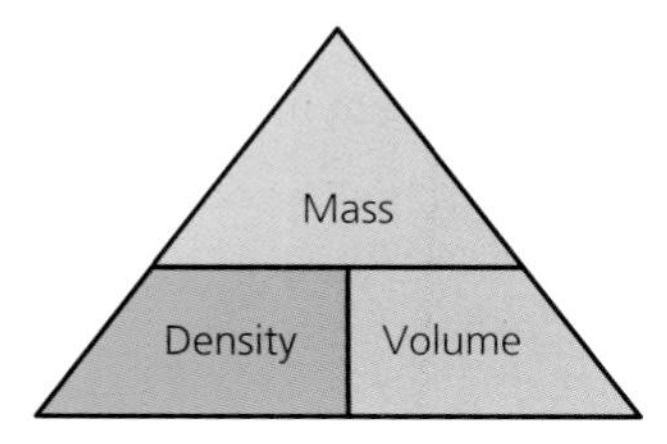

i.e. $\text{Density} = \frac{\text{Mass}}{\text{Volume}}$

$\text{Mass} = \text{Density} \times \text{Volume}$

$\text{Volume} = \frac{\text{Mass}}{\text{Density}}$

Population density is also a compound measure as it is a measure of a population per unit of area.

$$\text{Population Density} = \frac{\text{Population}}{\text{Area}}$$

An example of its units is the number of people per square kilometre (people/km^2). Again, the relationship between population density, population and area can be represented in a triangular diagram:

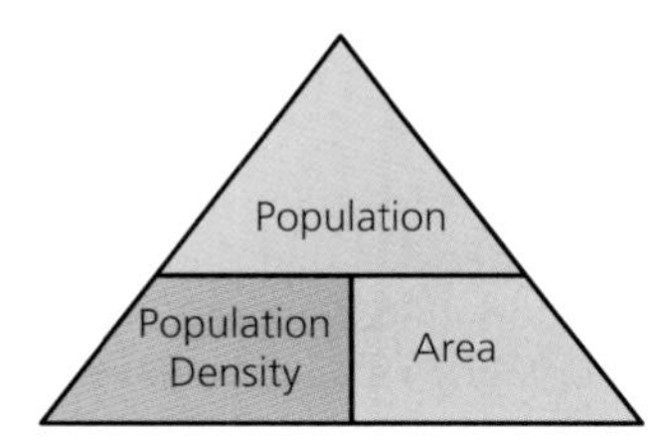

i.e. $\text{Population Density} = \frac{\text{Population}}{\text{Area}}$

$\text{Population} = \text{Population Density} \times \text{Area}$

$\text{Area} = \frac{\text{Population}}{\text{Population Density}}$

Worked examples

1 A train travels a total distance of 140 km in $1\frac{1}{2}$ hours.

a Calculate the average speed of the train during the journey.

$$\begin{aligned}\text{Average Speed} &= \frac{\text{Total Distance}}{\text{Total Time}}\\ &= \frac{140}{1\frac{1}{2}}\\ &= 93\tfrac{1}{3}\text{ km/h}\end{aligned}$$

b During the journey, the train spent 15 minutes stopped at stations. Calculate the average speed of the train whilst it was moving.

Notice that the original time was given in hours, whilst the time spent stopped at stations is given in minutes. To proceed with the calculation, the units have to be consistent, i.e. either both in hours or both in minutes.

The time spent travelling is $1\frac{1}{2} - \frac{1}{4} = 1\frac{1}{4}$ hours

$$\text{Therefore average speed} = \frac{140}{1\frac{1}{4}}$$
$$= 112\,\text{km/h}$$

c If the average speed was 120 km/h, calculate how long the journey took.

$$\text{Total Time} = \frac{\text{Total Distance}}{\text{Average Speed}}$$
$$= \frac{140}{120} = 1.1\dot{6}\text{ hours}$$

Note, it may be necessary to convert a decimal answer to hours and minutes. To convert a decimal time to minutes, multiply by 60.

$$0.1\dot{6} \times 60 = 10$$

Therefore total time is 1 hr 10 mins or 70 mins.

2 A village has a population of 540. Its total area is $8\,\text{km}^2$.

a Calculate the population density of the village.

$$\text{Population Density} = \frac{\text{Population}}{\text{Area}}$$
$$= \frac{540}{8} = 67.5\text{ people/km}^2$$

b A building company wants to build some new houses in the existing area of the village. It is decided that the maximum desirable population density of the village should not exceed 110 people/km^2. Calculate the extra number of people the village can have.

$$\text{Population} = \text{Population Density} \times \text{Area}$$
$$= 110 \times 8$$
$$= 880\text{ people}$$

Therefore the maximum number of extra people that will need housing is $880 - 540 = 340$.

Exercise 6.6

1 Aluminium has a density of 2900 kg/m^3. A construction company needs four cubic metres of aluminium. Calculate the mass of the aluminium needed.

2 A marathon race is 42 195 m in length. The world record in 2016 was 2 hours, 2 minutes and 57 seconds held by Dennis Kimetto of Kenya.
  a How many seconds in total did Kimetto take to complete the race?
  b Calculate his average speed in m/s for the race, giving your answer to 2 decimal places.
  c What average speed would the runner need to maintain to complete the marathon in under two hours?

3 The approximate densities of four metals in g/cm^3 are given below:

Aluminium	2.9 g/cm^3	Copper	9.3 g/cm^3
Brass	8.8 g/cm^3	Steel	8.2 g/cm^3

A cube of an unknown metal has side lengths of 5 cm. The mass of the cube is 1.1 kg.

  a By calculating the cube's density, determine which metal the cube is likely to be made from.
  b Another cube made of steel has a mass of 4.0 kg. Calculate the length of each of the sides of the steel cube, giving your answer to 1 d.p.

4 Singapore is the country with the highest population density in the world. Its population is 5 535 000 and it has a total area of 719 km^2.
  a Calculate Singapore's population density.

China is the country with the largest population.
  b Explain why China has not got the world's highest population density.
  c Find the area and population of your own country. Calculate your country's population density.

5 A farmer has a rectangular field measuring 600 m × 800 m. He uses the field for grazing his sheep.
  a Calculate the area of the field in km^2.
  b 40 sheep graze in the field. Calculate the population density of sheep in the field, giving your answer in sheep/km^2.
  c Guidelines for keeping sheep state that the maximum population density for grazing sheep is 180/km^2. Calculate the number of sheep the farmer is allowed to graze in his field.

6 The formula linking pressure (P N/m^2), force (F N) and surface area (A m^2) is given as $P = \frac{F}{A}$. A square-based box exerts a force of 160 N on a floor. If the pressure on the floor is 1000 N/m^2, calculate the length, in cm, of each side of the base of the box.

Student assessment 1

1 A piece of wood is cut in the ratio 3 : 7.
  a What fraction of the whole is the longer piece?
  b If the wood is 1.5 m long, how long is the shorter piece?

2 A recipe for two people requires $\frac{1}{4}$ kg of rice to 150 g of meat.
  a How much meat would be needed for five people?
  b How much rice would there be in 1 kg of the final dish?

3 The scale of a map is 1 : 10 000.
  a Two rivers are 4.5 cm apart on the map. How far apart are they in real life? Give your answer in metres.
  b Two towns are 8 km apart in real life. How far apart are they on the map? Give your answer in centimetres.

4 A model train is a $\frac{1}{25}$ scale model.
  a Express this as a ratio.
  b If the length of the model engine is 7 cm, what is its true length?

5 Divide 3 tonnes in the ratio 2 : 5 : 13.

6 The ratio of the angles of a quadrilateral is 2 : 3 : 3 : 4. Calculate the size of each of the angles.

The angles of a quadrilateral add up to 360°.

7 The ratio of the interior angles of a pentagon is 2 : 3 : 4 : 4 : 5. Calculate the size of the largest angle.

The interior angles of a pentagon add up to 540°.

8 A large swimming pool takes 36 hours to fill using three identical pumps.
  a How long would it take to fill using eight identical pumps?
  b If the pool needs to be filled in 9 hours, how many of these pumps will be needed?

9 The first triangle is an enlargement of the second. Calculate the size of the missing sides and angles.

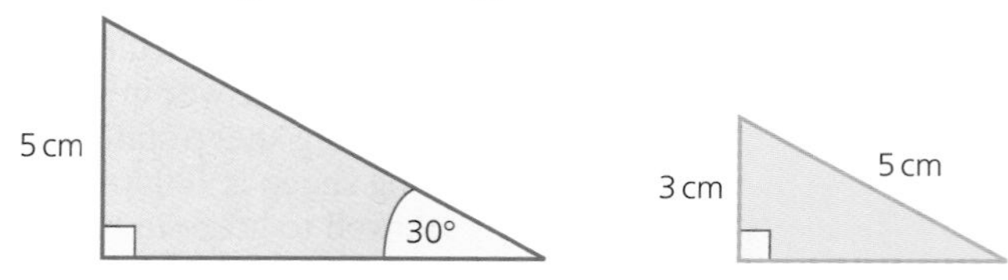

10 A tap issuing water at a rate of 1.2 litres per minute fills a container in 4 minutes.
  a How long would it take to fill the same container if the rate was decreased to 1 litre per minute? Give your answer in minutes and seconds.
  b If the container is to be filled in 3 minutes, calculate the rate at which the water should flow.

11 The population density of a small village increases over time from 18 people/km^2 to 26.5 people/km^2. If the area of the village remains unchanged at 14 km^2 during that time, calculate the increase in the number of people in the village.

7 Indices and standard form

Indices

Indices is the plural of index.

The index refers to the **power** to which a number is raised. In the example 5^3 the number 5 is raised to the power 3. The 3 is known as the **index**.

Worked examples

1 $5^3 = 5 \times 5 \times 5 = 125$

2 $7^4 = 7 \times 7 \times 7 \times 7 = 2401$

3 $3^1 = 3$

Exercise 7.1

1 Using indices, simplify these expressions:
- a $3 \times 3 \times 3$
- b $2 \times 2 \times 2 \times 2 \times 2$
- c 4×4
- d $6 \times 6 \times 6 \times 6$
- e $8 \times 8 \times 8 \times 8 \times 8 \times 8$
- f 5

2 Simplify the following using indices:
- a $2 \times 2 \times 2 \times 3 \times 3$
- b $4 \times 4 \times 4 \times 4 \times 4 \times 5 \times 5$
- c $3 \times 3 \times 4 \times 4 \times 4 \times 5 \times 5$
- d $2 \times 7 \times 7 \times 7 \times 7$
- e $1 \times 1 \times 6 \times 6$
- f $3 \times 3 \times 3 \times 4 \times 4 \times 6 \times 6 \times 6 \times 6 \times 6$

3 Write out the following in full:
- a 4^2
- b 5^7
- c 3^5
- d $4^3 \times 6^3$
- e $7^2 \times 2^7$
- f $3^2 \times 4^3 \times 2^4$

4 Without a calculator work out the value of:
- a 2^5
- b 3^4
- c 8^2
- d 6^3
- e 10^6
- f 4^4
- g $2^3 \times 3^2$
- h $10^3 \times 5^3$

Laws of indices

When working with numbers involving indices there are three basic laws which can be applied. These are:

1 $a^m \times a^n = a^{m+n}$

2 $a^m \div a^n$ or $\frac{a^m}{a^n} = a^{m-n}$

3 $(a^m)^n = a^{mn}$

Positive indices

Worked examples

1 Simplify $4^3 \times 4^2$.

$4^3 \times 4^2 = 4^{(3+2)}$

$= 4^5$

2 Simplify $2^5 \div 2^3$.

$2^5 \div 2^3 = 2^{(5-3)}$

$= 2^2$

3 Evaluate $3^3 \times 3^4$.

$3^3 \times 3^4 = 3^{(3+4)}$

$= 3^7$

$= 2187$

4 Evaluate $(4^2)^3$.

$(4^2)^3 = 4^{(2\times3)}$

$= 4^6$

$= 4096$

Exercise 7.2

1 Simplify the following using indices:

a $3^2 \times 3^4$
b $8^5 \times 8^2$
c $5^2 \times 5^4 \times 5^3$
d $4^3 \times 4^5 \times 4^2$
e $2^1 \times 2^3$
f $6^2 \times 3^2 \times 3^3 \times 6^4$
g $4^5 \times 4^3 \times 5^5 \times 5^4 \times 6^2$
h $2^4 \times 5^7 \times 5^3 \times 6^2 \times 6^6$

2 Simplify:

a $4^6 \div 4^2$
b $5^7 \div 5^4$
c $2^5 \div 2^4$
d $6^5 \div 6^2$
e $\frac{6^5}{6^2}$
f $\frac{8^6}{8^5}$
g $\frac{4^8}{4^5}$
h $\frac{3^9}{3^2}$

3 Simplify:

a $(5^2)^2$
b $(4^3)^4$
c $(10^2)^5$
d $(3^3)^5$
e $(6^2)^4$
f $(8^2)^3$

4 Simplify:

a $\frac{2^2 \times 2^4}{2^3}$
b $\frac{3^4 \times 3^2}{3^5}$
c $\frac{5^6 \times 5^7}{5^2 \times 5^8}$
d $\frac{(4^2)^5 \times 4^2}{4^7}$
e $\frac{4^4 \times 2^5 \times 4^2}{4^3 \times 2^3}$
f $\frac{6^3 \times 6^3 \times 8^5 \times 8^6}{8^6 \times 6^2}$
g $\frac{(5^2)^2 \times (4^4)^3}{5^8 \times 4^9}$
h $\frac{(6^3)^4 \times 6^3 \times 4^9}{6^8 \times (4^2)^4}$

The zero index

The zero index indicates that a number is raised to the power 0.
A number raised to the power 0 is equal to 1. This can be explained by applying the laws of indices.

$a^m \div a^n = a^{m-n}$ therefore $\frac{a^m}{a^m} = a^{m-m}$

$= a^0$

However, $\frac{a^m}{a^m} = 1$

therefore $a^0 = 1$

Exercise 7.3

Without using a calculator, evaluate:

a $2^3 \times 2^0$ b $5^2 \div 6^0$ c $5^2 \times 5^{-2}$

d $6^3 \times 6^{-3}$ e $(4^0)^2$ f $4^0 \div 2^2$

Negative indices

A negative index indicates that a number is being raised to a negative power, e.g. 4^{-3}.

Another law of indices states that $a^{-m} = \frac{1}{a^m}$. This can be proved as follows.

$a^{-m} = a^{0-m}$

$= \frac{a^0}{a^m}$ (from the second law of indices)

$= \frac{1}{a^m}$

therefore $a^{-m} = \frac{1}{a^m}$

Exercise 7.4

Without using a calculator, evaluate:

1 a 4^{-1} b 3^{-2} c 6×10^{-2}
d 5×10^{-3} e 100×10^{-2} f 10^{-3}

2 a 9×3^{-2} b 16×2^{-3} c 64×2^{-4}
d 4×2^{-3} e 36×6^{-3} f 100×10^{-1}

3 a $\frac{3}{2^{-2}}$ b $\frac{4}{2^{-3}}$ c $\frac{9}{5^{-2}}$

d $\frac{5}{4^{-2}}$ e $\frac{7^{-3}}{7^{-4}}$ f $\frac{8^{-6}}{8^{-8}}$

Fractional indices

$16^{\frac{1}{2}}$ can be written as $(4^2)^{\frac{1}{2}}$.

$(4^2)^{\frac{1}{2}} = 4^{\left(2\times\frac{1}{2}\right)}$
$= 4^1$
$= 4$

Therefore $16^{\frac{1}{2}} = 4$

but $\sqrt{16} = 4$

therefore $16^{\frac{1}{2}} = \sqrt{16}$

Similarly:

$125^{\frac{1}{3}}$ can be written as $\left(5^3\right)^{\frac{1}{3}}$

$$\left(5^3\right)^{\frac{1}{3}} = 5^{(3 \times \frac{1}{3})}$$
$$= 5^1$$
$$= 5$$

Therefore $125^{\frac{1}{3}} = 5$

but $\sqrt[3]{125} = 5$

therefore $125^{\frac{1}{3}} = \sqrt[3]{125}$

In general:

$$a^{\frac{1}{n}} = \sqrt[n]{a}$$

$$a^{\frac{m}{n}} = \sqrt[n]{\left(a^m\right)} \text{ or } \left(\sqrt[n]{a}\right)^m$$

Worked examples

1 Evaluate $16^{\frac{1}{4}}$ without the use of a calculator.

$$16^{\frac{1}{4}} = \sqrt[4]{16}$$
$$= \sqrt[4]{(2^4)}$$
$$= 2$$

Alternatively:

$$16^{\frac{1}{4}} = (2^4)^{\frac{1}{4}}$$
$$= 2^1$$
$$= 2$$

2 Evaluate $25^{\frac{3}{2}}$ without the use of a calculator.

$$25^{\frac{3}{2}} = (25^{\frac{1}{2}})^3$$
$$= \sqrt{25}^3$$
$$= 5^3$$
$$= 125$$

Alternatively:

$$25^{\frac{3}{2}} = (5^2)^{\frac{3}{2}}$$
$$= 5^3$$
$$= 125$$

Exercise 7.5

Evaluate the following without the use of a calculator:

1 a $16^{\frac{1}{2}}$ b $25^{\frac{1}{2}}$ c $100^{\frac{1}{2}}$

d $27^{\frac{1}{3}}$ e $81^{\frac{1}{2}}$ f $1000^{\frac{1}{3}}$

2 a $16^{\frac{1}{4}}$ b $81^{\frac{1}{4}}$ c $32^{\frac{1}{5}}$

d $64^{\frac{1}{6}}$ e $216^{\frac{1}{3}}$ f $256^{\frac{1}{4}}$

3 **a** $4^{\frac{3}{2}}$ **b** $4^{\frac{5}{2}}$ **c** $9^{\frac{3}{2}}$

d $16^{\frac{3}{2}}$ **e** $1^{\frac{5}{2}}$ **f** $27^{\frac{2}{3}}$

4 **a** $125^{\frac{2}{3}}$ **b** $32^{\frac{3}{5}}$ **c** $64^{\frac{5}{6}}$

d $1000^{\frac{2}{3}}$ **e** $16^{\frac{5}{4}}$ **f** $81^{\frac{3}{4}}$

Exercise 7.6 Evaluate the following without the use of a calculator:

1 **a** $\frac{27^{\frac{2}{3}}}{3^2}$ **b** $\frac{7^{\frac{3}{2}}}{\sqrt{7}}$ **c** $\frac{4^{\frac{5}{2}}}{4^2}$

d $\frac{16^{\frac{3}{2}}}{2^6}$ **e** $\frac{27^{\frac{5}{3}}}{\sqrt{9}}$ **f** $\frac{6^{\frac{4}{3}}}{6^{\frac{1}{3}}}$

2 **a** $5^{\frac{2}{3}} \times 5^{\frac{4}{3}}$ **b** $4^{\frac{1}{4}} \times 4^{\frac{1}{4}}$ **c** 8×2^{-2}

d $3^{\frac{4}{3}} \times 3^{\frac{5}{3}}$ **e** $2^{-2} \times 16$ **f** $8^{\frac{5}{3}} \times 8^{-\frac{4}{3}}$

3 **a** $\frac{2^{\frac{1}{2}} \times 2^{\frac{5}{2}}}{2}$ **b** $\frac{4^{\frac{5}{6}} \times 4^{\frac{1}{6}}}{4^{\frac{1}{2}}}$ **c** $\frac{2^3 \times 8^{\frac{3}{2}}}{\sqrt{8}}$

d $\frac{(3^2)^{\frac{3}{2}} \times 3^{-\frac{1}{2}}}{3^{\frac{1}{2}}}$ **e** $\frac{8^{\frac{1}{3}} + 7}{27^{\frac{1}{3}}}$ **f** $\frac{9^{\frac{1}{2}} \times 3^{\frac{5}{2}}}{3^{\frac{2}{3}} \times 3^{-\frac{1}{6}}}$

Standard form

Standard form is also known as standard index form or sometimes as scientific notation. It involves writing large numbers or very small numbers in terms of powers of 10.

Positive indices and large numbers

$$100 = 1 \times 10^2$$

$$1000 = 1 \times 10^3$$

$$10\,000 = 1 \times 10^4$$

$$3000 = 3 \times 10^3$$

For a number to be in standard form it must take the form $A \times 10^n$ where the index n is a positive or negative integer and A must lie in the range $1 \leqslant A < 10$.

e.g. 3100 can be written in many different ways:

3.1×10^3 31×10^2 0.31×10^4 etc.

However, only 3.1×10^3 satisfies the above conditions and therefore is the only one which is written in standard form.

Worked examples

1 Write 72 000 in standard form.
7.2×10^4

2 Write 4×10^4 as an ordinary number.
$4 \times 10^4 = 4 \times 10000$
$= 40000$

3 Multiply 600×4000 and write your answer in standard form.
600×4000
$= 2400000$
$= 2.4 \times 10^6$

Exercise 7.7

1 Deduce the value of n in the following:
a $79000 = 7.9 \times 10^n$
b $53000 = 5.3 \times 10^n$
c $4160000 = 4.16 \times 10^n$
d 8 million $= 8 \times 10^n$
e 247 million $= 2.47 \times 10^n$
f $24000000 = 2.4 \times 10^n$

2 Write the following numbers in standard form:
a 65 000
b 41 000
c 723 000
d 18 million
e 950 000
f 760 million
g 720 000
h $\frac{1}{4}$ million

3 Write the numbers below which are written in standard form:
a 26.3×10^5
b 2.6×10^7
c 0.5×10^3
d 8×10^8
e 0.85×10^9
f 8.3×10^{10}
g 1.8×10^7
h 18×10^5
i 3.6×10^6
j 6.0×10^1

4 Write the following as ordinary numbers:
a 3.8×10^3
b 4.25×10^6
c 9.003×10^7
d 1.01×10^5

5 Multiply the following and write your answers in standard form:
a 400×2000
b 6000×5000
c 75000×200
d 33000×6000
e 8 million $\times 250$
f 95000×3000
g 7.5 million $\times 2$
h 8.2 million $\times 50$
i $300 \times 200 \times 400$
j $(7000)^2$

6 Which of the following are not in standard form?
a 6.2×10^5
b 7.834×10^{16}
c 8.0×10^5
d 0.46×10^7
e 82.3×10^6
f 6.75×10^1

7 Write the following numbers in standard form:
a 600 000
b 48 000 000
c 784 000 000 000
d 534 000
e 7 million
f 8.5 million

8 Write the following in standard form:
a 68×10^5
b 720×10^6
c 8×10^5
d 0.75×10^8
e 0.4×10^{10}
f 50×10^6

9 Multiply the following and write your answers in standard form:
a 200×3000
b 6000×4000
c 7 million $\times 20$
d 500×6 million
e 3 million $\times 4$ million
f 4500×4000

10 Light from the Sun takes approximately 8 minutes to reach Earth. If light travels at a speed of 3×10^8 m/s, calculate to three significant figures (s.f.) the distance from the Sun to the Earth.

Multiplying and dividing numbers in standard form

When you multiply or divide numbers in standard form, you work with the numbers and the powers of 10 separately. You use the laws of indices when working with the powers of 10.

Worked examples

1 Multiply the following and write your answer in standard form:
$(2.4 \times 10^4) \times (5 \times 10^7)$
$= 12 \times 10^{11}$
$= 1.2 \times 10^{12}$ when written in standard form

2 Divide the following and write your answer in standard form:
$(6.4 \times 10^7) \div (1.6 \times 10^3)$
$= 4 \times 10^4$

Exercise 7.8

1 Multiply the following and write your answers in standard form:

a $(4 \times 10^3) \times (2 \times 10^5)$
b $(2.5 \times 10^4) \times (3 \times 10^4)$
c $(1.8 \times 10^7) \times (5 \times 10^6)$
d $(2.1 \times 10^4) \times (4 \times 10^7)$
e $(3.5 \times 10^4) \times (4 \times 10^7)$
f $(4.2 \times 10^5) \times (3 \times 10^4)$
g $(2 \times 10^4)^2$
h $(4 \times 10^8)^2$

2 Find the value of the following and write your answers in standard form:

a $(8 \times 10^6) \div (2 \times 10^3)$
b $(8.8 \times 10^9) \div (2.2 \times 10^3)$
c $(7.6 \times 10^8) \div (4 \times 10^7)$
d $(6.5 \times 10^{14}) \div (1.3 \times 10^7)$
e $(5.2 \times 10^6) \div (1.3 \times 10^6)$
f $(3.8 \times 10^{11}) \div (1.9 \times 10^3)$

3 Find the value of the following and write your answers in standard form:

a $(3 \times 10^4) \times (6 \times 10^5) \div (9 \times 10^5)$
b $(6.5 \times 10^8) \div (1.3 \times 10^4) \times (5 \times 10^3)$
c $(18 \times 10^3) \div 900 \times 250$
d $27\,000 \div 3000 \times 8000$
e $4000 \times 8000 \div 640$
f $2500 \times 2500 \div 1250$

4 Find the value of the following and write your answers in standard form:

a $(4.4 \times 10^3) \times (2 \times 10^5)$
b $(6.8 \times 10^7) \times (3 \times 10^3)$
c $(4 \times 10^5) \times (8.3 \times 10^5)$
d $(5 \times 10^9) \times (8.4 \times 10^{12})$
e $(8.5 \times 10^6) \times (6 \times 10^{15})$
f $(5.0 \times 10^{12})^2$

5 Find the value of the following and write your answers in standard form:

a $(3.8 \times 10^8) \div (1.9 \times 10^6)$
b $(6.75 \times 10^9) \div (2.25 \times 10^4)$
c $(9.6 \times 10^{11}) \div (2.4 \times 10^5)$
d $(1.8 \times 10^{12}) \div (9.0 \times 10^7)$
e $(2.3 \times 10^{11}) \div (9.2 \times 10^4)$
f $(2.4 \times 10^8) \div (6.0 \times 10^3)$

Adding and subtracting numbers in standard form

You can only add and subtract numbers in standard form if the indices are the same. If the indices are different, you can change one of the numbers so that it has the same index as the other. It will not then be in standard form and you may need to change your answer back to standard form after doing the calculation.

Worked examples

1 Add the following and write your answer in standard form:

$(3.8 \times 10^6) + (8.7 \times 10^4)$

Changing the indices to the same value gives the sum:

$(380 \times 10^4) + (8.7 \times 10^4)$

$= 388.7 \times 10^4$

$= 3.887 \times 10^6$ when written in standard form

2 Subtract the following and write your answer in standard form:

$(6.5 \times 10^7) - (9.2 \times 10^5)$

Changing the indices to the same value gives the sum:

$(650 \times 10^5) - (9.2 \times 10^5)$

$= 640.8 \times 10^5$

$= 6.408 \times 10^7$ when written in standard form

Exercise 7.9

Find the value of the following and write your answers in standard form:

a $(3.8 \times 10^5) + (4.6 \times 10^4)$
b $(7.9 \times 10^7) + (5.8 \times 10^8)$
c $(6.3 \times 10^7) + (8.8 \times 10^5)$
d $(3.15 \times 10^9) + (7.0 \times 10^6)$
e $(5.3 \times 10^8) - (8.0 \times 10^7)$
f $(6.5 \times 10^7) - (4.9 \times 10^6)$
g $(8.93 \times 10^{10}) - (7.8 \times 10^9)$
h $(4.07 \times 10^7) - (5.1 \times 10^6)$

Negative indices and small numbers

A negative index is used when writing a number between 0 and 1 in standard form.

e.g. $100 = 1 \times 10^2$
$10 = 1 \times 10^1$
$1 = 1 \times 10^0$
$0.1 = 1 \times 10^{-1}$
$0.01 = 1 \times 10^{-2}$
$0.001 = 1 \times 10^{-3}$
$0.0001 = 1 \times 10^{-4}$

Note that A must still lie within the range $1 \leqslant A < 10$.

Worked examples

1 Write 0.0032 in standard form.

3.2×10^{-3}

2 Write 1.8×10^{-4} as an ordinary number.

$1.8 \times 10^{-4} = 1.8 \div 10^4$

$= 1.8 \div 10000$

$= 0.00018$

3 Write the following numbers in order of magnitude, starting with the largest:

3.6×10^{-3} 5.2×10^{-5} 1×10^{-2} 8.35×10^{-2} 6.08×10^{-8}

8.35×10^{-2} 1×10^{-2} 3.6×10^{-3} 5.2×10^{-5} 6.08×10^{-8}

Exercise 7.10

1 Copy and complete the following so that the answers are correct (the first question is done for you):

a 0.0048 $= 4.8 \times 10^{-3}$
b 0.0079 $= 7.9 \times$ ...
c 0.00081 $= 8.1 \times$ ...
d 0.000009 $= 9 \times$ ...
e 0.00000045 $= 4.5 \times$ ...
f 0.000000003 24 $= 3.24 \times$ ...
g 0.00000842 $= 8.42 \times$ ...
h 0.000000000403 $= 4.03 \times$ ...

2 Write these numbers in standard form:

a 0.0006
b 0.000053
c 0.000864
d 0.000000088
e 0.0000007
f 0.0004145

3 Write the following as ordinary numbers:

a 8×10^{-3}
b 4.2×10^{-4}
c 9.03×10^{-2}
d 1.01×10^{-5}

4 Write the following numbers in standard form:

a 68×10^{-5}
b 750×10^{-9}
c 42×10^{-11}
d 0.08×10^{-7}
e 0.057×10^{-9}
f 0.4×10^{-10}

5 Deduce the value of n in each of the following:

a $0.00025 = 2.5 \times 10^n$
b $0.00357 = 3.57 \times 10^n$
c $0.00000006 = 6 \times 10^n$
d $0.004^2 = 1.6 \times 10^n$
e $0.00065^2 = 4.225 \times 10^n$
f $0.0002^n = 8 \times 10^{-12}$

6 Write these numbers in order of magnitude, starting with the largest:

3.2×10^{-4} 6.8×10^5 5.57×10^{-9} 6.2×10^3

5.8×10^{-7} 6.741×10^{-4} 8.414×10^2

Student assessment 1

1 Simplify the following using indices:
   a $2 \times 2 \times 2 \times 5 \times 5$
   b $2 \times 2 \times 3 \times 3 \times 3 \times 3 \times 3$

2 Write out in full:
   a 4^3
   b 6^4

3 Work out the value of the following without using a calculator:
   a $2^3 \times 10^2$
   b $1^4 \times 3^3$

4 Simplify the following using indices:
   a $3^4 \times 3^3$
   b $6^3 \times 6^2 \times 3^4 \times 3^5$
   c $\frac{4^5}{2^3}$
   d $\frac{(6^2)^3}{6^5}$
   e $\frac{3^5 \times 4^2}{3^3 \times 4^0}$
   f $\frac{4^{-2} \times 2^6}{2^2}$

5 Without using a calculator, evaluate:
   a $2^4 \times 2^{-2}$
   b $\frac{3^5}{3^3}$
   c $\frac{5^{-5}}{5^{-6}}$
   d $\frac{2^5 \times 4^{-3}}{2^{-1}}$

6 Evaluate the following without the use of a calculator:
   a $81^{\frac{1}{2}}$
   b $27^{\frac{1}{3}}$
   c $9^{\frac{1}{2}}$
   d $625^{\frac{3}{4}}$
   e $343^{\frac{2}{3}}$
   f $16^{-\frac{1}{4}}$
   g $\frac{1}{25^{-\frac{1}{2}}}$
   h $\frac{2}{16^{-\frac{3}{4}}}$

7 Evaluate the following without the use of a calculator:
   a $\frac{16^{\frac{1}{2}}}{2^2}$
   b $\frac{9^{\frac{5}{2}}}{3^3}$
   c $\frac{8^{\frac{4}{3}}}{8^{\frac{2}{3}}}$
   d $5^{\frac{6}{5}} \times 5^{\frac{4}{5}}$
   e $4^{\frac{3}{2}} \times 2^{-2}$
   f $\frac{27^{\frac{2}{3}} \times 3^{-2}}{4^{-\frac{3}{2}}}$
   g $\frac{(4^3)^{-\frac{1}{2}} \times 2^{\frac{3}{2}}}{2^{-\frac{3}{2}}}$
   h $\frac{(5^{\frac{2}{3}})^{\frac{1}{2}} \times 5^{\frac{2}{3}}}{3^{-2}}$

Student assessment 2

1 Simplify the following using indices:
a $3 \times 2 \times 2 \times 3 \times 27$
b $2 \times 2 \times 4 \times 4 \times 4 \times 2 \times 32$

2 Write out in full:
a 6^5
b 2^{-5}

3 Work out the value of the following without using a calculator:
a $3^3 \times 10^3$
b $1^{-4} \times 5^3$

4 Simplify the following using indices:
a $2^4 \times 2^3$
b $7^5 \times 7^2 \times 3^4 \times 3^8$
c $\frac{4^8}{2^{10}}$
d $\frac{(3^3)^4}{27^3}$
e $\frac{7^6 \times 4^2}{4^3 \times 7^6}$
f $\frac{8^{-2} \times 2^6}{2^{-2}}$

5 Without using a calculator, evaluate:
a $5^2 \times 5^{-1}$
b $\frac{4^5}{4^3}$
c $\frac{7^{-5}}{7^{-7}}$
d $\frac{3^{-5} \times 4^2}{3^{-6}}$

6 Evaluate the following without the use of a calculator:
a $64^{\frac{1}{6}}$
b $27^{\frac{4}{3}}$
c $9^{-\frac{1}{2}}$
d $512^{\frac{2}{3}}$
e $\sqrt[3]{27}$
f $\sqrt[4]{16}$
g $\frac{1}{36^{-\frac{1}{2}}}$
h $\frac{2}{64^{-\frac{2}{3}}}$

7 Evaluate the following without the use of a calculator:
a $\frac{25^{\frac{1}{2}}}{9^{-\frac{1}{2}}}$
b $\frac{4^{\frac{5}{2}}}{2^3}$
c $\frac{27^{\frac{4}{3}}}{3^3}$
d $25^{\frac{3}{2}} \times 5^2$
e $4^{\frac{3}{2}} \times 4^{-\frac{1}{2}}$
f $\frac{27^{\frac{2}{3}} \times 3^{-3}}{9^{-\frac{1}{2}}}$
g $\frac{(4^2)^{-\frac{1}{4}} \times 9^{\frac{3}{2}}}{(\frac{1}{4})^{\frac{1}{2}}}$
h $\frac{(5^{\frac{1}{3}})^{\frac{1}{2}} \times 5^{\frac{5}{6}}}{4^{-\frac{1}{2}}}$

? Student assessment 3

1 Write the following numbers in standard form:
   a 6 million
   b 0.0045
   c 3 800 000 000
   d 0.000 000 361
   e 460 million
   f 3

2 Write the following as ordinary numbers:
   a 8.112×10^6
   b 3.05×10^{-4}

3 Write the following numbers in order of magnitude, starting with the largest:

   3.6×10^2 2.1×10^{-3} 9×10^1 4.05×10^8 1.5×10^{-2} 7.2×10^{-3}

4 Write the following numbers:
   a in standard form
   b in order of magnitude, starting with the smallest.

   15 million   430 000   0.000 435   4.8   0.0085

5 Deduce the value of n in each of the following:
   a $4750 = 4.75 \times 10^n$
   b $6\,440\,000\,000 = 6.44 \times 10^n$
   c $0.0040 = 4.0 \times 10^n$
   d $1000^2 = 1 \times 10^n$
   e $0.9^3 = 7.29 \times 10^n$
   f $800^3 = 5.12 \times 10^n$

6 Write the answers to the following calculations in standard form:
   a $50\,000 \times 2400$
   b $(3.7 \times 10^6) \times (4.0 \times 10^4)$
   c $(5.8 \times 10^7) + (9.3 \times 10^6)$
   d $(4.7 \times 10^6) - (8.2 \times 10^5)$

7 The speed of light is 3×10^8 m/s. Jupiter is 778 million km from the Sun. Calculate the number of minutes it takes for sunlight to reach Jupiter.

8 A star is 300 light years away from Earth. The speed of light is 3×10^5 km/s. Calculate the distance from the star to Earth. Give your answer in kilometres and written in standard form.

8 Money and finance

Currency conversion

In 2017, 1 euro could be exchanged for 1.50 Australian dollars (A$).

The **conversion graph** from euros to dollars and from dollars to euros is:

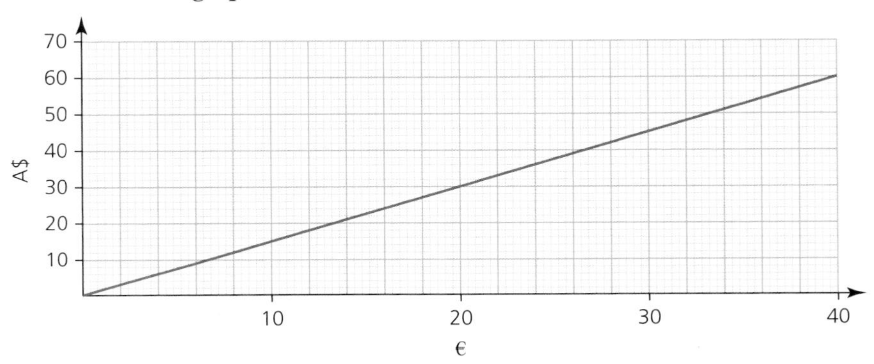

Exercise 8.1

1 Use the conversion graph to convert the following to Australian dollars:

a €20	b €30	c €5
d €25	e €35	f €15

2 Use the conversion graph to convert the following to euros:

a A$20	b A$30	c A$40
d A$35	e A$25	f A$48

3 1 euro can be exchanged for 75 Indian rupees. Draw a conversion graph. Use an appropriate scale with the horizontal scale up to €100. Use your graph to convert the following to rupees:

a €10	b €40	c €50
d €90	e €25	f €1000

4 Use your graph from Q.3 to convert the following to euros:

a 140 rupees	b 770 rupees	c 630 rupees
d 490 rupees	e 5600 rupees	f 2730 rupees

The table shows the exchange rate for €1 into various currencies. If necessary, draw conversion graphs for the exchange rates shown to answer Q.5–9.

Zimbabwe	412.8 Zimbabwe dollars
South Africa	15 rand
Turkey	4.0 Turkish lira
Japan	130 yen
Kuwait	0.35 dinar
United States of America	1.15 dollars

Exercise 8.1 (cont)

5 How many Zimbabwe dollars would you receive for the following?
  a €20 b €50 c €75
  d €30 e €25

6 How many euros would you receive for the following numbers of South African rand?
  a 225 rand b 420 rand
  c 1132.50 rand d 105 rand

7 In the Grand Bazaar in Istanbul, a visitor sees three carpets priced at 1200 Turkish lira, 400 Turkish lira and 880 Turkish lira. Draw and use a conversion graph to find the prices in euros.

8 €1 can be exchanged for US$1.15.
  €1 can also be exchanged for 130 yen.
  Draw a conversion graph for US dollars to Japanese yen, and answer the questions below.
  a How many yen would you receive for:
    i $300 ii $750 iii $1000?
  b How many US dollars would you receive for:
    i 5000 yen ii 8500 yen iii 100 yen?

9 Use the currency table (on page 75) to draw a conversion graph for Kuwaiti dinars to South African rand. Use the graph to find the number of rand you would receive for:
  a 35 dinars b 140 dinars
  c 41.30 dinars d 297.50 dinars

Personal and household finance

Net pay is what is left after deductions such as tax, insurance and pension contributions are taken from **gross earnings**. That is,

Net pay = Gross pay – Deductions

A **bonus** is an extra payment sometimes added to an employee's **basic pay**.

In many companies, there is a fixed number of hours that an employee is expected to work. Any work done in excess of this **basic week** is paid at a higher rate, referred to as **overtime**. Overtime may be 1.5 times basic pay, called **time and a half**, or twice basic pay, called **double time**.

Exercise 8.2

1 Copy the table and find the net pay for each employee:

	Gross pay ($)	Deductions ($)	Net pay ($)
a A Ahmet	162.00	23.50	
b B Martinez	205.50	41.36	
c C Stein	188.25	33.43	
d D Wong	225.18	60.12	

2 Copy and complete the table for each employee:

	Basic pay ($)	Overtime ($)	Bonus ($)	Gross pay ($)
a P Small	144	62	23	
b B Smith	152		31	208
c A Chang		38	12	173
d U Zafer	115	43		213
e M Said	128	36	18	

3 Copy and complete the table for each employee:

	Gross pay ($)	Tax ($)	Pension ($)	Net pay ($)
a A Hafar	203	54	18	
b K Zyeb		65	23	218
c H Such	345		41	232
d K Donald	185	23		147

4 Copy and complete the table to find the basic pay in each case:

	No. of hours worked	Basic rate per hour ($)	Basic pay ($)
a	40	3.15	
b	44	4.88	
c	38	5.02	
d	35	8.30	
e	48	7.25	

5 Copy and complete the table, which shows basic pay and overtime at time and a half:

	Basic hours worked	Rate per hour ($)	Basic pay ($)	Overtime hours worked	Overtime pay ($)	Total gross pay ($)
a	40	3.60		8		
b	35		203.00	4		
c	38	4.15		6		
d		6.10	256.20	5		
e	44	5.25		4		
f		4.87	180.19	3		
g	36	6.68		6		
h	45	7.10	319.50	7		

6 In Q.5, deductions amount to 32% of the total gross pay. Calculate the net pay for each employee.

Piece work is a method of payment where an employee is paid for the number of articles made, not for time taken.

Exercise 8.3

1 Four people help to pick grapes in a vineyard. They are paid $5.50 for each basket of grapes. Copy and complete the table:

	Mon	Tue	Wed	Thur	Fri	Total	Gross pay
Pepe	4	5	7	6	6		
Felicia	3	4	4	5	5		
Delores	5	6	6	5	6		
Juan	3	4	6	6	6		

2 Five people work in a pottery factory, making plates. They are paid $5 for every 12 plates made. Copy and complete the table, which shows the number of plates that each person produces:

	Mon	Tue	Wed	Thur	Fri	Total	Gross pay
Maria	240	360	288	192	180		
Ben	168	192	312	180	168		
Joe	288	156	192	204	180		
Bianca	228	144	108	180	120		
Selina	192	204	156	228	144		

3 A group of five people work at home making clothes. The patterns and material are provided by the company, and for each article produced they are paid:

Jacket $25 Shirt $13 Trousers $11 Dress $12

The five people make the numbers of articles of clothing shown in the table below.

a Find each person's gross pay.
b If the deductions amount to 15% of gross earnings, calculate each person's net pay.

	Jackets	Shirts	Trousers	Dresses
Neo	3	12	7	0
Keletso	8	5	2	9
Ditshele	0	14	12	2
Mpho	6	8	3	12
Kefilwe	4	9	16	5

4 A school organises a sponsored walk. The table shows how far students walked, the amount they were sponsored per mile, and the total each raised.

a Copy and complete the table.

Distance walked (km)	Amount per km ($)	Total raised ($)
10	0.80	
	0.65	9.10
18	0.38	
	0.72	7.31
12		7.92
	1.20	15.60
15	1	
	0.88	15.84
18		10.44
17		16.15

b How much was raised in total?

c This total was divided between three children's charities in the ratio of 2 : 3 : 5. How much did each charity receive?

Simple interest

Interest can be defined as money added by a bank to sums deposited by customers. The money deposited is called the **principal**. The **percentage interest** is the given rate and the money is left for a fixed period of time.

A formula can be obtained for **simple interest**:

$$SI = \frac{Ptr}{100}$$

where SI = simple interest, i.e. the interest paid
P = the principal
t = time in years
r = rate per cent

Worked example

Find the simple interest earned on $250 deposited for 6 years at 8% p.a.

$$SI = \frac{Ptr}{100}$$

$$SI = \frac{250 \times 6 \times 8}{100}$$

$$SI = 120$$

So the interest paid is $120.

p.a. stands for per annum, which means each year.

Exercise 8.4

All rates of interest given here are annual rates. Find the simple interest paid in the following cases:

a	Principal \$300	rate 6%	time 4 years
b	Principal \$750	rate 8%	time 7 years
c	Principal \$425	rate 6%	time 4 years
d	Principal \$2800	rate 4.5%	time 2 years
e	Principal \$6500	rate 6.25%	time 8 years
f	Principal \$880	rate 6%	time 7 years

Worked example

How long will it take for a sum of \$250 invested at 8% to earn interest of \$80?

$$SI = \frac{Ptr}{100}$$

$$80 = \frac{250 \times t \times 8}{100}$$

$$80 = 20t$$

$$4 = t$$

It will take 4 years.

Exercise 8.5

Calculate how long it will take for the following amounts of interest to be earned at the given rate.

a	$P = \$500$	$r = 6\%$	$SI = \$150$
b	$P = \$5800$	$r = 4\%$	$SI = \$96$
c	$P = \$4000$	$r = 7.5\%$	$SI = \$1500$
d	$P = \$2800$	$r = 8.5\%$	$SI = \$1904$
e	$P = \$900$	$r = 4.5\%$	$SI = \$243$
f	$P = \$400$	$r = 9\%$	$SI = \$252$

Worked example

What rate per year must be paid for a principal of \$750 to earn interest of \$180 in 4 years?

$$SI = \frac{Ptr}{100}$$

$$180 = \frac{750 \times 4 \times r}{100}$$

$$180 = 30r$$

$$6 = r$$

The rate must be 6% per year.

5 At an auction a company sells 150 television sets for an average of \$65 each. The production cost was \$10 000. How much loss did the company make?

6 A market trader sells tools and electrical goods. Find the profit or loss if he sells each of the following:
 a 15 torches: cost price \$2 each, selling price \$2.30 each
 b 60 plugs: cost price \$10 for 12, selling price \$1.10 each
 c 200 DVDs: cost price \$9 for 10, selling price \$1.30 each
 d 5 MP3 players: cost price \$82, selling price \$19 each
 e 96 batteries costing \$1 for 6, selling price 59c for 3
 f 3 clock radios costing \$65, sold for \$14 each

Percentage profit and loss

Most profits or losses are expressed as a percentage.

$$\text{Percentage profit or loss} = \frac{\text{Profit or Loss}}{\text{Cost price}} \times 100$$

Worked example

A woman buys a car for \$7500 and sells it two years later for \$4500. Calculate her loss over two years as a percentage of the cost price.

Cost price = \$7500 Selling price = \$4500 Loss = \$3000

$$\text{Percentage loss} = \frac{3000}{7500} \times 100 = 40$$

Her loss is 40%.

When something becomes worth less over a period of time, it is said to **depreciate**.

Exercise 8.11

1 Find the depreciation of each car as a percentage of the cost price. (C.P. = Cost price, S.P. = Selling price)

a	VW	C.P. \$4500	S.P. \$4005
b	Peugeot	C.P. \$9200	S.P. \$6900
c	Mercedes	C.P. \$11 000	S.P. \$5500
d	Toyota	C.P. \$4350	S.P. \$3480
e	Fiat	C.P. \$6850	S.P. \$4795
f	Ford	C.P. \$7800	S.P. \$2600

2 A company manufactures electrical items for the kitchen. Find the percentage profit on each appliance:

a	Cooker	C.P. \$240	S.P. \$300
b	Fridge	C.P. \$50	S.P. \$65
c	Freezer	C.P. \$80	S.P. \$96
d	Microwave	C.P. \$120	S.P. \$180
e	Washing machine	C.P. \$260	S.P. \$340
f	Dryer	C.P. \$70	S.P. \$91

Exercise 8.11 (cont)

3 A developer builds a number of different kinds of house on a village site. Given the cost prices and the selling prices in the table, which type of house gives the developer the largest percentage profit?

	Cost price ($)	Selling price ($)
Type A	40000	52000
Type B	65000	75000
Type C	81000	108000
Type D	110000	144000
Type E	132000	196000

4 Students in a school organise a disco. The disco company charges $350 hire charge. The students sell 280 tickets at $2.25. What is the percentage profit?

5 A shop sells second-hand boats. Calculate the percentage profit on each of the following:

	Cost price ($)	Selling price ($)
Mirror	400	520
Wayfarer	1100	1540
Laser	900	1305
Fireball	1250	1720

Student assessment 1

1 1 Australian dollar can be exchanged for €0.8. Draw a conversion graph to find the number of Australian dollars you would get for:
a €50 b €80 c €70

2 Use your graph from Q.1 to find the number of euros you could receive for:
a A$54 b A$81 c A$320

The currency conversion table shows the amounts of foreign currency received for €1. Draw appropriate conversion graphs to answer Q.3–5.

Nigeria	203 nairas
Malaysia	3.9 ringgits
Jordan	0.9 Jordanian dinars

3 Convert the following numbers of Malaysian ringgits into Jordanian dinars:
a 100 ringgits b 1200 ringgits c 150 ringgits

4 Convert the following numbers of Jordanian dinars into Nigerian nairas:
 a 1 dinar b 6 dinars c 4 dinars

5 Convert the following numbers of Nigerian nairas into Malaysian ringgits:
 a 1000 nairas b 5000 nairas c 7500 nairas

? Student assessment 2

1 A man worked 3 hours a day Monday to Friday for \$3.75 per hour. What was his 4-weekly gross payment?

2 A woman works at home making curtains. In one week she makes 4 pairs of long curtains and 13 pairs of short curtains. What is her gross pay if she receives \$2.10 for each long curtain, and \$1.85 for each short curtain?

3 Calculate the missing numbers from the simple interest table:

Principal (\$)	Rate (%)	Time (years)	Interest (\$)
200	9	3	a
350	7	b	98
520	c	5	169
d	3.75	6	189

4 A car cost \$7200 new and sold for \$5400 after two years. What was the percentage average annual depreciation?

5 A farmer sold eight cows at market at an average sale price of \$48 each. If his total costs for rearing all the animals were \$432, what was his percentage loss on each animal?

? Student assessment 3

1 A girl works in a shop on Saturdays for 8.5 hours. She is paid \$3.60 per hour. What is her gross pay for 4 weeks' work?

2 A potter makes cups and saucers in a factory. He is paid \$1.44 per batch of cups and \$1.20 per batch of saucers. What is his gross pay if he makes 9 batches of cups and 11 batches of saucers in one day?

3 Calculate the missing numbers from the simple interest table:

Principal ($)	Rate (%)	Time (years)	Interest ($)
300	6	4	a
250	b	3	60
480	5	c	96
650	d	8	390
e	3.75	4	187.50

4 A house was bought for $48000 12 years ago. It is now valued at $120000. What is the average annual increase in the value of the house?

5 An electrician bought five broken washing machines for $550. He repaired them and sold them for $143 each. What was his percentage profit?

9 Time

Times may be given in terms of the 12-hour clock. We tend to say, 'I get up at seven o'clock in the morning, play football at half past two in the afternoon, and go to bed before eleven o'clock'.

These times can be written as 7 a.m., 2.30 p.m. and 11 p.m.

In order to save confusion, most timetables are written using the 24-hour clock.

7 a.m. is written as 0700

2.30 p.m. is written as 1430

11 p.m. is written as 2300

To change p.m. times to 24-hour clock times, add 12 hours. To change 24-hour clock times later than 12.00 noon to 12-hour clock times, subtract 12 hours.

Exercise 9.1

1 These clocks show times in the morning. Write down the times using both the 12-hour and the 24-hour clock.

a

b

2 These clocks show times in the afternoon. Write down the times using both the 12-hour and the 24-hour clock.

a

b

3 Change these times into the 24-hour clock:

a 2.30 p.m. **b** 9 p.m. **c** 8.45 a.m. **d** 6 a.m.
e midday **f** 10.55 p.m. **g** 7.30 a.m. **h** 7.30 p.m.
i 1 a.m. **j** midnight

4 Change these times into the 24-hour clock:

a A quarter past seven in the morning
b Eight o'clock at night
c Ten past nine in the morning
d A quarter to nine in the morning
e A quarter to three in the afternoon
f Twenty to eight in the evening

Exercise 9.1 (cont)

5 These times are written for the 24-hour clock. Rewrite them using a.m. or p.m.

a 0720	**b** 0900	**c** 1430	**d** 1825
e 2340	**f** 0115	**g** 0005	**h** 1135
i 1750	**j** 2359	**k** 0410	**l** 0545

6 A journey to work takes a woman three quarters of an hour. If she catches the bus at the following times, when does she arrive?

a 0720 **b** 0755 **c** 0820 **d** 0845

7 The journey home for the same woman takes 55 minutes. If she catches the bus at these times, when does she arrive?

a 1725 **b** 1750 **c** 1805 **d** 1820

8 A boy cycles to school each day. His journey takes 70 minutes. When will he arrive if he leaves home at:

a 0715 **b** 0825 **c** 0840 **d** 0855?

9 The train into the city from a village takes 1 hour and 40 minutes. Copy and complete the train timetable.

Depart	0615		0925		1318		1854	
Arrive		0810		1200		1628		2105

10 The same journey by bus takes 2 hours and 5 minutes. Copy and complete the bus timetable.

Depart	0600		0855		1348		2125	
Arrive		0850		1114		1622		0010

11 A coach runs from Cambridge to the airports at Stansted, Gatwick and Heathrow. The time taken for the journey remains constant. Copy and complete the timetables for the outward and return journeys.

Cambridge	0400	0835	1250	1945	2110
Stansted	0515				
Gatwick	0650				
Heathrow	0735				

Heathrow	0625	0940	1435	1810	2215
Gatwick	0812				
Stansted	1003				
Cambridge	1100				

12 The flight time from London to Johannesburg is 11 hours and 20 minutes. Copy and complete the timetable.

	London	**Jo'burg**	**London**	**Jo'burg**
Sunday	0615		1420	
Monday		1845		0525
Tuesday	0720		1513	
Wednesday		1912		0730
Thursday	0610		1627	
Friday		1725		0815
Saturday	0955		1850	

13 The flight time from London to Kuala Lumpur is 13 hours and 45 minutes. Copy and complete the timetable.

	London	**Kuala Lumpur**	**London**	**Kuala Lumpur**	**London**	**Kuala Lumpur**
Sunday	0828		1400		1830	
Monday		2200		0315		0950
Tuesday	0915		1525		1755	
Wednesday		2135		0400		0822
Thursday	0700		1345		1840	
Friday		0010		0445		0738
Saturday	1012		1420		1908	

Average speed

Worked example

A train covers the 480 km journey from Paris to Lyon at an average speed of 100 km/h. If the train leaves Paris at 0835, when does it arrive in Lyon?

Time taken = $\frac{\text{distance}}{\text{speed}}$

Paris to Lyon: $\frac{480}{100}$ hours, that is, 4.8 hours.

4.8 hours is 4 hours and (0.8×60 minutes), that is, 4 hours and 48 minutes.

Departure 0835; arrival 0835 + 0448

Arrival time is 1323.

Exercise 9.2

1 Find the time in hours and minutes for the following journeys of the given distance at the average speed stated:

a 240 km at 60 km/h
b 340 km at 40 km/h
c 270 km at 80 km/h
d 100 km at 60 km/h
e 70 km at 30 km/h
f 560 km at 90 km/h
g 230 km at 100 km/h
h 70 km at 50 km/h
i 4500 km at 750 km/h
j 6000 km at 800 km/h

2 Grand Prix racing cars cover a 120 km race at the following average speeds. How long do the first five cars take to complete the race? Give your answer in minutes and seconds.

First 240 km/h
Second 220 km/h
Third 210 km/h
Fourth 205 km/h
Fifth 200 km/h

3 A train covers the 1500 km distance from Amsterdam to Barcelona at an average speed of 100 km/h. If the train leaves Amsterdam at 09 30, when does it arrive in Barcelona?

4 A plane takes off at 16 25 for the 3200 km journey from Moscow to Athens. If the plane flies at an average speed of 600 km/h, when will it land in Athens?

5 A plane leaves London for Boston, a distance of 5200 km, at 09 45. The plane travels at an average speed of 800 km/h. If Boston time is five hours behind British time, what is the time in Boston when the aircraft lands?

Student assessment 1

1 The clock shows a time in the afternoon. Write down the time using:

a the 12-hour clock
b the 24-hour clock.

2 Change these times into the 24-hour clock:

a 4.35 a.m.
b 6.30 p.m.
c a quarter to eight in the morning
d half past seven in the evening

3 These times are written using the 24-hour clock. Rewrite them using a.m. or p.m.

a 08 45
b 18 35
c 21 12
d 00 15

4 A journey to school takes a girl 25 minutes. What time does she arrive if she leaves home at the following times?

a 07 45
b 08 15
c 08 38

5 A bus service visits the towns on this timetable. Copy the timetable and fill in the missing times, given that the journey from:
Alphaville to Betatown takes 37 minutes
Betatown to Gammatown takes 18 minutes
Gammatown to Deltaville takes 42 minutes.

Alphaville	07 50		
Betatown		11 38	
Gammatown			16 48
Deltaville			

6 Find the times for the following journeys of given distance at the average speed stated. Give your answers in hours and minutes.

a 250 km at 50 km/h
b 375 km at 100 km/h
c 80 km at 60 km/h
d 200 km at 120 km/h
e 70 km at 30 km/h
f 300 km at 80 km/h

Student assessment 2

1 The clock shows a time in the morning. Write down the time using:

a the 12-hour clock
b the 24-hour clock.

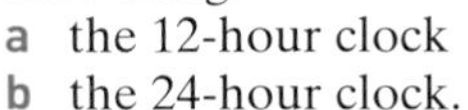

2 Change these times to the 24-hour clock:

a 5.20 a.m.
b 8.15 p.m.
c ten to nine in the morning
d half past eleven at night

3 These times are written using the 24-hour clock. Rewrite them using a.m. or p.m.

a 07 15
b 16 43
c 19 30
d 00 35

4 A journey to school takes a boy 22 minutes. When does he arrive if he leaves home at the following times?

a 07 48
b 08 17
c 08 38

5 A train stops at the following stations. Copy the timetable and fill in the times, given that the journey from:
Apple to Peach is 1 hr 38 minutes
Peach to Pear is 2 hrs 4 minutes
Pear to Plum is 1 hr 53 minutes.

Apple	10 14		
Peach		17 20	
Pear			23 15
Plum			

6 Find the time for the following journeys of given distance at the average speed stated. Give your answers in hours and minutes.

a 350 km at 70 km/h
b 425 km at 100 km/h
c 160 km at 60 km/h
d 450 km at 120 km/h
e 600 km at 160 km/h

10 Set notation and Venn diagrams

Sets

A **set** is a well-defined group of objects or symbols. The objects or symbols are called the **elements** of the set.

Worked examples

1 A particular set consists of the following elements:

{South Africa, Namibia, Egypt, Angola, ...}

a Describe the set.

The elements of the set are countries of Africa.

b Add another two elements to the set.

e.g. Zimbabwe, Ghana

c Is the set finite or infinite?

Finite. There is a finite number of countries in Africa.

2 Consider the set $A = \{x: x \text{ is a natural number}\}$

a Describe the set.

The elements of the set are the natural numbers.

b Write down two elements of the set.

e.g. 3 and 15

Exercise 10.1

1 In the following questions:

i describe the set in words

ii write down another two elements of the set.

a {Asia, Africa, Europe, ...}

b {2, 4, 6, 8, ...}

c {Sunday, Monday, Tuesday, ...}

d {January, March, July, ...}

e {1, 3, 6, 10, ...}

f {Mehmet, Michael, Mustapha, Matthew, ...}

g {11, 13, 17, 19, ...}

h {a, e, i, ...}

i {Earth, Mars, Venus, ...}

2 The number of elements in a set A is written as $n(A)$.
Give the value of $n(A)$ for the finite sets in questions 1a–i above.

Universal set

The **universal set** ($\mathscr{E}$) for any particular problem is the set which contains all the possible elements for that problem.

Worked examples

1 If $\mathscr{E}$ = {Integers from 1 to 10} state the numbers which form part of $\mathscr{E}$.

Therefore $\mathscr{E} = \{1, 2, 3, 4, 5, 6, 7, 8, 9, 10\}$

2 If $\mathscr{E}$ = {all 3D shapes} state three elements of $\mathscr{E}$.

e.g. sphere, cube, cylinder

Set notation and Venn diagrams

Venn diagrams are the principal way of showing sets diagrammatically. The method consists primarily of entering the elements of a set into a circle or circles. Some examples of the uses of Venn diagrams are shown.

$A = \{2, 4, 6, 8, 10\}$ can be represented as:

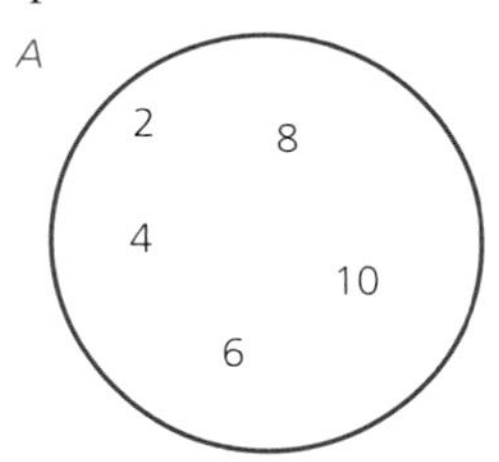

Elements which are in more than one set can also be represented using a Venn diagram.

$P = \{3, 6, 9, 12, 15, 18\}$ and $Q = \{2, 4, 6, 8, 10, 12\}$ can be represented as:

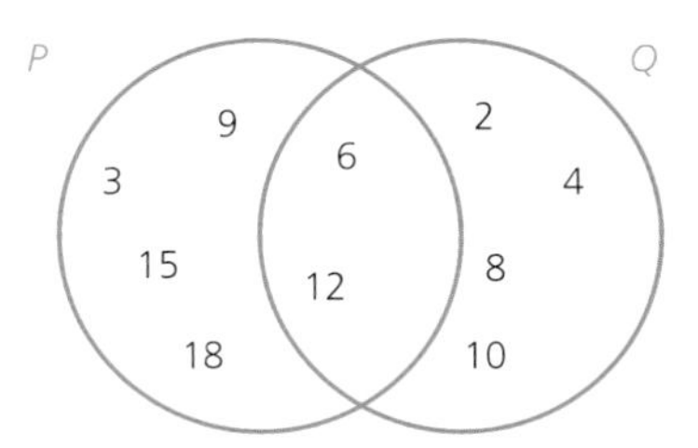

In the previous diagram, it can be seen that those elements which belong to both sets are placed in the region of overlap of the two circles.

When two sets P and Q overlap, as they do above, the notation $P \cap Q$ is used to denote the set of elements in the **intersection**, i.e. $P \cap Q = \{6, 12\}$.

Note that 6 belongs to the intersection of $P \cap Q$; 8 does not belong to the intersection of $P \cap Q$.

$J = \{10, 20, 30, 40, 50, 60, 70, 80, 90, 100\}$ and

$K = \{60, 70, 80\}$; as all the elements of K belong to J as well, this can be shown as:

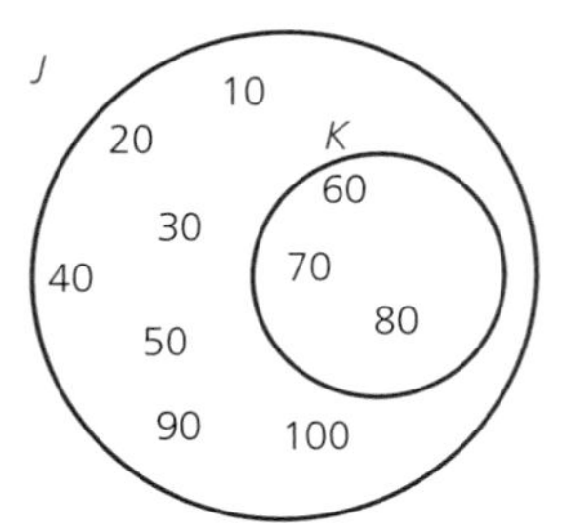

$X = \{1, 3, 6, 7, 14\}$ and $Y = \{3, 9, 13, 14, 18\}$ are represented as:

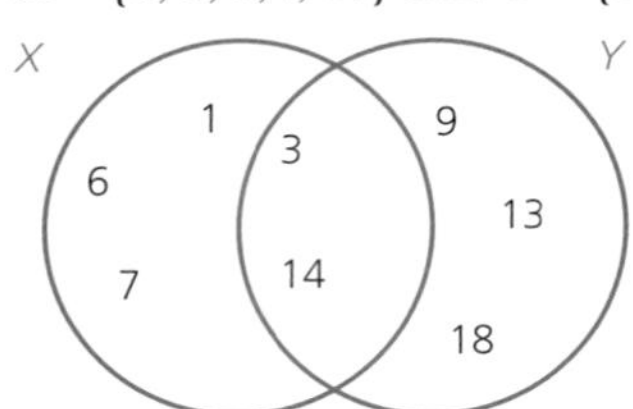

The **union** of two sets is everything which belongs to either or both sets and is represented by the symbol $\cup$.

Therefore, in the previous example, $X \cup Y = \{1, 3, 6, 7, 9, 13, 14, 18\}$.

Exercise 10.2

1 Using the Venn diagram indicate whether the following statements are true or false.

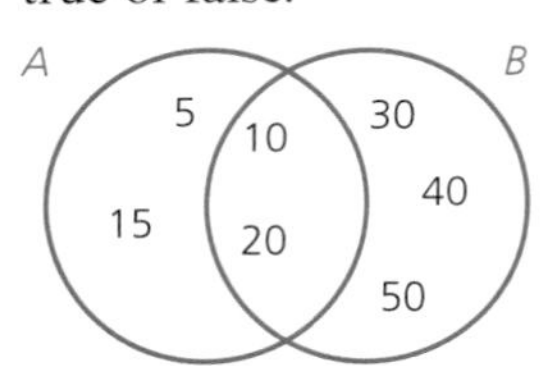

a 5 is an element of A
b 20 is an element of B
c 20 is not an element of A
d 50 is an element of A
e 50 is not an element of B
f $A \cap B = \{10, 20\}$

2 Complete the statement $A \cap B = \{...\}$ for each of the Venn diagrams below.

a

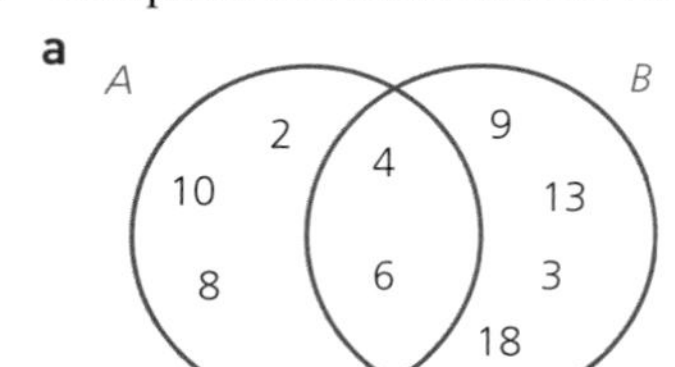

b

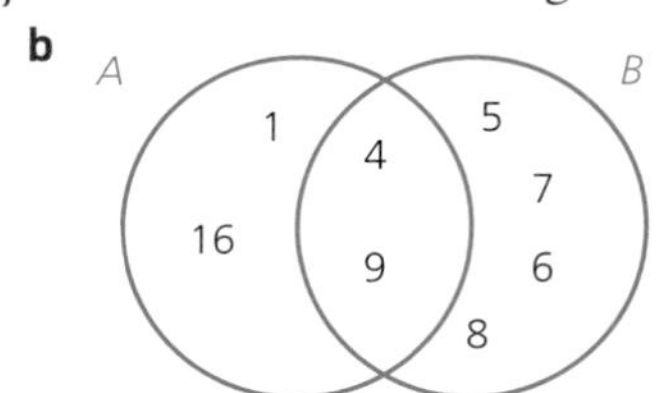

c

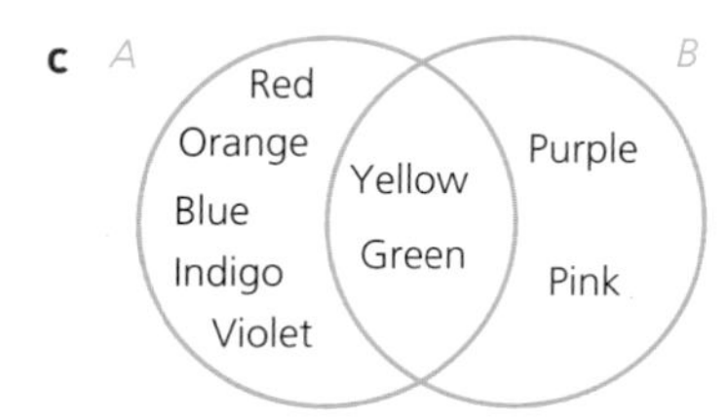

3 Complete the statement $A \cup B = \{...\}$ for each of the Venn diagrams in question 2.

4

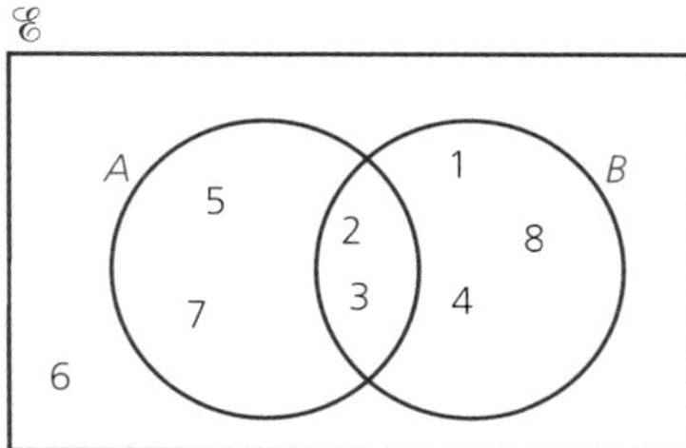

Copy and complete the following statements:

a $\mathscr{E} = \{...\}$
b $A \cap B = \{...\}$
c $A \cup B = \{...\}$

5

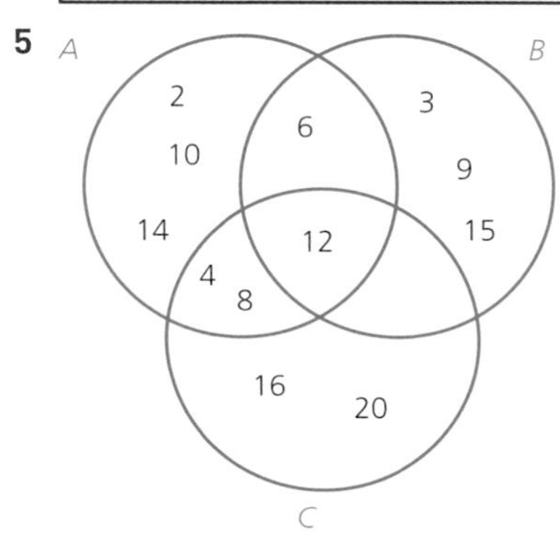

a Describe in words the elements of:

i set A **ii** set B **iii** set C

b Copy and complete the following statements:

i $A \cap B = \{...\}$ **ii** $A \cap C = \{...\}$ **iii** $B \cap C = \{...\}$
iv $A \cap B \cap C = \{...\}$ **v** $A \cup B = \{...\}$ **vi** $C \cup B = \{...\}$

6

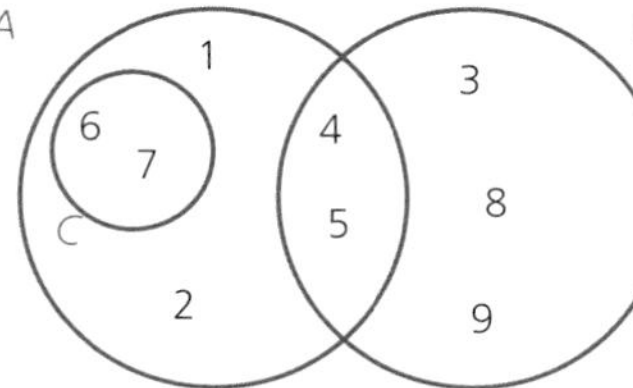

Copy and complete the following statements:

a $A = \{...\}$
b $B = \{...\}$
c $A \cap B = \{...\}$
d $A \cup B = \{...\}$

7

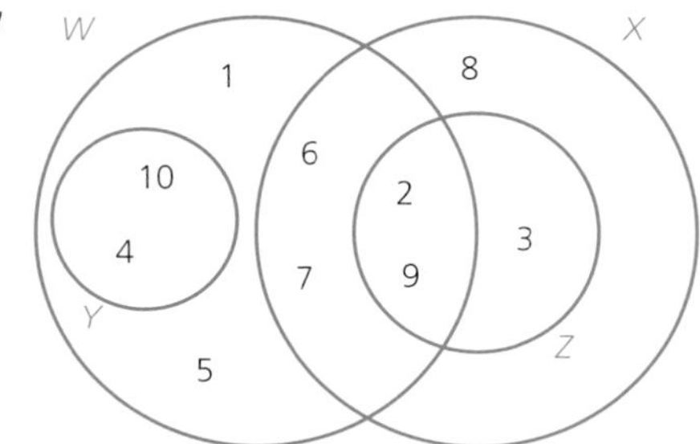

a Copy and complete the following statements:

i $W = \{...\}$ **ii** $X = \{...\}$
iii $W \cap Z = \{...\}$ **iv** $W \cap X = \{...\}$
v $Y \cap Z = \{...\}$

b The elements of which whole set also belong to the set X?

Exercise 10.3

1 $A = \{\text{Egypt, Libya, Morocco, Chad}\}$
$B = \{\text{Iran, Iraq, Turkey, Egypt}\}$
  a Draw a Venn diagram to illustrate the above information.
  b Copy and complete the following statements:
    i $A \cap B = \{...\}$ ii $A \cup B = \{...\}$

2 $P = \{2, 3, 5, 7, 11, 13, 17\}$
$Q = \{11, 13, 15, 17, 19\}$
  a Draw a Venn diagram to illustrate the above information.
  b Copy and complete the following statements:
    i $P \cap Q = \{...\}$ ii $P \cup Q = \{...\}$

3 $B = \{2, 4, 6, 8, 10\}$
$A \cup B = \{1, 2, 3, 4, 6, 8, 10\}$
$A \cap B = \{2, 4\}$
Represent the above information on a Venn diagram.

4 $X = \{a, c, d, e, f, g, l\}$
$Y = \{b, c, d, e, h, i, k, l, m\}$
$Z = \{c, f, i, j, m\}$
Represent the above information on a Venn diagram.

5 $P = \{1, 4, 7, 9, 11, 15\}$
$Q = \{5, 10, 15\}$
$R = \{1, 4, 9\}$
Represent the above information on a Venn diagram.

Problem solving involving sets

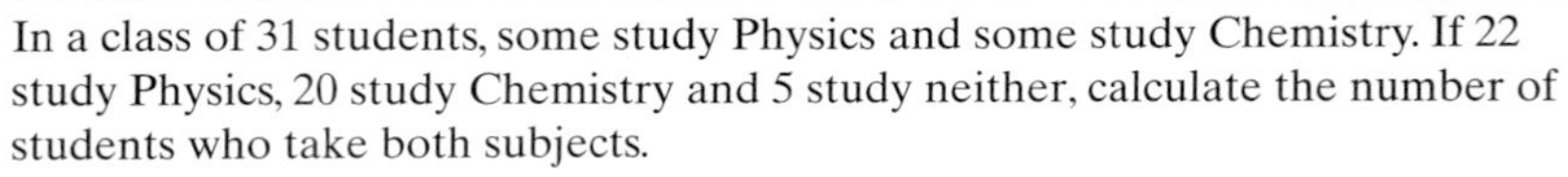

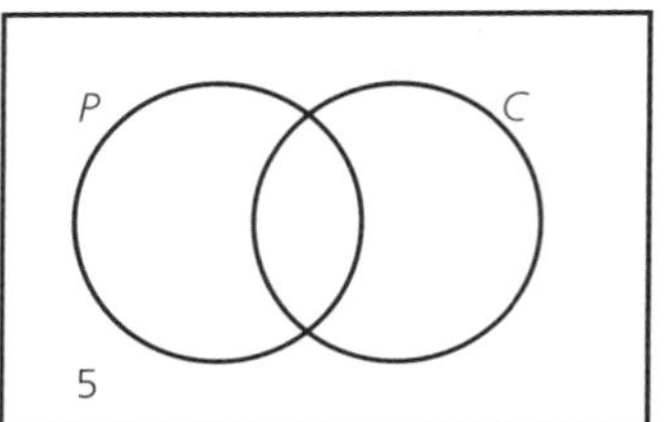

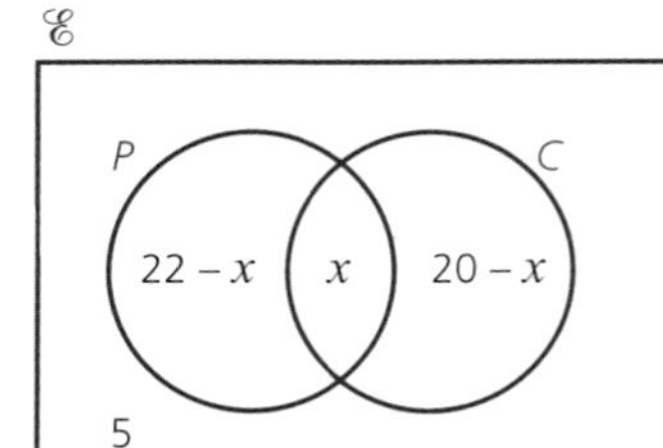

In a class of 31 students, some study Physics and some study Chemistry. If 22 study Physics, 20 study Chemistry and 5 study neither, calculate the number of students who take both subjects.

The information can be entered in a Venn diagram in stages.
The students taking neither Physics nor Chemistry can be put in first (as shown top left). This leaves 26 students to be entered into the set circles.

If x students take both subjects then

$n(P) = 22 - x + x$

$n(C) = 20 - x + x$

$P \cup C = 31 - 5 = 26$

Therefore $22 - x + x + 20 - x = 26$

$42 - x = 26$

$x = 16$

Substituting the value of x into the Venn diagram gives:

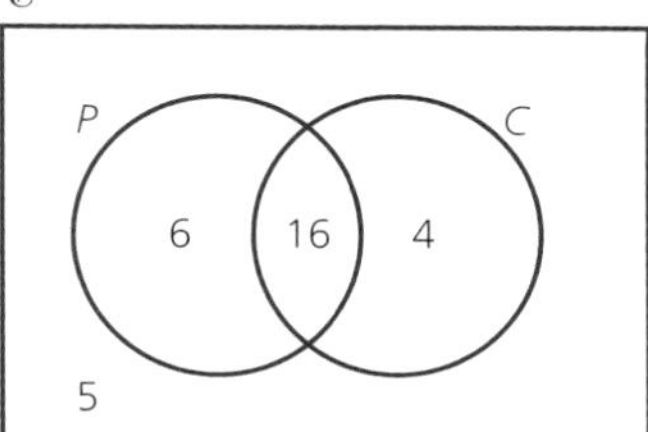

Therefore the number of students taking both Physics and Chemistry is 16.

Exercise 10.4

1 In a class of 35 students, 19 take Spanish, 18 take French and 3 take neither. Calculate how many take:
 a both French and Spanish
 b just Spanish
 c just French.

2 In a year group of 108 students, 60 liked football, 53 liked tennis and 10 liked neither. Calculate the number of students who liked football but not tennis.

3 In a year group of 113 students, 60 liked hockey, 45 liked rugby and 18 liked neither. Calculate the number of students who:
 a liked both hockey and rugby
 b liked only hockey.

4 One year 37 students sat an examination in Physics, 48 sat Chemistry and 45 sat Biology. 15 students sat Physics and Chemistry, 13 sat Chemistry and Biology, 7 sat Physics and Biology and 5 students sat all three.
 a Draw a Venn diagram to represent this information.
 b Calculate $n(P \cup C \cup B)$.

Student assessment 1

1 Describe the following sets in words:
 a $\{2, 4, 6, 8\}$
 b $\{2, 4, 6, 8, \ ...\}$
 c $\{1, 4, 9, 16, 25, \ ...\}$
 d {Arctic, Atlantic, Indian, Pacific}

2 Calculate the value of $n(A)$ for each of the sets shown below:
 a A = {days of the week}
 b A = {prime numbers between 50 and 60}
 c $A = \{x: x \text{ is an integer and } 5 \leqslant x \leqslant 10\}$
 d A = {days in a leap year}

3 Copy out the Venn diagram twice.
 a On one copy shade and label the region which represents $A \cap B$.
 b On the other copy shade and label the region which represents $A \cup B$.

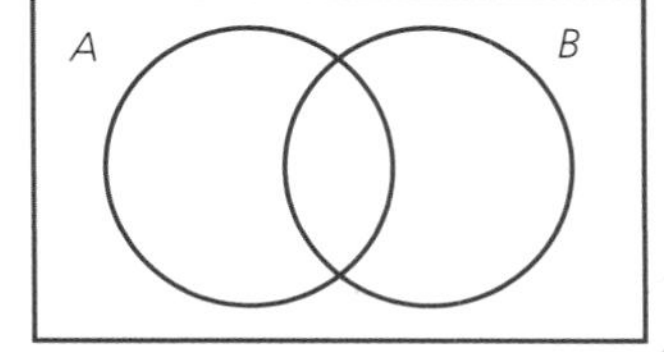

Student assessment 2

1 J = {London, Paris, Rome, Washington, Canberra, Ankara, Cairo}
K = {Cairo, Nairobi, Pretoria, Ankara}
   a Draw a Venn diagram to represent the above information.
   b Copy and complete the statement $J \cap K = \{...\}$.

2 $\mathscr{E}$ = {natural numbers}, M = {even numbers} and N = {multiples of 5}.
   a Draw a Venn diagram and place the numbers 1, 2, 3, 4, 5, 6, 7, 8, 9, 10 in the appropriate places in it.
   b If $X = M \cap N$, describe set X in words.

3 A group of 40 people were asked whether they like cricket (C) and football (F). The number liking both cricket and football was three times the number liking only cricket. Adding three to the number liking only cricket and doubling the answer equals the number of people liking only football. Four said they did not like sport at all.
   a Draw a Venn diagram to represent this information.
   b Calculate $n(C \cap F)$.

4 The Venn diagram below shows the number of elements in three sets P, Q and R.

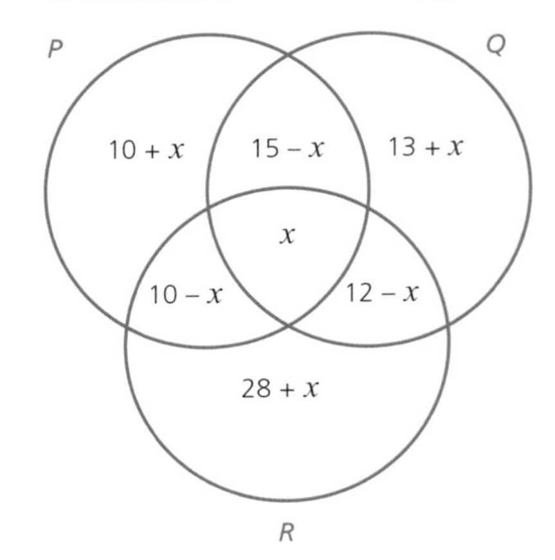

If $n(P \cup Q \cup R) = 93$ calculate:
   a x
   b $n(P)$
   c $n(Q)$
   d $n(R)$
   e $n(P \cap Q)$
   f $n(Q \cap R)$
   g $n(P \cap R)$
   h $n(R \cup Q)$

TOPIC 1

Mathematical investigations and ICT

Investigations are an important part of mathematical learning. All mathematical discoveries stem from an idea that a mathematician has and then investigates.

Sometimes when faced with a mathematical investigation, it can seem difficult to know how to start. The structure and example below may help you.

1. Read the question carefully and start with simple cases.
2. Draw simple diagrams to help.
3. Put the results from simple cases in a table.
4. Look for a pattern in your results.
5. Try to find a general rule in words.
6. Express your rule algebraically.
7. Test the rule for a new example.
8. Check that the original question has been answered.

Worked example

A mystic rose is created by placing a number of points evenly spaced on the circumference of a circle. Straight lines are then drawn from each point to every other point. The diagram shows a mystic rose with 20 points.

a How many straight lines are there?

b How many straight lines would there be on a mystic rose with 100 points?

To answer these questions, you are not expected to draw either of the shapes and count the number of lines.

1/2 Try simple cases:

By drawing some simple cases and counting the lines, some results can be found:

Mystic rose with 2 points

Number of lines = 1

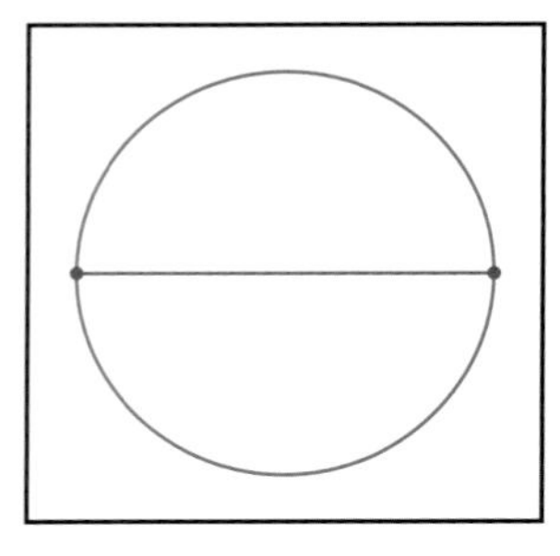

Mystic rose with 3 points

Number of lines = 3

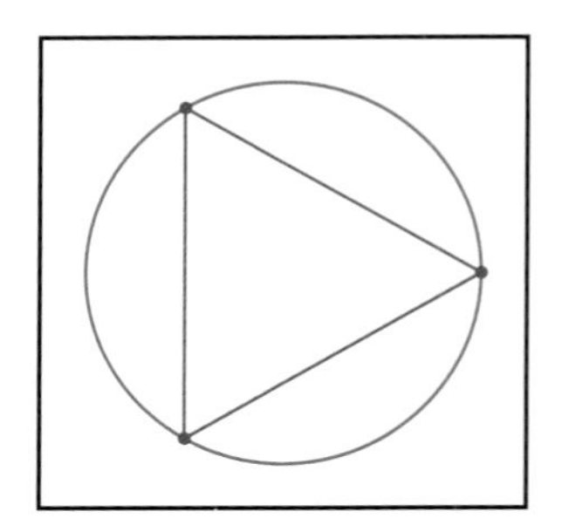

Mystic rose with 4 points

Number of lines = 6

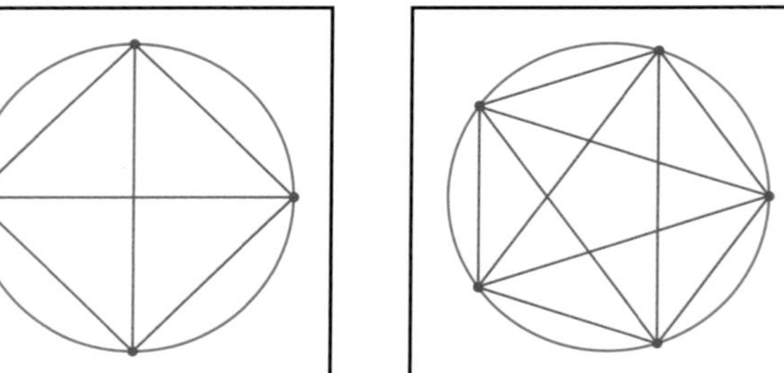

Mystic rose with 5 points

Number of lines = 10

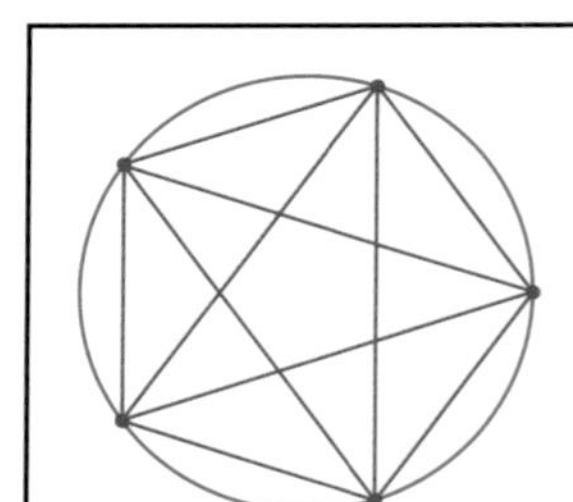

3 Enter the results in an ordered table:

Number of points	2	3	4	5
Number of lines	1	3	6	10

4/5 Look for a pattern in the results:

There are two patterns.
The first pattern shows how the values change.

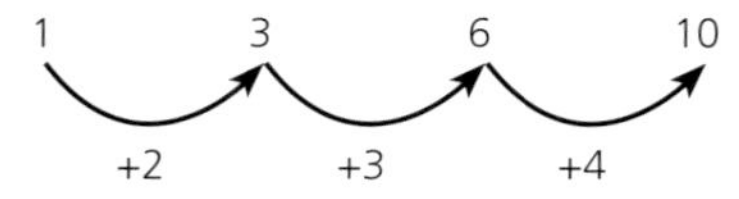

It can be seen that the difference between successive terms is increasing by one each time.
The problem with this pattern is that to find the 20th or 100th term, it would be necessary to continue this pattern and find all the terms leading up to the 20th or 100th term.

The second pattern is the relationship between the number of points and the number of lines.

Number of points	2	3	4	5
Number of lines	1	3	6	10

It is important to find a relationship that works for all values; for example, subtracting one from the number of points gives the number of lines in the first example only, so is not useful. However, halving the number of points and multiplying this by one less than the number of points works each time, i.e. Number of lines = (half the number of points) × (one less than the number of points).

6 Express the rule algebraically:

The rule expressed in words above can be written more elegantly using algebra. Let the number of lines be l and the number of points be p.

$$l = \frac{1}{2}p(p-1)$$

Note: Any letters can be used to represent the number of lines and the number of points, not just l and p.

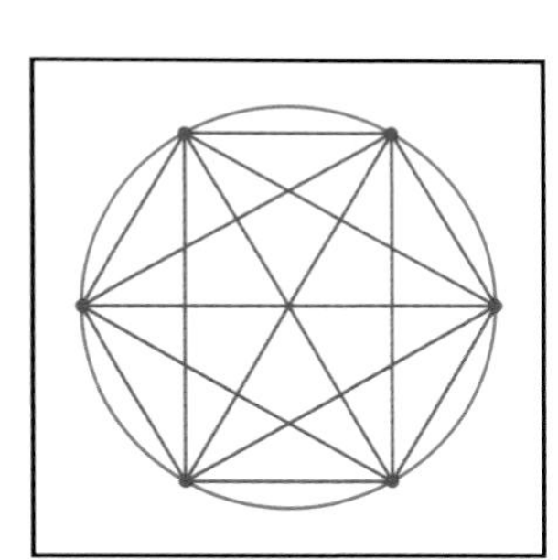

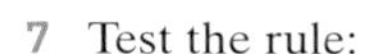

7 Test the rule:

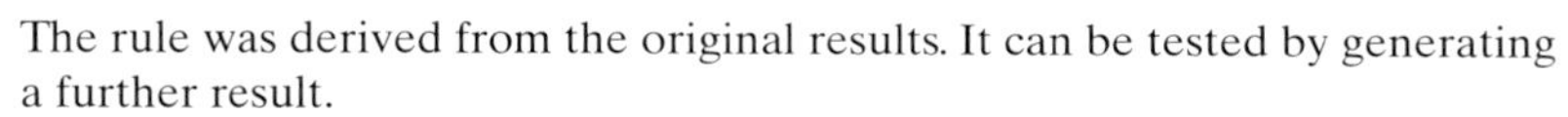

The rule was derived from the original results. It can be tested by generating a further result.
If the number of points $p = 6$, then the number of lines l is:

$$l = \frac{1}{2} \times 6(6-1)$$
$$= 3 \times 5$$
$$= 15$$

From the diagram to the left, the number of lines can also be counted as 15.

8 Check that the original questions have been answered:

Using the formula, the number of lines in a mystic rose with 20 points is:

$$l = \frac{1}{2} \times 20(20-1)$$
$$= 10 \times 19$$
$$= 190$$

The number of lines in a mystic rose with 100 points is:

$$l = \frac{1}{2} \times 100(100-1)$$
$$= 50 \times 99$$
$$= 4950$$

Primes and squares

13, 41 and 73 are prime numbers.

Two different square numbers can be added together to make these prime numbers, e.g. $3^2 + 8^2 = 73$.

1 Find the two square numbers that can be added to make 13 and 41.
2 List the prime numbers less than 100.
3 Which of the prime numbers less than 100 can be shown to be the sum of two different square numbers?
4 Is there a rule to the numbers in Q.3?
5 Your rule is a predictive rule not a formula. Discuss the difference.

Football leagues

There are 18 teams in a football league.

1 If each team plays the other teams twice, once at home and once away, then how many matches are played in a season?
2 If there are t teams in a league, how many matches are played in a season?

ICT activity 1

The step patterns follow a rule.

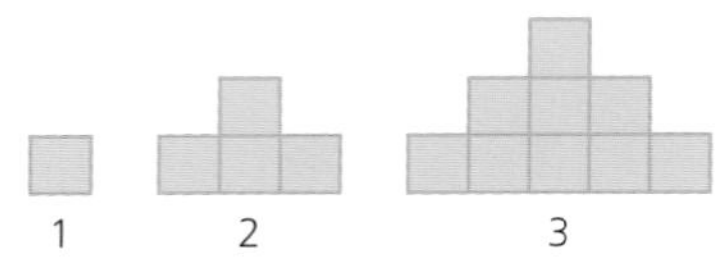

1 On squared paper, draw the next two patterns in this sequence.
2 Count the number of squares used in each of the first five patterns. Enter the results into a table on a spreadsheet, similar to the one shown.

	A	B
1	Pattern	Number of squares
2	1	
3	2	
4	3	
5	4	
6	5	
7	10	
8	20	
9	50	

3 The number of squares needed for each pattern follows a rule. Describe the rule.
4 By writing a formula in cell B7 and copying it down to B9, use the spreadsheet to generate the results for the 10th, 20th and 50th patterns.

5 Repeat Q.1–4 for the following patterns:

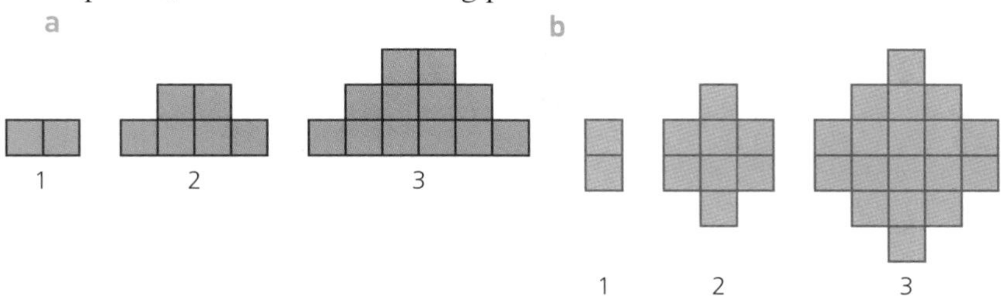

ICT activity 2

In this activity, you will be using both the internet and a spreadsheet in order to produce currency conversions.

1 Log on to the internet and search for a website that shows the exchange rates between different currencies.
2 Compare your own currency with another currency of your choice. Write down the exchange rate, e.g. $1 = €1.29.
3 Use a spreadsheet to construct a currency converter. Like this:

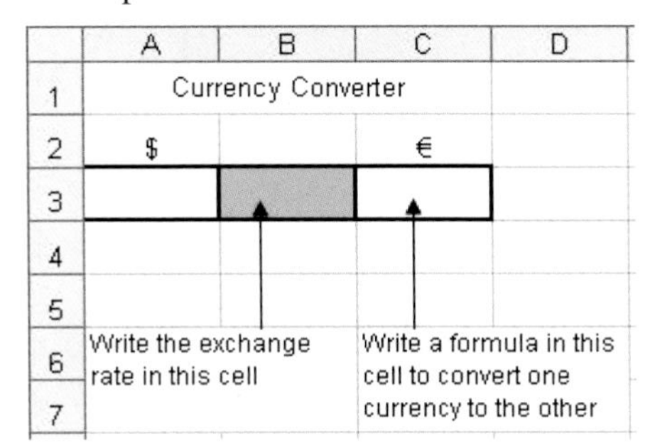

4 By entering different amounts of your own currency, use the currency converter to calculate the correct conversion. Record your answers in a table.
5 Repeat Q.1–4 for five different currencies of your choice.

TOPIC 2

Algebra and graphs

Contents

Course

C2.1

Use letters to express generalised numbers and express basic arithmetic processes algebraically.

Substitute numbers for words and letters in formulae.

Rearrange simple formulae.

Construct simple expressions and set up simple equations.

C2.2

Manipulate directed numbers.

Use brackets and extract common factors.

Expand products of algebraic expressions.

C2.3

Extended curriculum only.

C2.4

Use and interpret positive, negative and zero indices.

Use the rules of indices.

C2.5

Derive and solve simple linear equations in one unknown.

Derive and solve simultaneous linear equations in two unknowns.

C2.6

Extended curriculum only.

C2.7

Continue a given number sequence.

Recognise patterns in sequences including the term to term rule and relationships between different sequences.

Find and use the nth term of sequences.

C2.8

Extended curriculum only.

C2.9

Extended curriculum only.

C2.10

Interpret and use graphs in practical situations including travel graphs and conversion graphs.

Draw graphs from given data.

C2.11

Construct tables of values for functions of the form $ax + b$, $\pm x^2 + ax + b$, $\frac{a}{x}$ $(x \neq 0)$, where a and b are integer constants.

Draw and interpret these graphs.

Solve linear and quadratic equations approximately, including finding and interpreting roots by graphical methods.

Recognise, sketch and interpret graphs of functions.

C2.12

Extended curriculum only.

C2.13

Extended curriculum only.

The development of algebra

The roots of algebra can be traced to the ancient Babylonians, who used formulae for solving problems. However, the word *algebra* comes from the Arabic language. Muhammad ibn Musa al-Khwarizmi (AD790–850) wrote *Kitab al-Jabr* (*The Compendious Book on Calculation by Completion and Balancing*), which established algebra as a mathematical subject. He is known as the father of algebra.

Persian mathematician Omar Khayyam (1048–1131), who studied in Bukhara (now in Uzbekistan), discovered algebraic geometry and found the general solution of a cubic equation.

In 1545, Italian mathematician Girolamo Cardano published *Ars Magna* (*The Great Art*), a 40-chapter book in which he gave, for the first time, a method for solving a quartic equation.

al-Khwarizmi (790–850)

11 Algebraic representation and manipulation

Expanding brackets

When removing brackets, every term inside the bracket must be multiplied by whatever is outside the bracket.

Worked example

Expand:

a $3(x+4)$
$3x + 12$

b $5x(2y+3)$
$10xy + 15x$

c $2a(3a+2b-3c)$
$6a^2 + 4ab - 6ac$

d $-4p(2p-q+r^2)$
$-8p^2 + 4pq - 4pr^2$

e $-2x^2\left(x+3y-\frac{1}{x}\right)$
$-2x^3 - 6x^2y + 2x$

f $\frac{-2}{x}\left(-x+4y+\frac{1}{x}\right)$
$2 - \frac{8y}{x} - \frac{2}{x^2}$

Exercise 11.1 Expand:

1 **a** $2(a+3)$ **b** $4(b+7)$
c $5(2c+8)$ **d** $7(3d+9)$
e $9(8e-7)$ **f** $6(4f-3)$

2 **a** $3a(a+2b)$ **b** $4b(2a+3b)$
c $2c(a+b+c)$ **d** $3d(2b+3c+4d)$
e $e(3c-3d-e)$ **f** $f(3d-e-2f)$

3 **a** $2(2a^2+3b^2)$ **b** $4(3a^2+4b^2)$
c $-3(2c+3d)$ **d** $-(2c+3d)$
e $-4(c^2-2d^2+3e^2)$ **f** $-5(2e-3f^2)$

4 **a** $2a(a+b)$ **b** $3b(a-b)$
c $4c(b^2-c^2)$ **d** $3d^2(a^2-2b^2+c^2)$
e $-3e^2(4d-e)$ **f** $-2f(2d-3e^2-2f)$

Exercise 11.2

Expand:

1 a $4(x-3)$
b $5(2p-4)$
c $-6(7x-4y)$
d $3(2a-3b-4c)$
e $-7(2m-3n)$
f $-2(8x-3y)$

2 a $3x(x-3y)$
b $a(a+b+c)$
c $4m(2m-n)$
d $-5a(3a-4b)$
e $-4x(-x+y)$
f $-8p(-3p+q)$

3 a $-(2x^2-3y^2)$
b $-(-a+b)$
c $-(-7p+2q)$
d $\frac{1}{2}(6x-8y+4z)$
e $\frac{3}{4}(4x-2y)$
f $\frac{1}{5}x(10x-15y)$

4 a $3r(4r^2-5s+2t)$
b $a^2(a+b+c)$
c $3a^2(2a-3b)$
d $pq(p+q-pq)$
e $m^2(m-n+nm)$
f $a^3(a^3+a^2b)$

Exercise 11.3

Expand and simplify:

1 a $2a+2(3a+2)$
b $4(3b-2)-5b$
c $6(2c-1)-7c$
d $-4(d+2)+5d$
e $-3e+(e-1)$
f $5f-(2f-7)$

2 a $2(a+1)+3(b+2)$
b $4(a+5)-3(2b+6)$
c $3(c-1)+2(c-2)$
d $4(d-1)-3(d-2)$
e $-2(e-3)-(e-1)$
f $2(3f-3)+4(1-2f)$

3 a $2a(a+3)+2b(b-1)$
b $3a(a-4)-2b(b-3)$
c $2a(a+b+c)-2b(a+b-c)$
d $a^2(c^2+d^2)-c^2(a^2+d^2)$
e $a(b+c)-b(a-c)$
f $a(2d+3e)-2e(a-c)$

Exercise 11.4

Expand and simplify:

1 a $3a-2(2a+4)$
b $8x-4(x+5)$
c $3(p-4)-4$
d $7(3m-2n)+8n$
e $6x-3(2x-1)$
f $5p-3p(p+2)$

2 a $7m(m+4)+m^2+2$
b $3(x-4)+2(4-x)$
c $6(p+3)-4(p-1)$
d $5(m-8)-4(m-7)$
e $3a(a+2)-2(a^2-1)$
f $7a(b-2c)-c(2a-3)$

3 a $\frac{1}{2}(6x+4)+\frac{1}{3}(3x+6)$
b $\frac{1}{4}(2x+6y)+\frac{3}{4}(6x-4y)$
c $\frac{1}{8}(6x-12y)+\frac{1}{2}(3x-2y)$
d $\frac{1}{5}(15x+10y)+\frac{3}{10}(5x-5y)$
e $\frac{2}{3}(6x-9y)+\frac{1}{3}(9x+6y)$
f $\frac{x}{7}(14x-21y)-\frac{x}{2}(4x-6y)$

Expanding a pair of brackets

When multiplying together expressions in brackets, it is necessary to multiply all the terms in one bracket by all the terms in the other bracket.

Worked example

Expand:

a $(x+3)(x+5)$

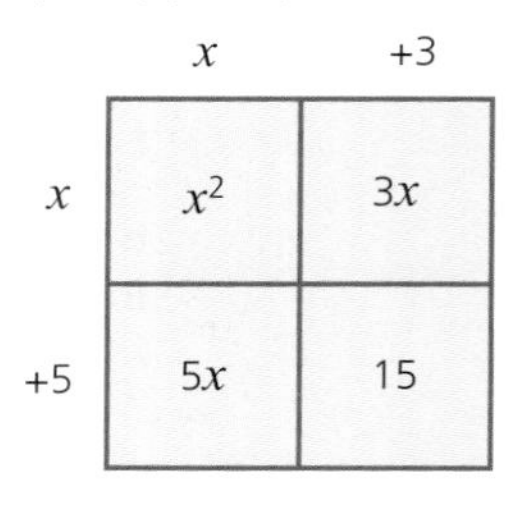

$= x^2 + 3x + 5x + 15$

$= x^2 + 8x + 15$

b $(2x-3)(3x-6)$.

	$2x$	-3
$3x$	$6x^2$	$-9x$
-6	$-12x$	18

$= 6x^2 - 9x - 12x + 18$

$= 6x^2 - 21x + 18$

Exercise 11.5

Expand and simplify:

1 a $(x+2)(x+3)$ b $(x+3)(x+4)$ c $(x+5)(x+2)$
d $(x+6)(x+1)$ e $(x-2)(x+3)$ f $(x+8)(x-3)$

2 a $(x-4)(x+6)$ b $(x-7)(x+4)$ c $(x+5)(x-7)$
d $(x+3)(x-5)$ e $(x+1)(x-3)$ f $(x-7)(x+9)$

3 a $(x-2)(x-3)$ b $(x-5)(x-2)$ c $(x-4)(x-8)$
d $(x+3)(x+3)$ e $(x-3)(x-3)$ f $(x-7)(x-5)$

4 a $(x+3)(x-3)$ b $(x+7)(x-7)$ c $(x-8)(x+8)$
d $(x+y)(x-y)$ e $(a+b)(a-b)$ f $(p-q)(p+q)$

5 a $(2x+1)(x+3)$ b $(3x-2)(2x+5)$ c $(4-3x)(x+2)$
d $(7-5y)^2$ e $(3+2x)(3-2x)$ f $(3+4x)(3-4x)$

Factorising

When factorising, the largest possible factor is removed from each of the terms and placed outside the brackets.

Worked example

Factorise the following expressions:

a $10x + 15$
$5(2x+3)$

b $8p - 6q + 10r$
$2(4p - 3q + 5r)$

c $-2q - 6p + 12$
$2(-q - 3p + 6)$

d $2a^2 + 3ab - 5ac$
$a(2a + 3b - 5c)$

e $6ax - 12ay - 18a^2$
$6a(x - 2y - 3a)$

f $3b + 9ba - 6bd$
$3b(1 + 3a - 2d)$

Exercise 11.6

Factorise:

1 a $4x - 6$ b $18 - 12p$
c $6y - 3$ d $4a + 6b$
e $3p - 3q$ f $8m + 12n + 16r$

2 a $3ab + 4ac - 5ad$ b $8pq + 6pr - 4ps$
c $a^2 - ab$ d $4x^2 - 6xy$
e $abc + abd + fab$ f $3m^2 + 9m$

3 a $3pqr - 9pqs$ b $5m^2 - 10mn$
c $8x^2y - 4xy^2$ d $2a^2b^2 - 3b^2c^2$
e $12p - 36$ f $42x - 54$

4 a $18 + 12y$ b $14a - 21b$
c $11x + 11xy$ d $4s - 16t + 20r$
e $5pq - 10qr + 15qs$ f $4xy + 8y^2$

5 a $m^2 + mn$ b $3p^2 - 6pq$
c $pqr + qrs$ d $ab + a^2b + ab^2$
e $3p^3 - 4p^4$ f $7b^3c + b^2c^2$

6 a $m^3 - m^2n + mn^2$ b $4r^3 - 6r^2 + 8r^2s$
c $56x^2y - 28xy^2$ d $72m^2n + 36mn^2 - 18m^2n^2$

7 a $3a^2 - 2ab + 4ac$ b $2ab - 3b^2 + 4bc$
c $2a^2c - 4b^2c + 6bc^2$ d $39cd^2 + 52c^2d$

8 a $12ac - 8ac^2 + 4a^2c$ b $34a^2b - 51ab^2$
c $33ac^2 + 121c^3 - 11b^2c^2$ d $38c^3d^2 - 57c^2d^3 + 95c^2d^2$

9 a $15\frac{a}{c} - 25\frac{b}{c} + 10\frac{d}{c}$ b $46\frac{a}{c^2} - 23\frac{b}{c^2}$
c $\frac{1}{2a} - \frac{1}{4a}$ d $\frac{3}{5d} - \frac{1}{10d} + \frac{4}{15d}$

10 a $\frac{5}{a^2} - \frac{3}{a}$ b $\frac{6}{b^2} - \frac{3}{b}$
c $\frac{2}{3a} - \frac{3}{3a^2}$ d $\frac{3}{5d^2} - \frac{4}{5d}$

Substitution

Worked example

Evaluate these expressions if $a = 3$, $b = 4$, $c = -5$:

a $2a + 3b - c$

$2 \times 3 + 3 \times 4 - (-5)$

$= 6 + 12 + 5$

$= 23$

b $3a - 4b + 2c$

$3 \times 3 - 4 \times 4 + 2 \times (-5)$

$= 9 - 16 - 10$

$= -17$

c $-2a + 2b - 3c$

$-2 \times 3 + 2 \times 4 - 3 \times (-5)$

$= -6 + 8 + 15$

$= 17$

d $a^2 + b^2 + c^2$

$3^2 + 4^2 + (-5)^2$

$= 9 + 16 + 25$

$= 50$

e $3a(2b - 3c)$

$3 \times 3 \times (2 \times 4 - 3 \times (-5))$

$= 9 \times (8 + 15)$

$= 9 \times 23$

$= 207$

f $-2c(-a + 2b)$

$-2 \times (-5) \times (-3 + 2 \times 4)$

$= 10 \times (-3 + 8)$

$= 10 \times 5$

$= 50$

Exercise 11.7

Evaluate the following expressions if $a = 2, b = 3$ and $c = 5$:

1 **a** $3a + 2b$ **b** $4a - 3b$
c $a - b - c$ **d** $3a - 2b + c$

2 **a** $-b(a + b)$ **b** $-2c(a - b)$
c $-3a(a - 3c)$ **d** $-4b(b - c)$

3 **a** $a^2 + b^2$ **b** $b^2 + c^2$
c $2a^2 - 3b^2$ **d** $3c^2 - 2b^2$

4 **a** $-a^2$ **b** $(-a)^2$
c $-b^3$ **d** $(-b)^3$

5 **a** $-c^3$ **b** $(-c)^3$
c $(-ac)^2$ **d** $-(ac)^2$

Exercise 11.8

Evaluate the following expressions if $p = 4, q = -2, r = 3$ and $s = -5$:

1 **a** $2p + 4q$ **b** $5r - 3s$
c $3q - 4s$ **d** $6p - 8q + 4s$
e $3r - 3p + 5q$ **f** $-p - q + r + s$

2 **a** $2p - 3q - 4r + s$ **b** $3s - 4p + r + q$
c $p^2 + q^2$ **d** $r^2 - s^2$
e $p(q - r + s)$ **f** $r(2p - 3q)$

3 **a** $2s(3p - 2q)$ **b** $pq + rs$
c $2pr - 3rq$ **d** $q^3 - r^2$
e $s^3 - p^3$ **f** $r^4 - q^5$

4 **a** $-2pqr$ **b** $-2p(q + r)$
c $-2rq + r$ **d** $(p + q)(r - s)$
e $(p + s)(r - q)$ **f** $(r + q)(p - s)$

5 **a** $(2p + 3q)(p - q)$ **b** $(q + r)(q - r)$
c $q^2 - r^2$ **d** $p^2 - r^2$
e $(p + r)(p - r)$ **f** $(-s + p)q^2$

Rearrangement of formulae

In the formula $a = 2b + c$, 'a' is the **subject**. In order to make either b or c the subject, the formula has to be rearranged.

Worked example

Rearrange the following formulae to make the red letter the subject:

a $a = 2b + c$

$a - 2b = c$

$c = a - 2b$

b $2r + p = q$

$p = q - 2r$

c $ab = cd$

$\frac{ab}{d} = c$

$c = \frac{ab}{d}$

d $\frac{a}{b} = \frac{c}{d}$

$ad = cb$

$d = \frac{cb}{a}$

Exercise 11.9

In the following questions, make the letter in red the subject of the formula:

1 a $a + b = c$ b $b + 2c = d$
c $2b + c = 4a$ d $3d + b = 2a$

2 a $ab = c$ b $ac = bd$
c $ab = c + 3$ d $ac = b - 4$

3 a $m + n = r$ b $m + n = p$
c $2m + n = 3p$ d $3x = 2p + q$
e $ab = cd$ f $ab = cd$

4 a $3xy = 4m$ b $7pq = 5r$
c $3x = c$ d $3x + 7 = y$
e $5y - 9 = 3r$ f $5y - 9 = 3x$

5 a $6b = 2a - 5$ b $6b = 2a - 5$
c $3x - 7y = 4z$ d $3x - 7y = 4z$
e $3x - 7y = 4z$ f $2pr - q = 8$

6 a $\frac{p}{4} = r$ b $\frac{4}{p} = 3r$
c $\frac{1}{5}n = 2p$ d $\frac{1}{5}n = 2p$
e $p(q + r) = 2t$ f $p(q + r) = 2t$

7 a $3m - n = rt(p + q)$ b $3m - n = rt(p + q)$
c $3m - n = rt(p + q)$ d $3m - n = rt(p + q)$
e $3m - n = rt(p + q)$ f $3m - n = rt(p + q)$

8 a $\frac{ab}{c} = de$ b $\frac{ab}{c} = de$
c $\frac{ab}{c} = de$ d $\frac{a + b}{c} = d$
e $\frac{a}{c} + b = d$ f $\frac{a}{c} + b = d$

Student assessment 1

1 Expand:
 a $4(a+2)$ b $5(2b-3)$ c $2c(c+2d)$
 d $3d(2c-4d)$ e $-5(3e-f)$ f $-(-f+2g)$

2 Expand and simplify where possible:
 a $2a+5(a+2)$ b $3(2b-3)-b$ c $-4c-(4-2c)$
 d $3(d+2)-2(d+4)$ e $-e(2e+3)+3(2+e^2)$ f $f(d-e-f)-e(e+f)$
 g $(x-7)(x+8)$ h $(3x+1)(x+2)$ i $(4x-3)(-x+2)$

3 Factorise:
 a $7a+14$ b $26b^2+39b$
 c $3cf-6df+9gf$ d $5d^2-10d^3$

4 If $a=2$, $b=3$ and $c=5$, evaluate the following:
 a $a-b-c$ b $2b-c$
 c $a^2-b^2+c^2$ d $(a+c)^2$

5 Rearrange the formulae to make the green letter the subject:
 a $a-b=c$ b $2c=b-3d$ c $ad=bc$
 d $e=5d-3c$ e $4a=e(f+g)$ f $4a=e(f+g)$

Student assessment 2

1 Expand and simplify where possible:
 a $3(2x-3y+5z)$ b $4p(2m-7)$
 c $-4m(2mn-n^2)$ d $4p^2(5pq-2q^2-2p)$
 e $4x-2(3x+1)$ f $4x(3x-2)+2(5x^2-3x)$
 g $\frac{1}{5}(15x-10)-\frac{1}{2}(9x-12)$ h $\frac{1}{2}(4x-6)+\frac{x}{4}(2x+8)$
 i $(10-x)(10+x)$ j $(c-d)(c+d)$
 k $(3x-5)^2$ l $(-2x+1)(\frac{1}{2}x+2)$

2 Factorise:
 a $16p-8q$ b p^2-6pq
 c $5p^2q-10pq^2$ d $9pq-6p^2q+12q^2p$

3 If $a=4$, $b=3$ and $c=-2$, evaluate the following:
 a $3a-2b+3c$ b $5a-3b^2$
 c $a^2+b^2+c^2$ d $(a+b)(a-b)$
 e a^2-b^2 f b^3-c^3

4 Rearrange the formulae to make the green letter the subject:
 a $p=4m+n$ b $4x-3y=5z$
 c $2x=\frac{3y}{5p}$ d $m(x+y)=3w$
 e $\frac{pq}{4r}=\frac{mn}{t}$ f $\frac{p+q}{r}=m-n$

12 Algebraic indices

In Chapter 7 you saw how numbers can be expressed using indices. For example, $5 \times 5 \times 5 = 125$, therefore $125 = 5^3$. The 3 is called the index.

Three laws of indices were introduced:

1 $a^m \times a^n = a^{m+n}$

2 $a^m \div a^n$ or $\frac{a^m}{a^n} = a^{m-n}$

3 $(a^m)^n = a^{mn}$

Positive indices

Worked examples

1 Simplify $d^3 \times d^4$.

$$d^3 \times d^4 = d^{(3+4)} = d^7$$

2 Simplify $\frac{(p^2)^4}{p^2 \times p^4}$.

$$\frac{(p^2)^4}{p^2 \times p^4} = \frac{p^{(2\times 4)}}{p^{(2+4)}} = \frac{p^8}{p^6} = p^{(8-6)} = p^2$$

Exercise 12.1

1 Simplify:

a $c^5 \times c^3$

b $m^4 \div m^2$

c $(b^3)^5 \div b^6$

d $\frac{m^4n^9}{mn^3}$

e $\frac{6a^6b^4}{3a^2b^3}$

f $\frac{12x^5y^7}{4x^2y^5}$

g $\frac{4u^3v^6}{8u^2v^3}$

h $\frac{3x^6y^5z^3}{9x^4y^2z}$

2 Simplify:

a $4a^2 \times 3a^3$

b $2a^2b \times 4a^3b^2$

c $(2p^2)^3$

d $(4m^2n^3)^2$

e $(5p^2)^2 \times (2p^3)^3$

f $(4m^2n^2) \times (2mn^3)^3$

g $\frac{(6x^2y^4)^2 \times (2xy)^3}{12x^6y^8}$

h $(ab)^d \times (ab)^e$

The zero index

As shown in Chapter 7, the zero index indicates that a number or algebraic term is raised to the power of zero. A term raised to the power of zero is always equal to 1. This is shown below:

$$a^m \div a^n = a^{m-n} \qquad \text{therefore } \frac{a^m}{a^m} = a^{m-m}$$

$$= a^0$$

However, $\dfrac{a^m}{a^m} = 1$

therefore $a^0 = 1$

Exercise 12.2 Simplify:

a $c^3 \times c^0$ **b** $g^{-2} \times g^3 \div g^0$ **c** $(p^0)^3(q^2)^{-1}$ **d** $(m^3)^3(m^{-2})^5$

Negative indices

A negative index indicates that a number or an algebraic term is being raised to a negative power, e.g. a^{-4}.

As shown in Chapter 7, one of the laws of indices states that:

$a^{-m} = \frac{1}{a^m}$. This is proved as follows:

$$a^{-m} = a^{0-m}$$

$$= \frac{a^0}{a^m} \text{ (from the second law of indices)}$$

$$= \frac{1}{a^m}$$

therefore $a^{-m} = \dfrac{1}{a^m}$

Exercise 12.3 Simplify:

a $\dfrac{a^{-3} \times a^5}{(a^2)^0}$ **b** $\dfrac{(r^3)^{-2}}{(p^{-2})^3}$ **c** $(t^3 \div t^{-5})^2$ **d** $\dfrac{m^0 \div m^{-6}}{(m^{-1})^3}$

Student assessment 1

1 Simplify the following using indices:
 a $a \times a \times a \times b \times b$
 b $d \times d \times e \times e \times e \times e \times e$

2 Write out in full:
 a m^3
 b r^4

3 Simplify the following using indices:
 a $a^4 \times a^3$
 b $p^3 \times p^2 \times q^4 \times q^5$
 c $\frac{b^7}{b^4}$
 d $\frac{(e^4)^5}{e^{14}}$

4 Simplify:
 a $r^4 \times t^0$
 b $\frac{(a^3)^0}{b^2}$
 c $\frac{(m^0)^5}{n^{-3}}$

5 Simplify:
 a $\frac{(p^2 \times p^{-5})^2}{p^3}$
 b $\frac{(h^{-2} \times h^{-5})^{-1}}{h^0}$

13 Equations

An **equation** is formed when the value of an unknown quantity is needed.

Deriving and solving linear equations in one unknown

Worked example

Solve the following linear equations:

a $3x + 8 = 14$

$3x = 6$

$x = 2$

b $12 = 20 + 2x$

$-8 = 2x$

$-4 = x$

c $3(p + 4) = 21$

$3p + 12 = 21$

$3p = 9$

$p = 3$

d $4(x - 5) = 7(2x - 5)$

$4x - 20 = 14x - 35$

$4x + 15 = 14x$

$15 = 10x$

$1.5 = x$

Exercise 13.1 Solve these linear equations:

1 **a** $5a - 2 = 18$ **b** $7b + 3 = 17$ **c** $9c - 12 = 60$
d $6d + 8 = 56$ **e** $4e - 7 = 33$ **f** $12f + 4 = 76$

2 **a** $4a = 3a + 7$ **b** $8b = 7b - 9$
c $7c + 5 = 8c$ **d** $5d - 8 = 6d$

3 **a** $3a - 4 = 2a + 7$ **b** $5b + 3 = 4b - 9$
c $8c - 9 = 7c + 4$ **d** $3d - 7 = 2d - 4$

4 **a** $6a - 3 = 4a + 7$ **b** $5b - 9 = 2b + 6$
c $7c - 8 = 3c + 4$ **d** $11d - 10 = 6d - 15$

5 **a** $\frac{a}{4} = 3$ **b** $\frac{1}{4}b = 2$ **c** $\frac{c}{5} = 2$
d $\frac{1}{5}d = 3$ **e** $4 = \frac{e}{3}$ **f** $-2 = \frac{1}{8}f$

6 **a** $\frac{a}{3}+1=4$ **b** $\frac{b}{5}+2=6$ **c** $8=2+\frac{c}{3}$

d $-4=3+\frac{d}{5}$ **e** $9=5+\frac{2e}{3}$ **f** $-7=\frac{3f}{2}-1$

7 **a** $\frac{2a}{3}=3$ **b** $5=\frac{3b}{2}$ **c** $\frac{4c}{5}=2$

d $7=\frac{5d}{8}$ **e** $1+\frac{3e}{8}=-5$ **f** $2=\frac{5f}{7}-8$

8 **a** $\frac{a+3}{2}=4$ **b** $\frac{b+5}{3}=2$ **c** $5=\frac{c-2}{3}$

d $2=\frac{d-5}{3}$ **e** $3=\frac{2e-1}{5}$ **f** $6=\frac{4f-2}{5}$

9 **a** $3(a+1)=9$ **b** $5(b-2)=25$ **c** $8=2(c-3)$

d $14=4(3-d)$ **e** $21=3(5-e)$ **f** $36=9(5-2f)$

10 **a** $\frac{a+2}{3}=\frac{a-3}{2}$ **b** $\frac{b-1}{4}=\frac{b+5}{3}$ **c** $\frac{2-c}{5}=\frac{7-c}{4}$

d $\frac{8+d}{7}=\frac{7+d}{6}$ **e** $\frac{3-e}{4}=\frac{5-e}{2}$ **f** $\frac{10+f}{3}=\frac{5-f}{2}$

Exercise 13.2

Solve the linear equations:

1 **a** $3x=2x-4$ **b** $5y=3y+10$ **c** $2y-5=3y$

d $p-8=3p$ **e** $3y-8=2y$ **f** $7x+11=5x$

2 **a** $3x-9=4$ **b** $4=3x-11$ **c** $6x-15=3x+3$

d $4y+5=3y-3$ **e** $8y-31=13-3y$ **f** $4m+2=5m-8$

3 **a** $7m-1=5m+1$ **b** $5p-3=3+3p$ **c** $12-2k=16+2k$

d $6x+9=3x-54$ **e** $8-3x=18-8x$ **f** $2-y=y-4$

4 **a** $\frac{x}{2}=3$ **b** $\frac{1}{2}y=7$ **c** $\frac{x}{4}=1$

d $\frac{1}{4}m=3$ **e** $7=\frac{x}{5}$ **f** $4=\frac{1}{5}p$

5 **a** $\frac{x}{3}-1=4$ **b** $\frac{x}{5}+2=1$ **c** $\frac{2}{3}x=5$

d $\frac{3}{4}x=6$ **e** $\frac{1}{5}x=\frac{1}{2}$ **f** $\frac{2x}{5}=4$

6 **a** $\frac{x+1}{2}=3$ **b** $4=\frac{x-2}{3}$ **c** $\frac{x-10}{3}=4$

d $8=\frac{5x-1}{3}$ **e** $\frac{2(x-5)}{3}=2$ **f** $\frac{3(x-2)}{4}=4x-8$

7 **a** $6=\frac{2(y-1)}{3}$ **b** $2(x+1)=3(x-5)$

c $5(x-4)=3(x+2)$ **d** $\frac{3+y}{2}=\frac{y+1}{4}$

e $\frac{7+2x}{3}=\frac{9x-1}{7}$ **f** $\frac{2x+3}{4}=\frac{4x-2}{6}$

Constructing equations

NB: All diagrams are not drawn to scale.

In many cases, when dealing with the practical applications of mathematics, equations need to be constructed first before they can be solved. Often the information is either given within the context of a problem or in a diagram.

Worked examples

1 Find the size of each of the angles in the triangle by constructing an equation and solving it to find the value of x.

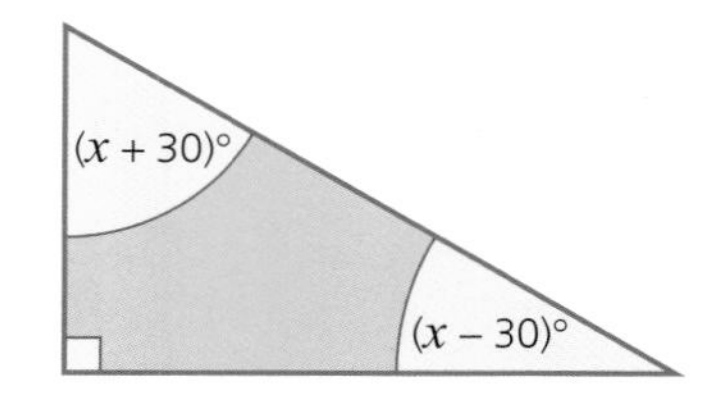

The sum of the angles of a triangle is 180°.

$$(x + 30) + (x - 30) + 90 = 180$$
$$2x + 90 = 180$$
$$2x = 90$$
$$x = 45$$

The three angles are therefore: 90°, $x + 30 = 75°$ and $x - 30 = 15°$.
Check: $90° + 75° + 15° = 180°$.

2 Find the size of each of the angles in the quadrilateral by constructing an equation and solving it to find the value of x.

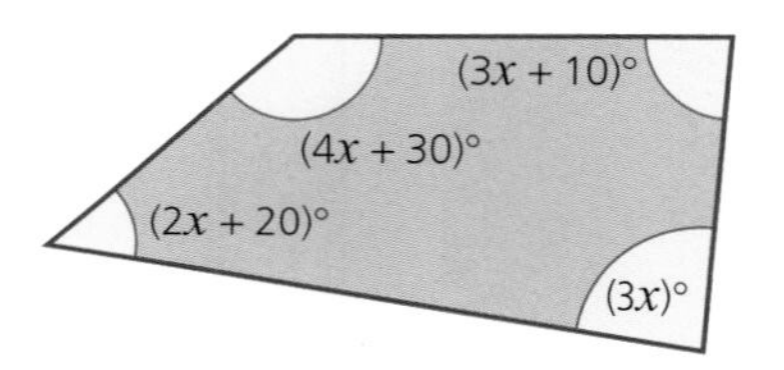

The sum of the angles of a quadrilateral is 360°.

$$4x + 30 + 3x + 10 + 3x + 2x + 20 = 360$$
$$12x + 60 = 360$$
$$12x = 300$$
$$x = 25$$

The angles are:

$$4x + 30 = (4 \times 25) + 30 = 130°$$
$$3x + 10 = (3 \times 25) + 10 = 85°$$
$$3x = 3 \times 25 = 75°$$
$$2x + 20 = (2 \times 25) + 20 = 70°$$
$$\text{Total} = 360°$$

3 Construct an equation and solve it to find the value of x in the diagram.

Area of rectangle = base × height

$$2(x + 3) = 16$$
$$2x + 6 = 16$$
$$2x = 10$$
$$x = 5$$

Exercise 13.4

In each case, find the number by forming an equation.

1 a If I multiply a number by 7 and add 1 the total is 22.
  b If I multiply a number by 9 and add 7 the total is 70.

2 a If I multiply a number by 8 and add 12 the total is 92.
  b If I add 2 to a number and then multiply by 3 the answer is 18.

3 a If I add 12 to a number and then multiply by 5 the answer is 100.
  b If I add 6 to a number and divide it by 4 the answer is 5.

4 a If I add 3 to a number and multiply by 4 the answer is the same as multiplying by 6 and adding 8.
  b If I add 1 to a number and then multiply by 3 the answer is the same as multiplying by 5 and subtracting 3.

Simultaneous equations

When the values of two unknowns are needed, two equations need to be formed and solved. The process of solving two equations and finding a common solution is known as **solving equations simultaneously**.

The two most common ways of solving simultaneous equations algebraically are by **elimination** and by **substitution**.

By elimination

The aim of this method is to eliminate one of the unknowns by either adding or subtracting the two equations.

Worked examples

Solve the following simultaneous equations by finding the values of x and y which satisfy both equations.

a $3x + y = 9$ (1)
  $5x - y = 7$ (2)

By adding equations (1) + (2) we eliminate the variable y:

$8x = 16$
$x = 2$

To find the value of y we substitute $x = 2$ into either equation (1) or (2). Substituting $x = 2$ into equation (1):

$3x + y = 9$

$6 + y = 9$

$y = 3$

To check that the solution is correct, the values of x and y are substituted into equation (2). If it is correct then the left-hand side of the equation will equal the right-hand side.

$$5x - y = 7$$

$$\text{LHS} = 10 - 3 = 7$$

$$= \text{RHS}$$

b $4x + y = 23$ (1)

$x + y = 8$ (2)

By subtracting the equations, i.e. (1) – (2), we eliminate the variable y:

$$3x = 15$$

$$x = 5$$

By substituting $x = 5$ into equation (2), y can be calculated:

$$x + y = 8$$

$$5 + y = 8$$

$$y = 3$$

Check by substituting both values into equation (1):

$$4x + y = 23$$

$$\text{LHS} = 20 + 3 = 23$$

$$= \text{RHS}$$

By substitution

The same equations can also be solved by the method known as substitution.

Worked examples

a $3x + y = 9$ (1)

$5x - y = 7$ (2)

Equation (2) can be rearranged to give: $y = 5x - 7$
This can now be substituted into equation (1):

$$3x + (5x - 7) = 9$$

$$3x + 5x - 7 = 9$$

$$8x - 7 = 9$$

$$8x = 16$$

$$x = 2$$

To find the value of y, $x = 2$ is substituted into either equation (1) or (2) as before, giving $y = 3$.

b $4x + y = 23$ (1)

$x + y = 8$ (2)

Equation (2) can be rearranged to give $y = 8 - x$.

This can be substituted into equation (1):

$$4x + (8 - x) = 23$$
$$4x + 8 - x = 23$$
$$3x + 8 = 23$$
$$3x = 15$$
$$x = 5$$

y can be found as before, giving the result $y = 3$.

Exercise 13.5

Solve the simultaneous equations either by elimination or by substitution:

1 **a** $x + y = 6$, $x - y = 2$
b $x + y = 11$, $x - y - 1 = 0$
c $x + y = 5$, $x - y = 7$
d $2x + y = 12$, $2x - y = 8$
e $3x + y = 17$, $3x - y = 13$
f $5x + y = 29$, $5x - y = 11$

2 **a** $3x + 2y = 13$, $4x = 2y + 8$
b $6x + 5y = 62$, $4x - 5y = 8$
c $x + 2y = 3$, $8x - 2y = 6$
d $9x + 3y = 24$, $x - 3y = -14$
e $7x - y = -3$, $4x + y = 14$
f $3x = 5y + 14$, $6x + 5y = 58$

3 **a** $2x + y = 14$, $x + y = 9$
b $5x + 3y = 29$, $x + 3y = 13$
c $4x + 2y = 50$, $x + 2y = 20$
d $x + y = 10$, $3x = -y + 22$
e $2x + 5y = 28$, $4x + 5y = 36$
f $x + 6y = -2$, $3x + 6y = 18$

4 **a** $x - y = 1$, $2x - y = 6$
b $3x - 2y = 8$, $2x - 2y = 4$
c $7x - 3y = 26$, $2x - 3y = 1$
d $x = y + 7$, $3x - y = 17$
e $8x - 2y = -2$, $3x - 2y = -7$
f $4x - y = -9$, $7x - y = -18$

5 **a** $x + y = -7$, $x - y = -3$
b $2x + 3y = -18$, $2x = 3y + 6$
c $5x - 3y = 9$, $2x + 3y = 19$
d $7x + 4y = 42$, $9x - 4y = -10$
e $4x - 4y = 0$, $8x + 4y = 12$
f $x - 3y = -25$, $5x - 3y = -17$

Exercise 13.5 (cont)

6 **a** $2x + 3y = 13$
$2x - 4y + 8 = 0$

b $2x + 4y = 50$
$2x + y = 20$

c $x + y = 10$
$3y = 22 - x$

d $5x + 2y = 28$
$5x + 4y = 36$

e $2x - 8y = 2$
$2x - 3y = 7$

f $x - 4y = 9$
$x - 7y = 18$

7 **a** $-4x = 4y$
$4x - 8y = 12$

b $3x = 19 + 2y$
$-3x + 5y = 5$

c $3x + 2y = 12$
$-3x + 9y = -12$

d $3x + 5y = 29$
$3x + y = 13$

e $-5x + 3y = 14$
$5x + 6y = 58$

f $-2x + 8y = 6$
$2x = 3 - y$

Further simultaneous equations

If neither x nor y can be eliminated by simply adding or subtracting the two equations then it is necessary to multiply one or both of the equations. The equations are multiplied by a number in order to make the coefficients of x (or y) numerically equal.

Worked examples

a $3x + 2y = 22$ (1)

$x + y = 9$ (2)

To eliminate y, equation (2) is multiplied by 2:

$3x + 2y = 22$ (1)

$2x + 2y = 18$ (3)

By subtracting (3) from (1), the variable y is eliminated:

$x = 4$

Substituting $x = 4$ into equation (2), we have:

$x + y = 9$
$4 + y = 9$
$y = 5$

Check by substituting both values into equation (1):

$3x + 2y = 22$
LHS $= 12 + 10 = 22$
$=$ RHS

b $5x - 3y = 1$ (1)

$3x + 4y = 18$ (2)

To eliminate the variable y, equation (1) is multiplied by 4, and equation (2) is multiplied by 3.

$20x - 12y = 4$ (3)

$9x + 12y = 54$ (4)

By adding equations (3) and (4) the variable y is eliminated:

$$29x = 58$$

$$x = 2$$

Substituting $x = 2$ into equation (2) gives:

$$3x + 4y = 18$$

$$6 + 4y = 18$$

$$4y = 12$$

$$y = 3$$

Check by substituting both values into equation (1):

$$5x - 3y = 1$$

$$\text{LHS} = 10 - 9 = 1$$

$$= \text{RHS}$$

Exercise 13.6

Solve the simultaneous equations either by elimination or by substitution.

1 **a** $2a + b = 5$
$3a - 2b = 4$

b $3b + 2c = 18$
$2b - c = 5$

c $4c - d = 18$
$2c + 2d = 14$

d $d + 5e = 17$
$2d - 8e = -2$

e $3e - f = 5$
$e + 2f = 11$

f $f + 3g = 5$
$2f - g = 3$

2 **a** $a + 2b = 8$
$3a - 5b = -9$

b $4b - 3c = 17$
$b + 5c = 10$

c $6c - 4d = -2$
$5c + d = 7$

d $5d + e = 18$
$2d + 3e = 15$

e $e + 2f = 14$
$3e - f = 7$

f $7f - 5g = 9$
$f + g = 3$

Exercise 13.6 (cont)

3 **a** $3a - 2b = -5$
$a + 5b = 4$

b $b + 2c = 3$
$3b - 5c = -13$

c $c - d = 4$
$3c + 4d = 5$

d $2d + 3e = 2$
$3d - e = -8$

e $e - 2f = -7$
$3e + 3f = -3$

f $f + g = -2$
$3f - 4g = 1$

4 **a** $2x + y = 7$
$3x + 2y = 12$

b $5x + 4y = 21$
$x + 2y = 9$

c $x + y = 7$
$3x + 4y = 23$

d $2x - 3y = -3$
$3x + 2y = 15$

e $4x = 4y + 8$
$x + 3y = 10$

f $x + 5y = 11$
$2x - 2y = 10$

5 **a** $x + y = 5$
$3x - 2y + 5 = 0$

b $2x - 2y = 6$
$x - 5y = -5$

c $2x + 3y = 15$
$2y = 15 - 3x$

d $x - 6y = 0$
$3x - 3y = 15$

e $2x - 5y = -11$
$3x + 4y = 18$

f $x + y = 5$
$2x - 2y = -2$

6 **a** $3y = 9 + 2x$
$3x + 2y = 6$

b $x + 4y = 13$
$3x - 3y = 9$

c $2x = 3y - 19$
$3x + 2y = 17$

d $2x - 5y = -8$
$-3x - 2y = -26$

e $5x - 2y = 0$
$2x + 5y = 29$

f $8y = 3 - x$
$3x - 2y = 9$

7 **a** $4x + 2y = 5$
$3x + 6y = 6$

b $4x + y = 14$
$6x - 3y = 3$

c $10x - y = -2$
$-15x + 3y = 9$

d $-2y = 0.5 - 2x$
$6x + 3y = 6$

e $x + 3y = 6$
$2x - 9y = 7$

f $5x - 3y = -0.5$
$3x + 2y = 3.5$

Exercise 13.7

1 The sum of two numbers is 17 and their difference is 3. Find the two numbers by forming two equations and solving them simultaneously.

2 The difference between two numbers is 7. If their sum is 25, find the two numbers by forming two equations and solving them simultaneously.

3 Find the values of x and y.

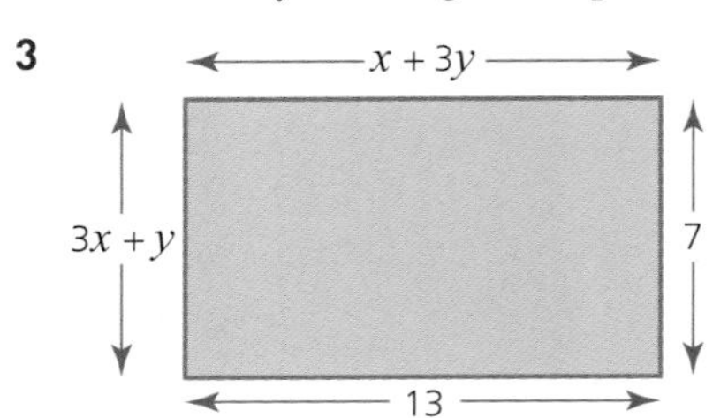

4 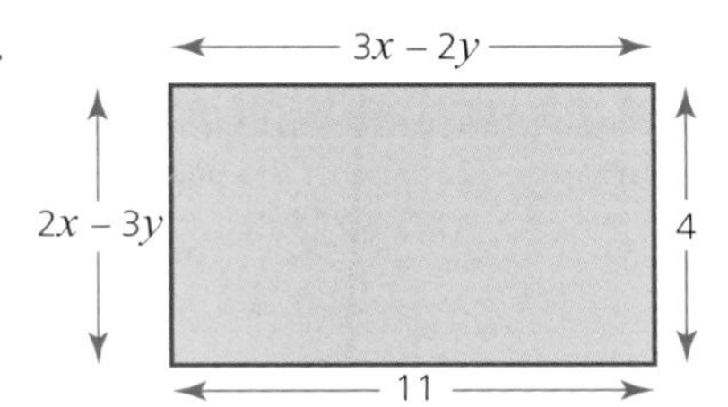

Find the values of x and y.

5 This year, a man's age, m, is three times his son's age, s. Ten years ago, the man's age was five times the age his son was then. By forming two equations and solving them simultaneously, find both of their ages.

6 A grandfather is ten times as old as his granddaughter. He is also 54 years older than her. How old is each of them?

Student assessment 1

Solve the equations:

1 a $a + 9 = 15$ b $3b + 7 = -14$
c $3 - 5c = 18$ d $4 - 7d = -24$

2 a $5a + 7 = 4a - 3$ b $8 - 3b = 4 - 2b$
c $6 - 3c = c + 8$ d $4d - 3 = d + 9$

3 a $\frac{a}{5} = 2$ b $\frac{b}{7} = 3$
c $4 = c - 2$ d $6 = \frac{1}{3}d$

4 a $\frac{a}{2} + 1 = 5$ b $\frac{b}{3} - 2 = 3$
c $7 = \frac{c}{3} - 1$ d $1 = \frac{1}{3}d - 2$

5 a $\frac{a-2}{3} = \frac{a+2}{2}$ b $\frac{b+5}{4} = \frac{2+b}{3}$
c $4(c - 5) = 3(c + 1)$ d $6(2 + 3d) = 5(4d - 2)$

Solve the simultaneous equations:

6 a $a + 2b = 4$
$3a + b = 7$
b $b - 2c = -2$
$3b + c = 15$
c $2c - 3d = -5$
$4c + d = -3$
d $4d + 5e = 0$
$d + e = -1$

7 Two students are buying school supplies. One buys a ruler and three pens, paying \$7.70. The other student buys a ruler and five pens, paying \$12.30. Calculate the cost of each of the items.

Student assessment 2

Solve the equations:

1 a $x + 7 = 16$ b $2x - 9 = 13$

c $8 - 4x = 24$ d $5 - 3x = -13$

2 a $7 - m = 4 + m$ b $5m - 3 = 3m + 11$

c $6m - 1 = 9m - 13$ d $18 - 3p = 6 + p$

3 a $\frac{x}{-5} = 2$ b $4 = \frac{1}{3}x$

c $\frac{x + 2}{3} = 4$ d $\frac{2x - 5}{7} = \frac{5}{2}$

4 a $\frac{2}{3}(x - 4) = 8$ b $4(x - 3) = 7(x + 2)$

c $4 = \frac{2}{7}(3x + 8)$ d $\frac{3}{4}(x - 1) = \frac{5}{8}(2x - 4)$

5 Solve the simultaneous equations:

a $2x + 3y = 16$
$2x - 3y = 4$

b $4x + 2y = 22$
$-2x + 2y = 2$

c $x + y = 9$
$2x + 4y = 26$

d $2x - 3y = 7$
$-3x + 4y = -11$

6 Two numbers added together equal 13. The difference between the two numbers is 6.5. Calculate each of the two numbers.

14 Sequences

Sequences

A **sequence** is an ordered set of numbers. Each number in a sequence is known as a **term**. The terms of a sequence form a pattern. For the sequence of numbers:

2, 5, 8, 11, 14, 17, ...

the difference between successive terms is +3. The **term-to-term rule** is therefore + 3.

Worked examples

1 Below is a sequence of numbers.

5, 9, 13, 17, 21, 25, ...

a What is the term-to-term rule for the sequence?

The term-to-term rule is + 4.

b Calculate the 10th term of the sequence.

Continuing the pattern gives:

5, 9, 13, 17, 21, 25, 29, 33, 37, 41, ...

Therefore the 10th term is 41.

2 Below is a sequence of numbers.

1, 2, 4, 8, 16, ...

a What is the term-to-term rule for the sequence?

The term-to-term rule is $\times 2$.

b Calculate the 10th term of the sequence.

Continuing the pattern gives:

1, 2, 4, 8, 16, 32, 64, 128, 256, 512, ...

Therefore the 10th term is 512.

Exercise 14.1

For each of the sequences:

i State a rule to describe the sequence.
ii Calculate the 10th term.

a 3, 6, 9, 12, 15, ...
b 8, 18, 28, 38, 48, ...
c 11, 33, 55, 77, 99, ...
d 0.7, 0.5, 0.3, 0.1, ...
e $\frac{1}{2}, \frac{1}{3}, \frac{1}{4}, \frac{1}{5}, \ldots$
f $\frac{1}{2}, \frac{2}{3}, \frac{3}{4}, \frac{4}{5}, \ldots$
g 1, 4, 9, 16, 25, ...
h 4, 7, 12, 19, 28, ...
i 1, 8, 27, 64, ...
j 5, 25, 125, 625, ...

Sometimes, the pattern in a sequence of numbers is not obvious. By looking at the differences between successive terms a pattern can often be found.

Worked examples

1 Calculate the 8th term in the sequence

8, 12, 20, 32, 48, ...

The pattern in this sequence is not immediately obvious, so a row for the differences between successive terms can be constructed.

	8		12		20		32		48
1st differences		4		8		12		16	

The pattern in the differences row is + 4 and this can be continued to complete the sequence to the 8th term.

	8		12		20		32		48		68		92		120
1st differences		4		8		12		16		20		24		28	

So the 8th term is 120.

2 Calculate the 8th term in the sequence

3, 6, 13, 28, 55, ...

1st differences	3	7	15	27

The row of first differences is not sufficient to spot the pattern, so a row of 2nd differences is constructed.

	3		6		13		28		55
1st differences		3		7		15		27	
2nd differences			4		8		12		

The pattern in the 2nd differences row can be seen to be + 4. This can now be used to complete the sequence.

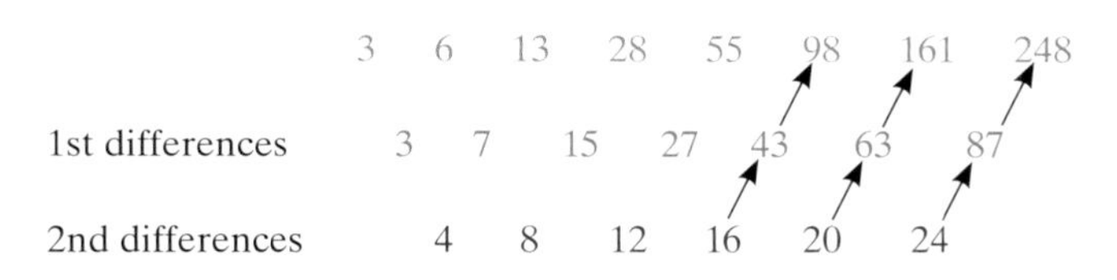

So the 8th term is 248.

Exercise 14.2

For each of the sequences calculate the next two terms:

a 8, 11, 17, 26, 38, ...
b 5, 7, 11, 19, 35, ...
c 9, 3, 3, 9, 21, ...
d –2, 5, 21, 51, 100, ...
e 11, 9, 10, 17, 36, 79, ...
f 4, 7, 11, 19, 36, 69, ...
g –3, 3, 8, 13, 17, 21, 24, ...

The *n*th term

So far, the method used for generating a sequence relies on knowing the previous term to work out the next one. This method works but can be a little cumbersome if the 100th term is needed and only the first five terms are given! A more efficient rule is one which is related to a term's position in a sequence.

Worked examples

1 For the sequence shown, give an expression for the *n*th term.

Position	1	2	3	4	5	n
Term	3	6	9	12	15	?

By looking at the sequence it can be seen that the term is always $3 \times$ position.

Therefore the *n*th term can be given by the expression $3n$.

2 For the sequence shown, give an expression for the *n*th term.

Position	1	2	3	4	5	n
Term	2	5	8	11	14	?

You will need to spot similarities between sequences. The terms of the above sequence are the same as the terms in example 1 above but with 1 subtracted each time.

The expression for the *n*th term is therefore $3n - 1$.

Exercise 14.3

1 For each of the sequences:

i Write down the next two terms.

ii Give an expression for the nth term.

a 5, 8, 11, 14, 17, ...

b 5, 9, 13, 17, 21, ...

c 4, 9, 14, 19, 24, ...

d 8, 10, 12, 14, 16, ...

e 1, 8, 15, 22, 29, ...

f 0, 4, 8, 12, 16, 20, ...

g 1, 10, 19, 28, 37, ...

h 15, 25, 35, 45, 55, ...

i 9, 20, 31, 42, 53, ...

j 1.5, 3.5, 5.5, 7.5, 9.5, 11.5, ...

k 0.25, 1.25, 2.25, 3.25, 4.25, ...

l 0, 1, 2, 3, 4, 5, ...

2 For each of the sequences:

i Write down the next two terms.

ii Give an expression for the nth term.

a 2, 5, 10, 17, 26, 37, ...

b 8, 11, 16, 23, 32, ...

c 0, 3, 8, 15, 24, 35, ...

d 1, 8, 27, 64, 125, ...

e 2, 9, 28, 65, 126, ...

f 11, 18, 37, 74, 135, ...

g –2, 5, 24, 61, 122, ...

h 2, 6, 12, 20, 30, 42, ...

Further sequences

To be able to spot rules for more complicated sequences it is important to be aware of some other common types of sequence.

Worked examples

1

Position	1	2	3	4	5	n
Term	1	4	9	16		

a Describe the sequence in words.

The terms form the sequence of square numbers.

i.e. $1 \times 1 = 1$, $2 \times 2 = 4$, $3 \times 3 = 9$, $4 \times 4 = 16$

b Predict the 5th term.

The 5th term is the fifth square number, i.e. $5 \times 5 = 25$

c Write the rule for the nth term.

The nth term is $n \times n = n^2$

2

Position	1	2	3	4	5	n
Term	1	8	27	64		

a Describe the sequence in words.

The terms form the sequence of **cube numbers**.

i.e. $1 \times 1 \times 1 = 1,\ 2 \times 2 \times 2 = 8,\ 3 \times 3 \times 3 = 27,\ 4 \times 4 \times 4 = 64$

b Predict the 5th term.

The 5th term is the fifth cube number, i.e. $5 \times 5 \times 5 = 125$

c Write the rule for the nth term.

The nth term is $n \times n \times n = n^3$

3 The table shows a sequence that is a pattern of growing triangles:

Position	1	2	3	4	5	n
Pattern						
Term	1	3	6	10		

a Predict the number of dots in the fifth position of the pattern.

The number of dots added each time is the same as the position number. Therefore the number of dots in the fifth position is the number of dots in the fourth position + 5, i.e. $10 + 5 = 15$

b Calculate the nth term.

The rule can be deduced by looking at the dot patterns themselves.

The second pattern can be doubled and arranged as a rectangle as shown (left).

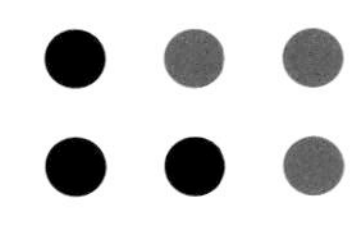

The total number of dots is $2 \times 3 = 6$

The total number of black dots is therefore $\frac{1}{2} \times 2 \times 3 = 3$

Notice that the height of the rectangle is equal to the position number (i.e. 2) and its length is one more than the position number (i.e. 3).

The third pattern can be doubled and arranged as a rectangle as shown.

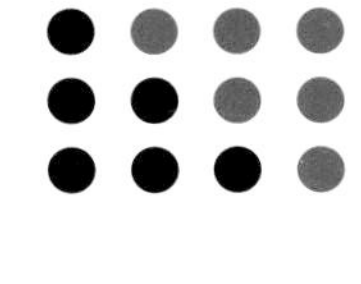

The total number of dots is $3 \times 4 = 12$

The total number of black dots is therefore $\frac{1}{2} \times 3 \times 4 = 6$

Once again the height of the rectangle is equal to the position number (i.e. 3) and its length one more than the position number (i.e. 4).

Therefore the nth pattern of dots can also be doubled and arranged into a rectangle with a height equal to n and a width equal to $n + 1$ as shown:

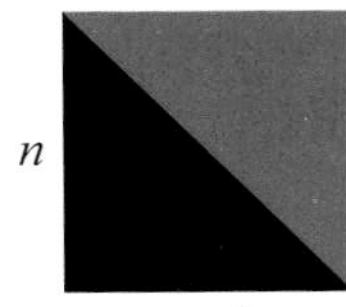

The total area is $n(n + 1)$

The black area is therefore $\frac{1}{2}n(n+1)$

Note

The sequence of numbers 1, 3, 6, 10, 15, etc. is known as the sequence of **triangular numbers**. The formula for the nth triangular number is $\frac{1}{2}n(n+1)$.

It is important to be able to identify these sequences as often they are used to form other sequences.

Worked examples

1 A sequence of numbers is: 3, 6, 11, 18, 27, ...

a Calculate the next two terms.

The difference between the terms goes up by two each time.

Therefore the next two terms are 38 and 51.

b Calculate the rule for the nth term.

This is not as difficult as it first seems if its similarity to the sequence of square numbers is noticed as shown:

Position	1	2	3	4	5	n
Square numbers	1	4	9	16	25	n^2
Term	3	6	11	18	27	

The numbers in the given sequence are always two more than the sequence of square numbers.

Therefore the rule for the nth term is $n^2 + 2$.

2 A sequence of numbers is: 0, 2, 5, 9, 14, ...

a Calculate the next two terms.

The difference between the terms goes up by one each time.

Therefore the next two terms are 20 and 27.

b Calculate the rule for the nth term.

This too is not as difficult as it first seems if its similarity to the sequence of triangular numbers is noticed as shown:

Position	1	2	3	4	5	n
Triangular numbers	1	3	6	10	15	$\frac{1}{2}n(n+1)$
Term	0	2	5	9	14	

The numbers in the given sequence are always one less than the sequence of triangular numbers. Therefore the rule for the nth term is $\frac{1}{2}n(n+1) - 1$

Exercise 14.4

For each of the sequences in questions 1–5, consider their relation to sequences of square, cube or triangular numbers and then:

i Write down the next two terms.
ii Write down the rule for the nth term.

1 2, 8, 18, 32, ...

2 2, 6, 12, 20, ...

3 –1, 6, 25, 62, ...

4 5, 8, 13, 20, ...

5 $\frac{1}{2}$, 4, $13\frac{1}{2}$, 32, ...

For each of the sequences in questions 6–8, consider their relation to the sequences above and write the rule for the nth term.

6 1, 7, 17, 31, ...

7 1, 5, 11, 19, ...

8 –2, 12, 50, 124, ...

Student assessment 1

1 For each of the sequences:
i Calculate the next two terms.
ii Explain the pattern in words.

a 9, 18, 27, 36, ...
b 54, 48, 42, 36, ...
c 18, 9, 4.5, ...
d 12, 6, 0, –6, ...
e 216, 125, 64, ...
f 1, 3, 9, 27, ...

2 For each of the sequences:
i Calculate the next two terms.
ii Explain the pattern in words.

a 6, 12, 18, 24, ...
b 24, 21, 18, 15, ...
c 10, 5, 0, ...
d 16, 25, 36, 49, ...
e 1, 10, 100, ...
f 1, $\frac{1}{2}$, $\frac{1}{4}$, $\frac{1}{8}$, ...

3 For each of the sequences, give an expression for the nth term:

a 6, 10, 14, 18, 22, ...
b 13, 19, 25, 31, ...
c 3, 9, 15, 21, 27, ...
d 4, 7, 12, 19, 28, ...
e 0, 10, 20, 30, 40, ...
f 0, 7, 26, 63, 124, ...

4 For each of the sequences, give an expression for the nth term:

a 3, 5, 7, 9, 11, ...
b 7, 13, 19, 25, 31, ...
c 8, 18, 28, 38, ...
d 1, 9, 17, 25, ...
e –4, 4, 12, 20, ...
f 2, 5, 10, 17, 26, ...

5 a Write down the first five terms of the sequence of square numbers.
b Write down the first five terms of the sequence of cube numbers.
c Write down the first five terms of the sequence of triangular numbers.

6 For each of the sequences:
i Write the next two terms.
ii Write the rule for the nth term.

a 4, 7, 12, 19, ...
b 2, 16, 54, 128, ...
c $\frac{1}{2}$, $1\frac{1}{2}$, 3, 5, ...
d 0, 4, 10, 18, ...

15 Graphs in practical situations

Conversion graphs

A straight line graph can be used to convert one set of units to another. Examples include converting from one currency to another, converting miles to kilometres and converting temperature from degrees Celsius to degrees Fahrenheit.

Worked example

The graph below converts US dollars into Chinese yuan based on an exchange rate of $1 = 8.80 yuan.

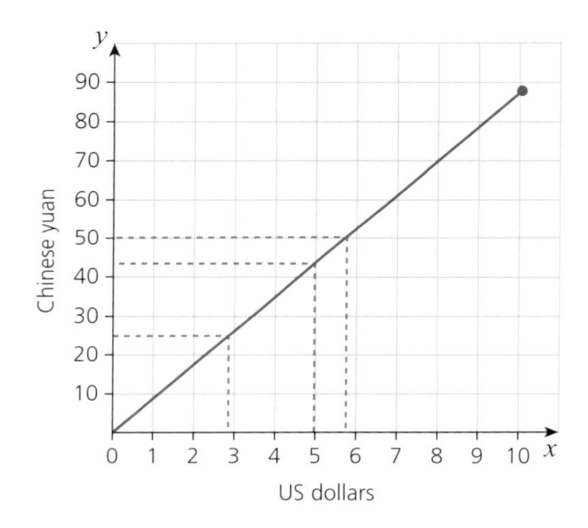

a Using the graph, estimate the number of yuan equivalent to $5.

A line is drawn up from $5 until it reaches the plotted line, then across to the y-axis. From the graph it can be seen that $5 ≈ 44 yuan.

≈ is the symbol for 'is approximately equal to'.

b Using the graph, what would be the cost in dollars of a drink costing 25 yuan?

A line is drawn across from 25 yuan until it reaches the plotted line, then down to the x-axis. From the graph it can be seen that the cost of the drink ≈ $2.80.

c If a meal costs 200 yuan, use the graph to estimate its cost in US dollars.

The graph does not go up to 200 yuan, therefore a factor of 200 needs to be used, e.g. 50 yuan. From the graph 50 yuan ≈ $5.70, therefore it can be deduced that 200 yuan ≈ $22.80 (i.e. 4 × $5.70).

Exercise 15.1

1 Given that 80 km = 50 miles, draw a conversion graph up to 100 km. Using your graph, estimate:
  a how many miles is 50 km
  b how many kilometres is 80 miles
  c the speed in miles per hour (mph) equivalent to 100 km/h
  d the speed in km/h equivalent to 40 mph.

2 You can roughly convert temperature in degrees Celsius to degrees Fahrenheit by doubling the degrees Celsius and adding 30.

  Draw a conversion graph up to 50 °C. Use your graph to estimate the following:
  a the temperature in °F equivalent to 25 °C
  b the temperature in °C equivalent to 100 °F
  c the temperature in °F equivalent to 0 °C.

3 Given that 0 °C = 32 °F and 50 °C = 122 °F, on the graph you drew for Q.2, draw a true conversion graph.
  a Use the true graph to calculate the conversions in Q.2.
  b Where would you say the rough conversion is most useful?

4 Long-distance calls from New York to Harare are priced at 85 cents/min off peak and $1.20/min at peak times.
  a Draw, on the same axes, conversion graphs for the two different rates.
  b From your graph estimate the cost of an 8-minute call made off peak.
  c Estimate the cost of the same call made at peak rate.
  d A call is to be made from a telephone box. If the caller has only $4 to spend, estimate how much more time he can talk for if he rings at off peak instead of at peak times.

5 A maths exam is marked out of 120. Draw a conversion graph and use it to change the following marks to percentages.
  a 80 b 110 c 54 d 72

Speed, distance and time

You may already be aware of the formula:

$$\text{distance} = \text{speed} \times \text{time}$$

Rearranging the formula gives:

$$\text{time} = \frac{\text{distance}}{\text{speed}} \text{ and speed} = \frac{\text{distance}}{\text{time}}$$

Where the speed is not constant:

$$\text{average speed} = \frac{\text{total distance}}{\text{total time}}$$

Exercise 15.2

1 Find the average speed of an object moving:
  a 30 m in 5 s b 48 m in 12 s
  c 78 km in 2 h d 50 km in 2.5 h
  e 400 km in 2 h 30 min f 110 km in 2 h 12 min

Exercise 15.2 (cont)

2 How far will an object travel during:

a 10 s at 40 m/s
b 7 s at 26 m/s
c 3 hours at 70 km/h
d 4 h 15 min at 60 km/h
e 10 min at 60 km/h
f 1 h 6 min at 20 m/s?

3 How long will it take to travel:

a 50 m at 10 m/s
b 1 km at 20 m/s
c 2 km at 30 km/h
d 5 km at 70 m/s
e 200 cm at 0.4 m/s
f 1 km at 15 km/h?

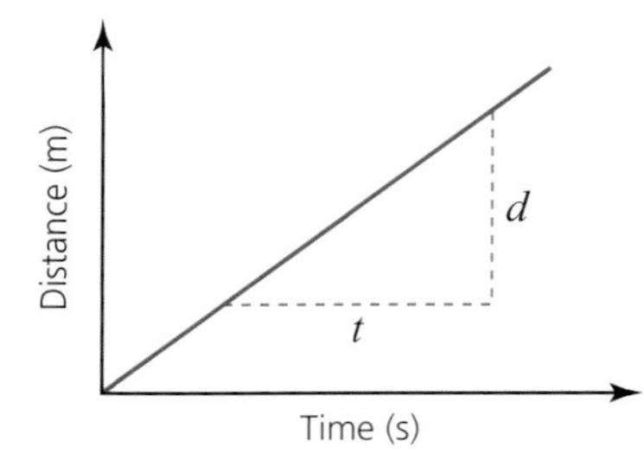

Travel graphs

The graph of an object travelling at a constant speed is a straight line as shown (left).

$$\text{Gradient} = \frac{d}{t}$$

The units of the gradient are m/s, hence the gradient of a distance–time graph represents the speed at which the object is travelling. Another way of interpreting the gradient of a distance–time graph is that it represents the rate of change of distance with time.

Worked example

The graph represents an object travelling at constant speed.

a From the graph calculate how long it took to cover a distance of 30 m.

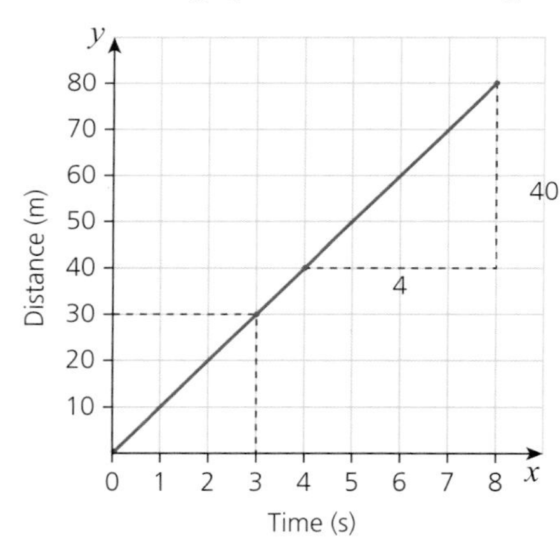

The time taken to travel 30 m was 3 seconds.

b Calculate the gradient of the graph.

Taking two points on the line,

$$\text{gradient} = \frac{40\,\text{m}}{4\,\text{s}}$$

$$= 10\,\text{m/s}.$$

c Calculate the speed at which the object was travelling.

Gradient of a distance–time graph = speed.
Therefore the speed is 10 m/s.

Exercise 15.3

1 Draw a distance–time graph for the first 10 seconds of an object travelling at 6 m/s.

2 Draw a distance–time graph for the first 10 seconds of an object travelling at 5 m/s. Use your graph to estimate:
  a the time taken to travel 25 m
  b how far the object travels in 3.5 seconds.

3 Two objects A and B set off from the same point and move in the same straight line. B sets off first, whilst A sets off 2 seconds later.

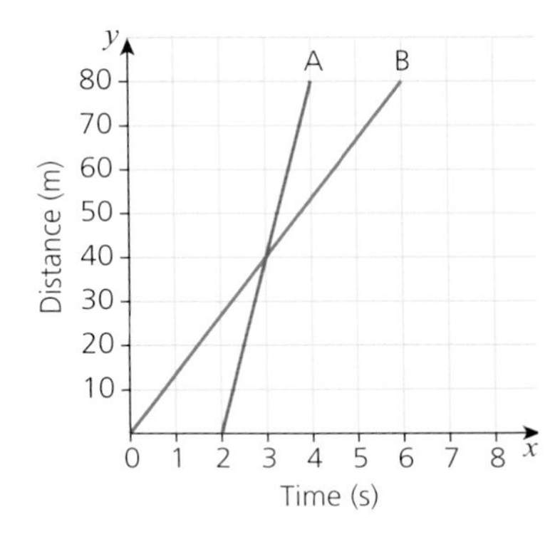

Using the distance–time graph estimate:
  a the rate of change of distance with time of each of the objects
  b how far apart the objects would be 20 seconds after the start.

4 Three objects A, B and C move in the same straight line away from a point X. Both A and C change their speed during the journey, whilst B travels at a constant speed throughout.

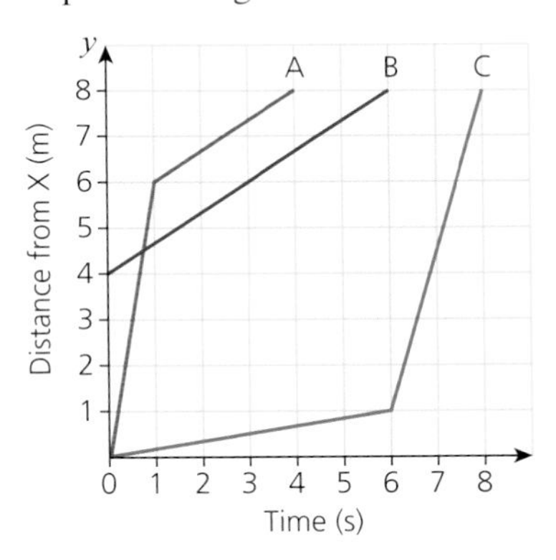

From the distance–time graph estimate:
  a the speed of object B
  b the two speeds of object A
  c the average rate of change of distance with time of object C
  d how far object C is from X, 3 seconds from the start
  e how far apart objects A and C are, 4 seconds from the start.

The graphs of two or more journeys can be shown on the same axes. The shape of the graph gives a clear picture of the movement of each of the objects.

Worked example

The journeys of two cars, X and Y, travelling between A and B, are represented on the distance–time graph. Car X and Car Y both reach point B 100 km from A at 11 00.

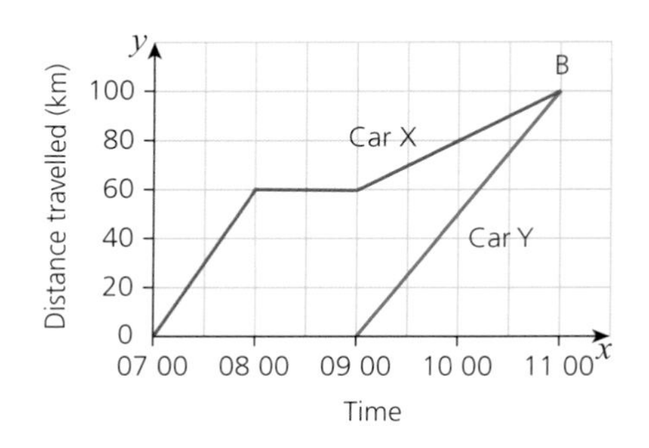

a Calculate the speed of Car X between 07 00 and 08 00.

$\text{speed} = \frac{\text{distance}}{\text{time}}$

$\frac{60}{1}\text{ km/h} = 60\text{ km/h}$

b Calculate the speed of Car Y between 09 00 and 11 00.

$\text{speed} = \frac{100}{2}\text{ km/h}$

$= 50\text{ km/h}$

c Explain what is happening to Car X between 08 00 and 09 00.

No distance has been travelled, therefore Car X is stationary.

Exercise 15.4

1 Two friends Paul and Helena arrange to meet for lunch at noon. They live 50 km apart and the restaurant is 30 km from Paul's home. The travel graph illustrates their journeys.

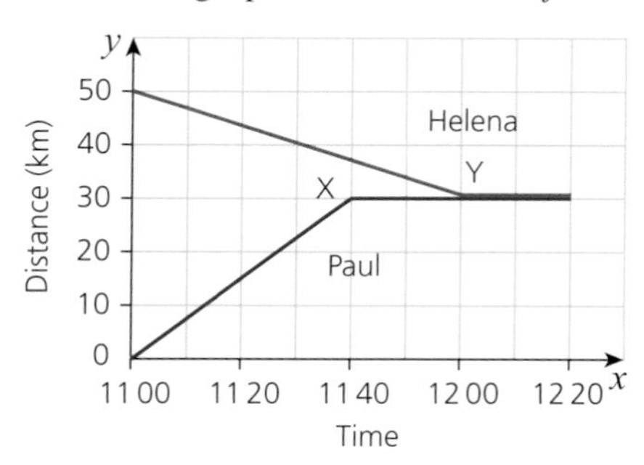

a What is Paul's average speed between 11 00 and 11 40?
b What is Helena's average speed between 11 00 and 12 00?
c What does the line XY represent?

2 A car travels at a speed of 60 km/h for 1 hour. It stops for 30 minutes, then continues at a constant speed of 80 km/h for a further 1.5 hours. Draw a distance–time graph for this journey.

3 A girl cycles for 1.5 hours at 10 km/h. She stops for an hour, then travels for a further 15 km in 1 hour. Draw a distance–time graph of the girl's journey.

4 Two friends leave their houses at 1600. The houses are 4 km apart and the friends travel towards each other on the same road. Fyodor walks at 7 km/h and Yin walks at 5 km/h.
  a On the same axes, draw a distance–time graph of their journeys.
  b From your graph estimate the time at which they meet.
  c Estimate the distance from Fyodor's house to the point where they meet.

5 A train leaves a station P at 1800 and travels to station Q 150 km away. It travels at a steady speed of 75 km/h. At 1810 another train leaves Q for P at a steady speed of 100 km/h.
  a On the same axes draw a distance–time graph to show both journeys.
  b From the graph estimate the time at which the trains pass each other.
  c At what distance from station Q do the trains pass each other?
  d Which train arrives at its destination first?

6 A train sets off from town P at 0915 and heads towards town Q 250 km away. Its journey is split into the three stages, a, b and c. At 0900 a second train leaves town Q heading for town P. Its journey is split into the two stages, d and e. Using the graph, calculate the following:
  a the speed of the first train during stages a, b and c
  b the speed of the second train during stages d and e.

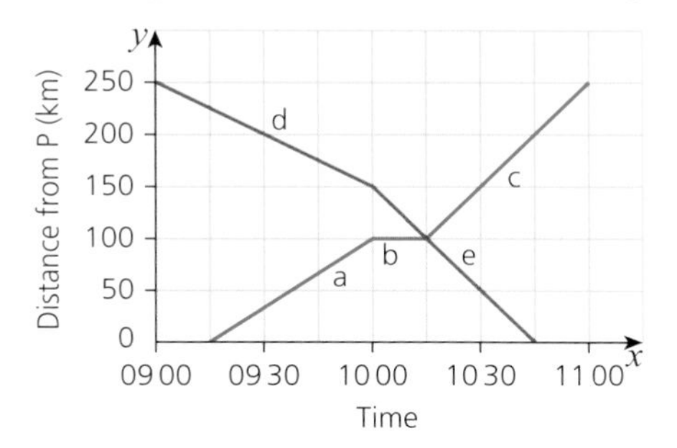

Student assessment 1

1 Absolute zero (0 K) is equivalent to –273 °C and 0 °C is equivalent to 273 K. Draw a conversion graph which will convert K into °C. Use your graph to estimate:
   a the temperature in K equivalent to –40 °C
   b the temperature in °C equivalent to 100 K.

2 A German plumber has a call-out charge of €70 and then charges a rate of €50 per hour.
   a Draw a conversion graph and estimate the cost of the following:
      i a job lasting $4\frac{1}{2}$ hours
      ii a job lasting $6\frac{3}{4}$ hours.
   b If a job cost €245, estimate from your graph how long it took to complete.

3 A boy lives 3.5 km from his school. He walks home at a constant speed of 9 km/h for the first 10 minutes. He then stops and talks to his friends for 5 minutes. He finally runs the rest of his journey home at a constant speed of 12 km/h.
   a Illustrate this information on a distance–time graph.
   b Use your graph to estimate the total time it took the boy to get home that day.

4 Look at the distance–time graphs A, B, C and D.
   a Two of the graphs are not possible.
      i Which two graphs are impossible?
      ii Explain why the two you have chosen are not possible.
   b Explain what the horizontal lines in the graphs say about the motion.

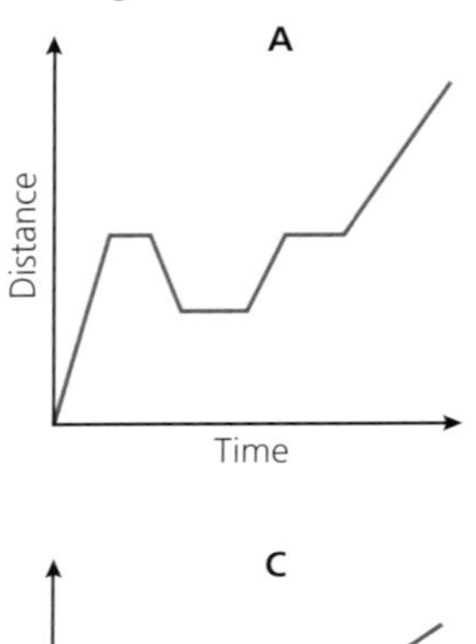

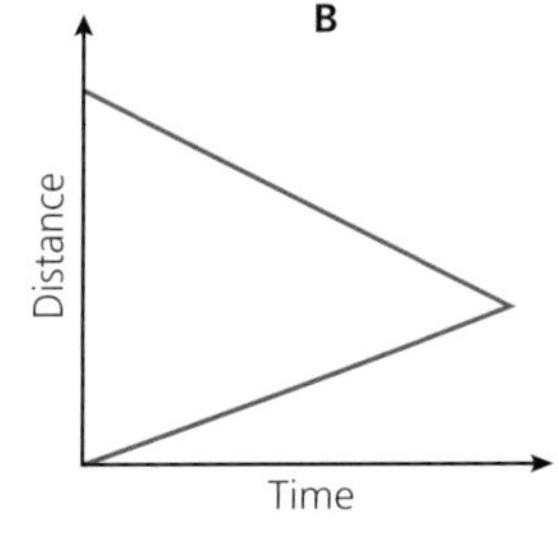

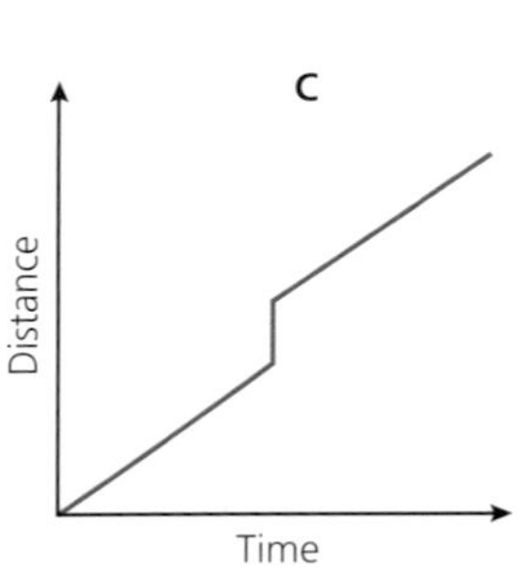

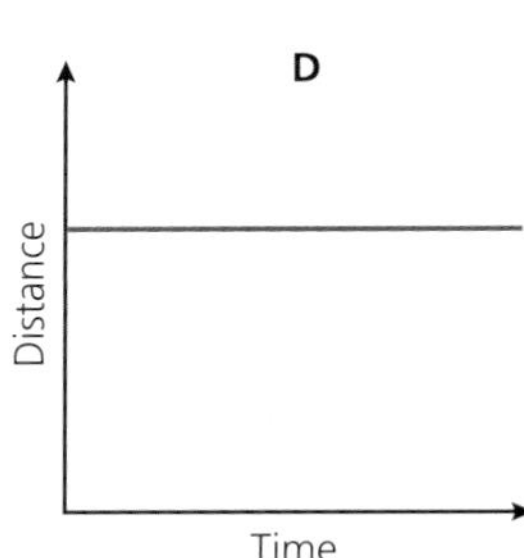

16 Graphs of functions

Linear functions

A **linear function** produces a straight line when plotted. A straight line consists of an infinite number of points. However, to plot a linear function, only two points on the line are needed. Once these have been plotted, the line can be drawn through them and extended, if necessary, at both ends.

Worked examples

1 Plot the line $y = x + 3$.

To identify two points simply choose two values of x, substitute these into the equation and calculate the corresponding y-values. Sometimes a small table of results is clearer.

x	y
0	3
4	7

Using the table, two points on the line are (0, 3) and (4, 7).
Plot the points on a pair of axes and draw a line through them:

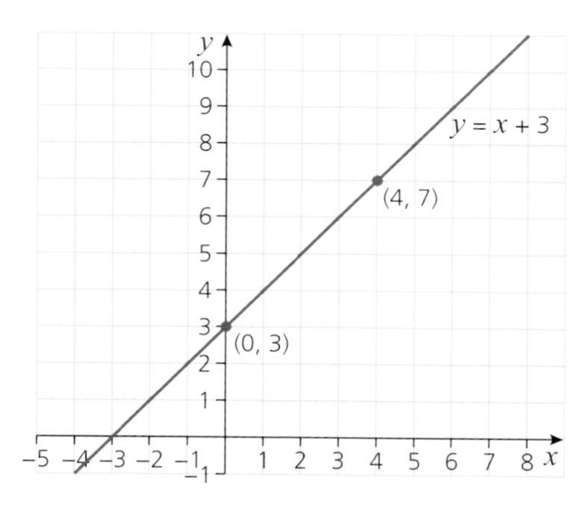

It is good practice to check with a third point:

Substituting $x = 2$ into the equation gives $y = 5$. As the point (2, 5) lies on the line, the line is drawn correctly.

2 Plot the line $2y+x=6$.

It is often easier to plot a line if the function is first written with y as the subject:

$$2y + x = 6$$
$$2y = -x + 6$$
$$y = -\frac{1}{2}x + 3$$

Choose two values of x and find the corresponding values of y:

x	y
0	3
6	0

From the table, two points on the line are (0, 3) and (6, 0).
Plot the points on a pair of axes and draw a line through them:

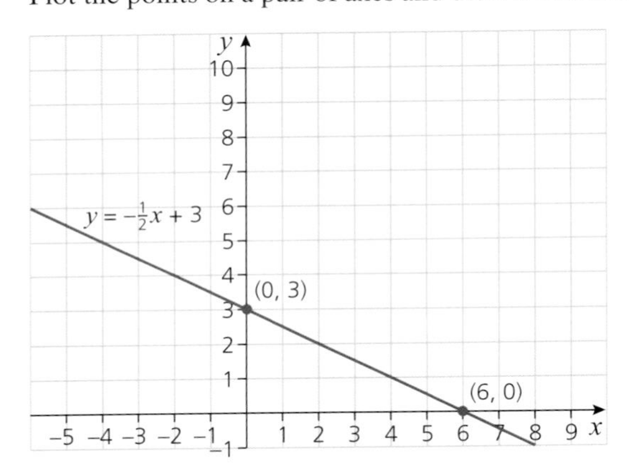

Check with a third point:
Substituting $x=4$ into the equation gives $y=1$. As the point (4, 1) lies on the line, the line is drawn correctly.

Exercise 16.1

1 Plot the following straight lines.

a $y=2x+4$ **b** $y=2x+3$ **c** $y=2x-1$

d $y=x-4$ **e** $y=x+1$ **f** $y=x+3$

g $y=1-x$ **h** $y=3-x$ **i** $y=-(x+2)$

2 Plot the following straight lines.

a $y=2x+3$ **b** $y=x-4$ **c** $y=3x-2$

d $y=-2x$ **e** $y=-x-1$ **f** $-y=x+1$

g $-y=3x-3$ **h** $2y=4x-2$ **i** $y-4=3x$

3 Plot the following straight lines.

a $-2x+y=4$ **b** $-4x+2y=12$ **c** $3y=6x-3$

d $2x=x+1$ **e** $3y-6x=9$ **f** $2y+x=8$

g $x+y+2=0$ **h** $3x+2y-4=0$ **i** $4=4y-2x$

Graphical solution of simultaneous equations

When solving two equations simultaneously, you need to find a solution that works for both equations. Chapter 13 shows how to arrive at the solution algebraically. It is, however, possible to arrive at the same solution graphically.

Worked example

a By plotting the graphs of both of the following equations on the same axes, find a common solution.

$x + y = 4$

$x - y = 2$

When both lines are plotted, the point at which they cross gives the common solution. This is because it is the only point which lies on both lines.

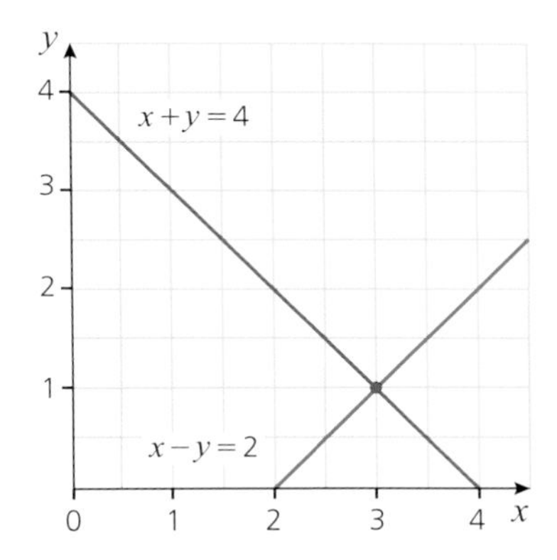

Therefore the common solution is (3, 1).

b Check your answer to part a by solving the equations algebraically.

$x + y = 4 \quad (1)$

$x - y = 2 \quad (2)$

$(1) + (2) \rightarrow 2x = 6$

$x = 3$

Substituting $x = 3$ into equation (1):

$3 + y = 4$

$y = 1$

Therefore the common solution occurs at (3, 1).

Exercise 16.2 Solve the simultaneous equations below:

i by graphical means ii by algebraic means.

1 **a** $x + y = 5$
$x - y = 1$

b $x + y = 7$
$x - y = 3$

c $2x + y = 5$
$x - y = 1$

d $2x + 2y = 6$
$2x - y = 3$

e $x + 3y = -1$
$x - 2y = -6$

f $x - y = 6$
$x + y = 2$

2 **a** $3x - 2y = 13$
$2x + y = 4$

b $4x - 5y = 1$
$2x + y = -3$

c $x + 5 = y$
$2x + 3y - 5 = 0$

d $x = y$
$x + y + 6 = 0$

e $2x + y = 4$
$4x + 2y = 8$

f $y - 3x = 1$
$y = 3x - 3$

Quadratic functions

The general expression for a quadratic function takes the form $ax^2 + bx + c$, where a, b and c are constants. Some examples of quadratic functions are:

$$y = 2x^2 + 3x - 12 \qquad y = x^2 - 5x + 6 \qquad y = 3x^2 + 2x - 3$$

If a graph of a quadratic function is plotted, the smooth curve produced is called a **parabola**. For example:

$y = x^2$

x	–4	–3	–2	–1	0	1	2	3	4
y	16	9	4	1	0	1	4	9	16

$y = -x^2$

x	–4	–3	–2	–1	0	1	2	3	4
y	–16	–9	–4	–1	0	–1	–4	–9	–16

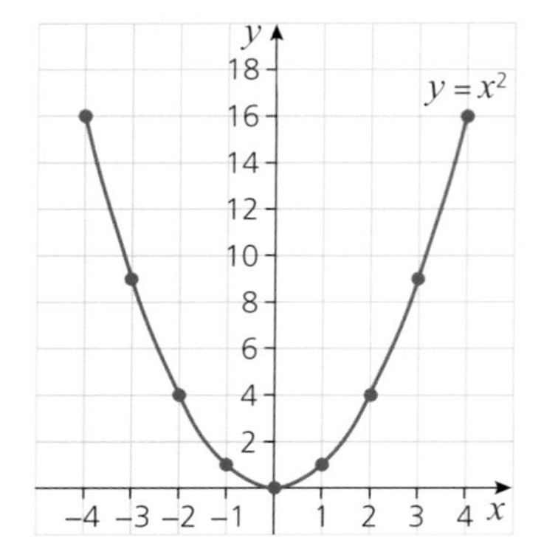

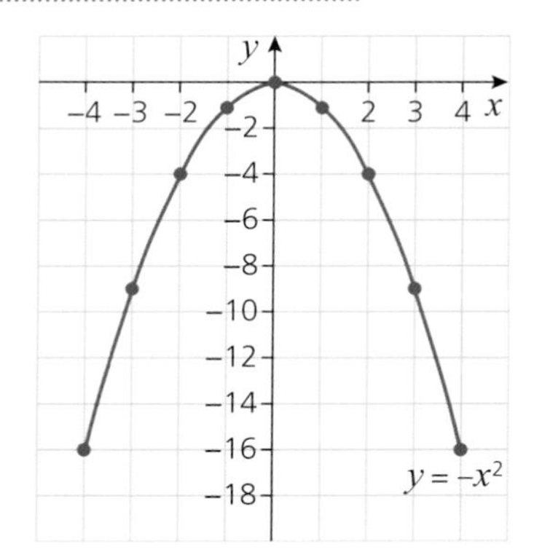

Worked examples

1 Plot a graph of the function $y = x^2 - 5x + 6$ for $0 \leqslant x \leqslant 5$.

First create a table of values for x and y:

x	0	1	2	3	4	5
$y = x^2 - 5x + 6$	6	2	0	0	2	6

These can then be plotted to give the graph:

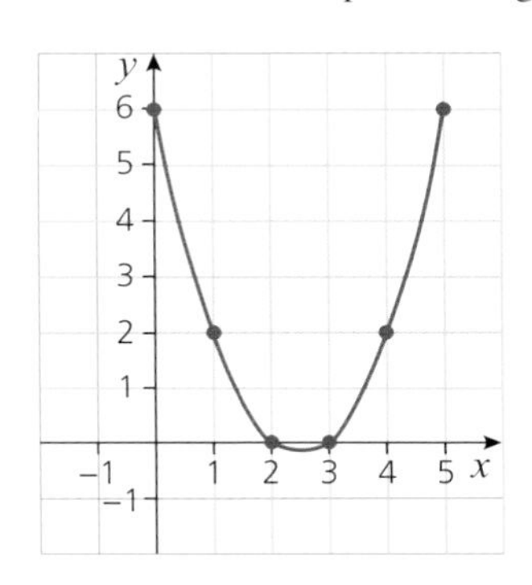

2 Plot a graph of the function $y = -x^2 + x + 2$ for $-3 \leqslant x \leqslant 4$.

Draw up a table of values:

x	−3	−2	−1	0	1	2	3	4
$y = -x^2 + x + 2$	−10	−4	0	2	2	0	−4	−10

Then plot the points and join them with a smooth curve:

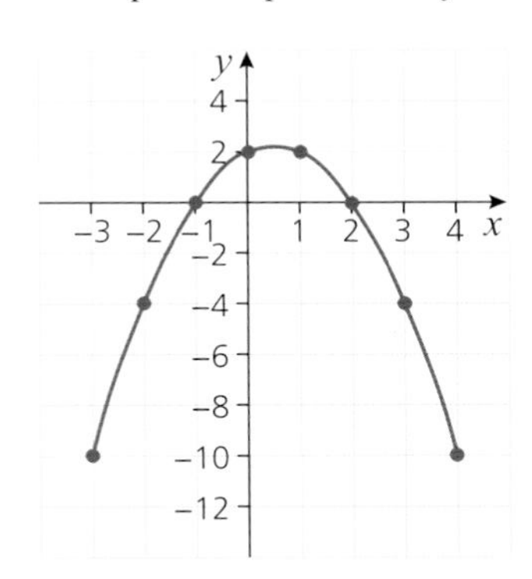

Exercise 16.3 For each of the following quadratic functions, construct a table of values for the stated range and then draw the graph.

a $y = x^2 + x - 2, \quad -4 \leqslant x \leqslant 3$
b $y = -x^2 + 2x + 3, \quad -3 \leqslant x \leqslant 5$
c $y = x^2 - 4x + 4, \quad -1 \leqslant x \leqslant 5$
d $y = -x^2 - 2x - 1, \quad -4 \leqslant x \leqslant 2$
e $y = x^2 - 2x - 15, \quad -4 \leqslant x \leqslant 6$

Graphical solution of a quadratic equation

To solve an equation, you need to find the values of x when $y = 0$. On a graph, these are the values of x where the curve crosses the x-axis. These are known as the **roots** of the equation.

Worked examples

1 Draw a graph of $y = x^2 - 4x + 3$ for $-2 \leqslant x \leqslant 5$.

x	−2	−1	0	1	2	3	4	5
y	15	8	3	0	−1	0	3	8

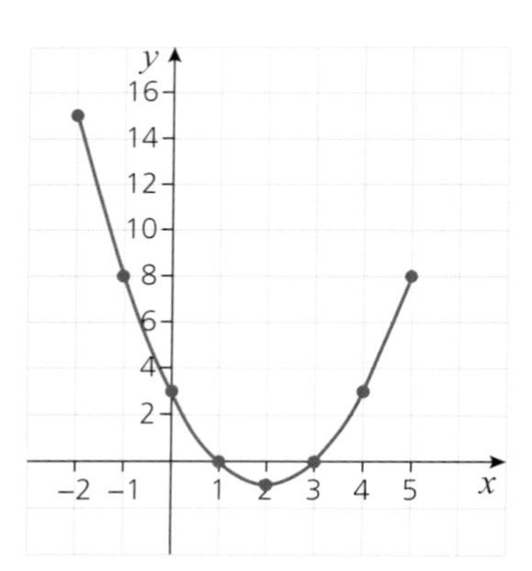

2 Use the graph to solve the equation $x^2 - 4x + 3 = 0$.

The solutions are $x = 1$ and $x = 3$

These are the values of x where the graph crosses the x-axis.

Exercise 16.4 Find the roots of each of the quadratic equations below by plotting a graph for the ranges of x stated.

a $x^2 - x - 6 = 0, \quad -4 \leqslant x \leqslant 4$
b $-x^2 + 1 = 0, \quad -4 \leqslant x \leqslant 4$
c $x^2 - 6x + 9 = 0, \quad 0 \leqslant x \leqslant 6$
d $-x^2 - x + 12 = 0, \quad -5 \leqslant x \leqslant 4$
e $x^2 - 4x + 4 = 0, \quad -2 \leqslant x \leqslant 6$

In the previous worked example in which $y = x^2 - 4x + 3$, a solution was found to the equation $x^2 - 4x + 3 = 0$ by reading the values of x where the graph crossed the x-axis. The graph can, however, also be used to solve other related quadratic equations.

Worked example

Look at how the given equation relates to the given graph.

Use the graph of $y = x^2 - 4x + 3$ to solve the equation $y = x^2 - 4x + 1 = 0$.

$x^2 - 4x + 1 = 0$ can be rearranged to give

$x^2 - 4x + 3 = 2$

Using the graph of $y = x^2 - 4x + 3$ and plotting the line $y = 2$ on the same axes gives the graph shown below.

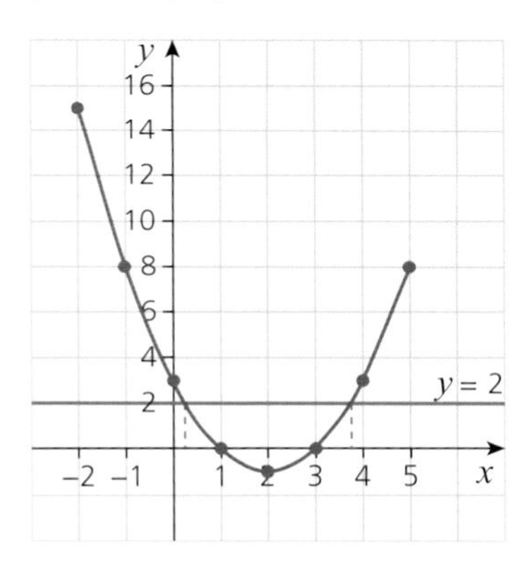

Where the curve and the line cross gives the solution to $x^2 - 4x + 3 = 2$, and hence also $x^2 - 4x + 1 = 0$.

Therefore the solutions to $x^2 - 4x + 1 = 0$ are $x \approx 0.3$ and $x \approx 3.7$.

Exercise 16.5

Using the graphs that you drew in Exercise 16.4, solve the following quadratic equations. Show your method clearly.

a $x^2 - x - 4 = 0$

b $-x^2 - 1 = 0$

c $x^2 - 6x + 8 = 0$

d $-x^2 - x + 9 = 0$

e $x^2 - 4x + 1 = 0$

The reciprocal function

If a graph of a **reciprocal function** is plotted, the curve produced is called a **hyperbola**.

Worked example

Draw the graph of $y = \frac{2}{x}$ for $-4 \leqslant x \leqslant 4$.

x	−4	−3	−2	−1	0	1	2	3	4
y	−0.5	−0.7	−1	−2	–	2	1	0.7	0.5

$y = \frac{2}{x}$ is a reciprocal function so the graph is a hyperbola.

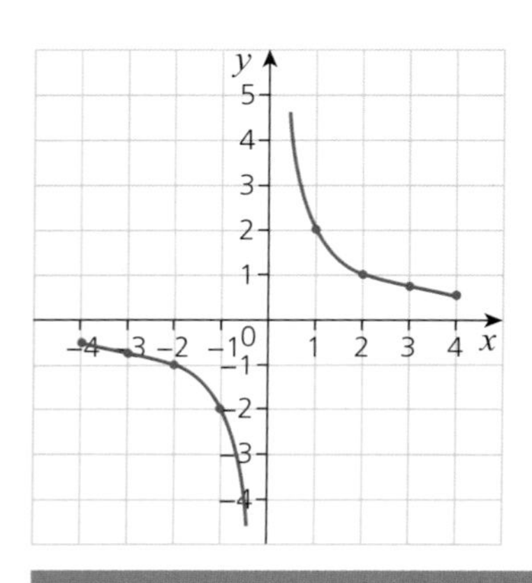

The graph gets closer and closer to both the x and y axes, without actually touching or crossing them. The axes are known as asymptotes to the graph.

Exercise 16.6

1 Plot the graph of the function $y = \frac{1}{x}$ for $-4 \leqslant x \leqslant 4$.

2 Plot the graph of the function $y = \frac{3}{x}$ for $-4 \leqslant x \leqslant 4$.

3 Plot the graph of the function $y = \frac{5}{2x}$ for $-4 \leqslant x \leqslant 4$.

Recognising and sketching functions

So far in this chapter, you have plotted graphs of functions. In other words, you have substituted values of x into the equation of a function, calculated the corresponding values of y, and plotted and joined the resulting (x, y) coordinates.

However, plotting an accurate graph is time-consuming and is not always necessary to answer a question. In many cases, a sketch of a graph is as useful and is considerably quicker.

When doing a sketch, certain key pieces of information need to be included. As a minimum, where the graph intersects both the x-axis and y-axis needs to be given.

Sketching linear functions

Straight line graphs can be sketched simply by working out where the line intersects both axes.

Worked example

Sketch the graph of $y = -3x + 5$.

The graph intersects the y-axis when $x = 0$.
This is substituted in to the equation:

$$y = -3(0) + 5$$
$$y = 5$$

The graph intersects the x-axis when $y = 0$.
This is then substituted in to the equation and solved:

$$0 = -3x + 5$$
$$3x = 5$$
$$x = \frac{5}{3} \text{ (or } 1\tfrac{2}{3}\text{)}$$

Mark the two points and join them with a straight line:

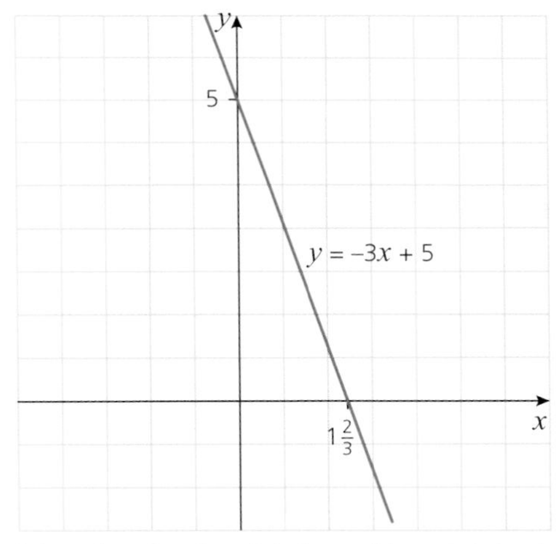

Note that the sketch below, although it looks very different from the one above, is also acceptable as it shows the same intersections with the axes.

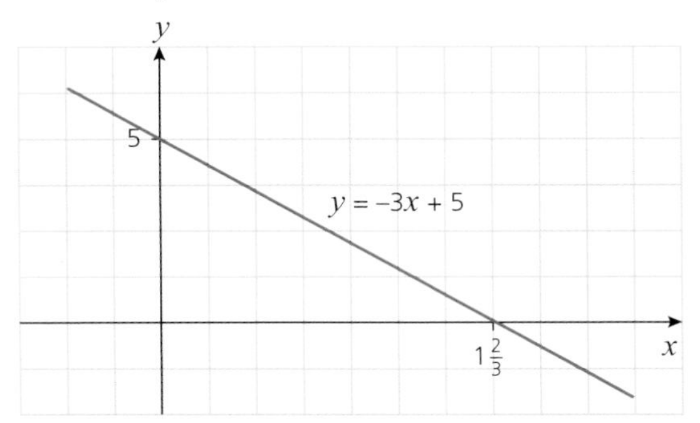

Exercise 16.7 Sketch the following linear functions, showing clearly where the lines intersect both axes.

a $y = 2x - 4$ **b** $y = \frac{1}{2}x + 6$ **c** $y = -2x - 3$

d $y = -\frac{1}{3}x + 9$ **e** $2y + x - 2 = 0$ **f** $x = \frac{2y+4}{3}$

Sketching quadratic functions

You will have seen that plotting a quadratic function produces either a ∪ or a ∩ shaped curve, depending on whether the x^2 term is positive or negative.

Therefore, a quadratic graph has either a highest point or a lowest point, depending on its shape. These points are known as **turning points**.

To sketch a quadratic, the key points that need to be included are the intersection with the y-axis and either the coordinates of the turning point or the intersection(s) with the x-axis.

Worked examples

1 The graph of the quadratic equation $y = x^2 - 6x + 11$ has a turning point at (3, 2). Sketch the graph, showing clearly where it intersects the y-axis.

As the x^2 term is positive, the graph is ∪-shaped.
To find where the graph intersects the y-axis, substitute $x = 0$ into the equation.

$y = (0)^2 - 6(0) + 11$
$y = 11$

Therefore the graph can be sketched as follows:

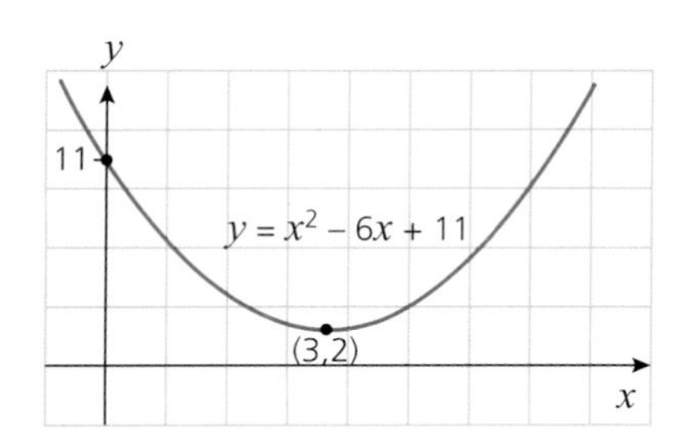

2 The coordinates of the turning point of a quadratic graph are (–4, 4). The equation of the function is $y = -x^2 - 8x - 12$. Sketch the graph.

As the x^2 term is negative, the graph is ∩-shaped.
To find where the graph intersects the y-axis, substitute $x = 0$ into the equation.

$y = -(0)^2 - 8(0) - 12$
$y = -12$

Therefore the graph of $y = -x^2 - 8x - 12$ can be sketched as follows:

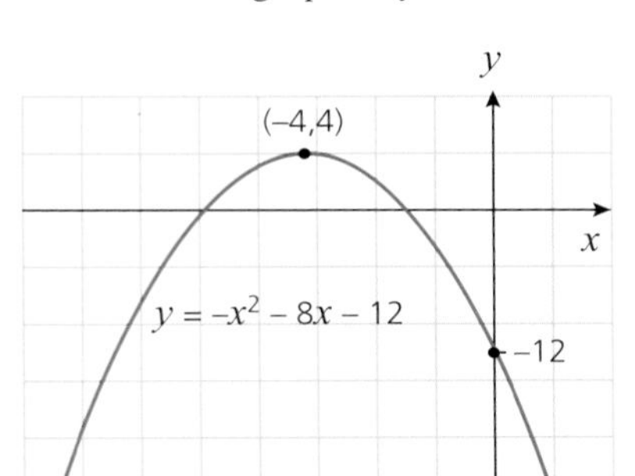

You will have seen in Chapter 11 that an expression such as $(x - 2)(x + 3)$ can be expanded to give $x^2 + x - 6$. Therefore $(x - 2)(x + 3)$ is also a quadratic expression. If a quadratic function is given in this form then the graph can still be sketched and, in particular, the points where it crosses the x-axis can be found too.

Worked examples

1 Show that $(x+4)(x+2) = x^2 + 6x + 8$.

$$(x+4)(x+2) = x^2 + 4x + 2x + 8$$
$$= x^2 + 6x + 8$$

2 Hence sketch the graph of $y = (x+4)(x+2)$.

The x^2 term is positive so the graph is ∪-shaped.
The graph intersects the y-axis when $x = 0$:

$$y = (0+4)(0+2)$$
$$= 4 \times 2$$
$$= 8$$

As the coordinates of the turning point are not given, the intersection with the x-axis needs to be calculated. At the x-axis, $y = 0$. Substituting $y = 0$ into the equation gives:

$$(x+4)(x+2) = 0$$

For the product of two terms to be zero, one of the terms must be zero.
If $(x+4) = 0$, then $x = -4$.
If $(x+2) = 0$, then $x = -2$.

$x = -4$ and $x = -2$ are therefore where the graph intersects the x-axis.
The graph can be sketched as:

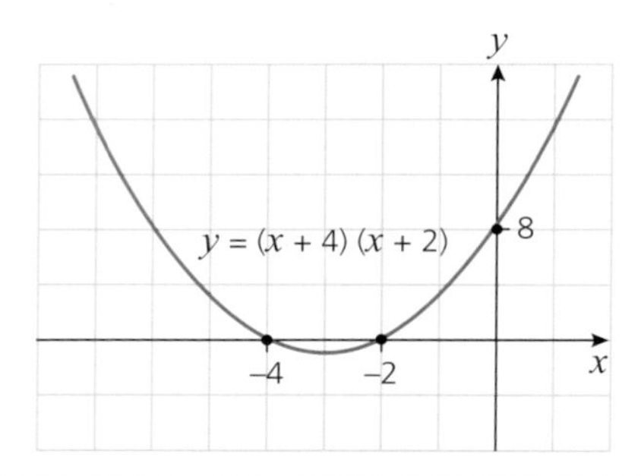

Exercise 16.8

1 For each part, the equation of a quadratic function and the coordinates of its turning point are given. Sketch a graph for each function.

a $y = x^2 - 10x + 27$; turning point at $(5, 2)$

b $y = x^2 + 2x - 5$; turning point at $(-1, -6)$

c $y = -x^2 + 4x - 3$; turning point at $(2, 1)$

d $y = x^2 - 12x + 36$; turning point at $(6, 0)$

e $y = 4x^2 - 20x$; turning point at $\left(\frac{5}{2}, -25\right)$

2 a Expand the brackets $(x + 3)(x - 3)$.

b Use your expansion in part a) to sketch the graph of $y = (x + 3)(x - 3)$. Label any point(s) of intersection with the axes.

3 a Expand the expression $-(x - 2)(x - 4)$.

b Use your expansion in part a) to sketch the graph of $y = -(x - 2)(x - 4)$. Label any point(s) of intersection with the axes.

4 a Expand the brackets $(-3x - 6)(x + 1)$.

b Use your expansion in part a) to sketch the graph of $y = (-3x - 6)(x + 1)$. Label any point(s) of intersection with the axes.

Student assessment 1

1 Plot these lines on the same pair of axes. Label each line clearly.

a $x = -2$ b $y = 3$ c $y = 2x$ d $y = -\frac{x}{2}$

2 Plot the graph of each linear equation.

a $y = x + 1$ b $y = 3 - 3x$

c $2x - y = 4$ d $2y - 5x = 8$

3 Solve each pair of simultaneous equations graphically.

a $x + y = 4$
$x - y = 0$

b $3x + y = 2$
$x - y = 2$

c $y + 4x + 4 = 0$
$x + y = 2$

d $x - y = -2$
$3x + 2y + 6 = 0$

4 a Copy and complete the table of values for the function $y = x^2 + 8x + 15$.

x	−7	−6	−5	−4	−3	−2	−1	0	1	2
y		3				3				

b Plot a graph of the function.

5 Plot the graph of each function for the given limits of x.

a $y = x^2 - 3; -4 \leqslant x \leqslant 4$

b $y = 3 - x^2; -4 \leqslant x \leqslant 4$

c $y = -x^2 - 2x - 1; -3 \leqslant x \leqslant 3$

d $y = x^2 + 2x - 7; -5 \leqslant x \leqslant 3$

6 a Plot the graph of the quadratic function $y = x^2 + 9x + 20$ for $-7 \leqslant x \leqslant -2$.

b Showing your method clearly, use information from your graph to solve the equation $x^2 = -9x - 14$.

7 Plot the graph of $y = \frac{1}{x}$ for $-4 \leqslant x \leqslant 4$.

8 Sketch the graph of $y = \frac{1}{3}x - 5$. Label clearly any points of intersection with the axes.

9 The quadratic equation $y = 4x^2 - 8x - 4$ has its turning point at $(1, -8)$. Sketch the graph of the function.

Student assessment 2

1 Plot these lines on the same pair of axes. Label each line clearly.

a $x = 3$

b $y = -2$

c $y = -3x$

d $y = \frac{x}{4} + 4$

2 Plot the graph of each linear equation.

a $y = 2x + 3$

b $y = 4 - x$

c $2x - y = 3$

d $-3x + 2y = 5$

3 Solve each pair of simultaneous equations graphically.

a $x + y = 6$
$x - y = 0$

b $x + 2y = 8$
$x - y = -1$

c $2x - y = -5$
$x - 3y = 0$

d $4x - 2y = -2$
$3x - y + 2 = 0$

4 a Copy and complete the table of values for the function $y = -x^2 - 7x - 12$.

x	−7	−6	−5	−4	−3	−2	−1	0	1	2
y		−6				−2				

b Plot a graph of the function.

5 Plot the graph of each function for the given limits of x.

a $y = x^2 - 5; -4 \leq x \leq 4$

b $y = 1 - x^2; -4 \leq x \leq 4$

c $y = x^2 - 3x - 10; -3 \leq x \leq 6$

d $y = -x^2 - 4x - 4; -5 \leq x \leq 1$

6 a Plot the graph of the quadratic equation $y = -x^2 - x + 15$ for $-6 \leq x \leq 4$.

b Showing your method clearly, use your graph to help you solve these equations.

i $10 = x^2 + x$

ii $x^2 = -x + 5$

7 Plot the graph of $y = \frac{2}{x}$ for $-4 \leq x \leq 4$.

8 Sketch the graph of $y = -\frac{5}{2}x + 10$. Label clearly any points of intersection with the axes.

9 Sketch the graph of the quadratic equation $y = -(2x - 2)(x - 7)$. Show clearly any points of intersection with the axes.

TOPIC 2

Mathematical investigations and ICT

House of cards

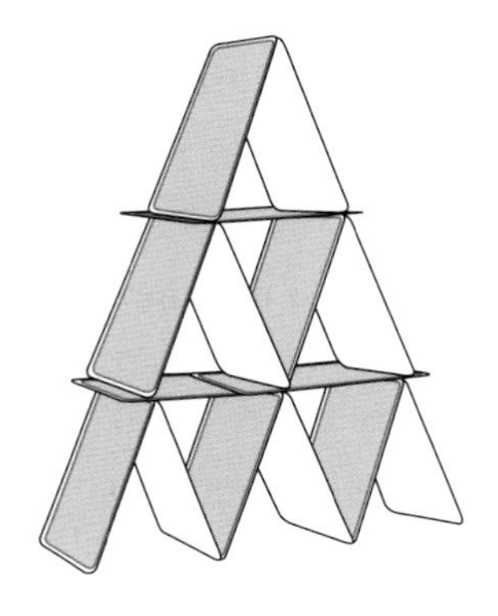

The drawing shows a house of cards 3 layers high. 15 cards are needed to construct it.

1 How many cards are needed to construct a house 10 layers high?
2 The largest house of cards ever constructed was 75 layers high. How many cards were needed?
3 Show that the general formula for the number of cards, c, needed to construct a house of cards n layers high is:

$$c = \frac{1}{2}n(3n+1)$$

Chequered boards

A chessboard is an 8×8 square grid with alternating black and white squares.

There are 64 unit squares of which 32 are black and 32 are white.

Consider boards of different sizes consisting of alternating black and white unit squares.

For example:

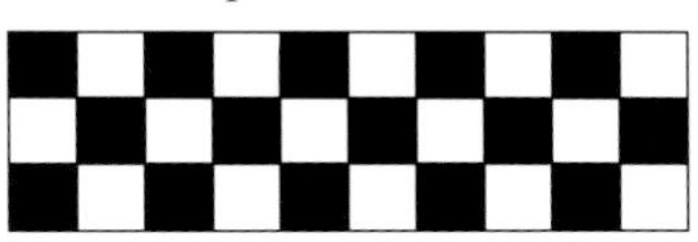

Total number of unit squares: 30
Number of black squares: 15
Number of white squares: 15

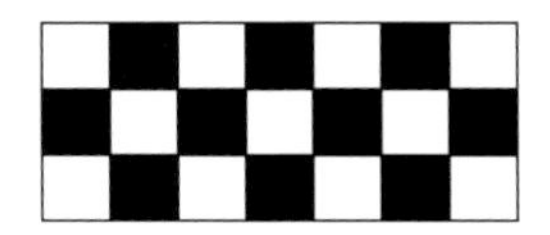

Total number of unit squares: 21
Number of black squares: 10
Number of white squares: 11

1 Investigate the number of black and white unit squares on different rectangular boards. Note: For consistency you may find it helpful to always keep the bottom right-hand square the same colour.
2 What are the numbers of black and white squares on a board $m \times n$ units?

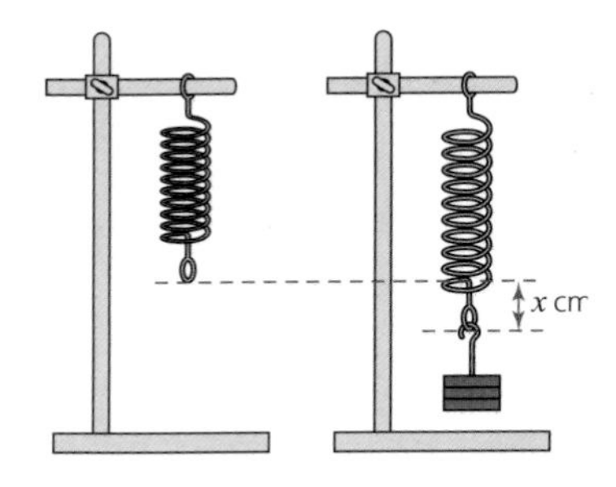

Modelling: Stretching a spring

A spring is attached to a clamp stand as shown.

Different weights are attached to the end of the spring. The table shows the mass (m) in grams of each weight and the amount by which the spring stretches (x) in centimetres.

Mass (g)	50	100	150	200	250	300	350	400	450	500
Extension (cm)	3.1	6.3	9.5	12.8	15.4	18.9	21.7	25.0	28.2	31.2

1 Plot a graph of mass against extension.
2 Describe the approximate relationship between the mass and the extension.
3 Draw a line of best fit through the data.
4 Calculate the equation of the line of best fit.
5 Use your equation to predict the extension of the spring for a mass of 275 g.
6 Explain why it is unlikely that the equation would be useful to find the extension when a mass of 5 kg is added to the spring.

ICT activity 1

You have seen that the solution of two simultaneous equations gives the coordinates of the point that satisfies both equations. If the simultaneous equations were plotted, the point at which the lines cross would correspond to the solution of the simultaneous equations.

For example:

Solving $x + y = 6$ and $x - y = 2$ produces the result $x = 4$ and $y = 2$, i.e. coordinates $(4, 2)$.

Plotting the graphs of both lines confirms that the point of intersection occurs at $(4, 2)$.

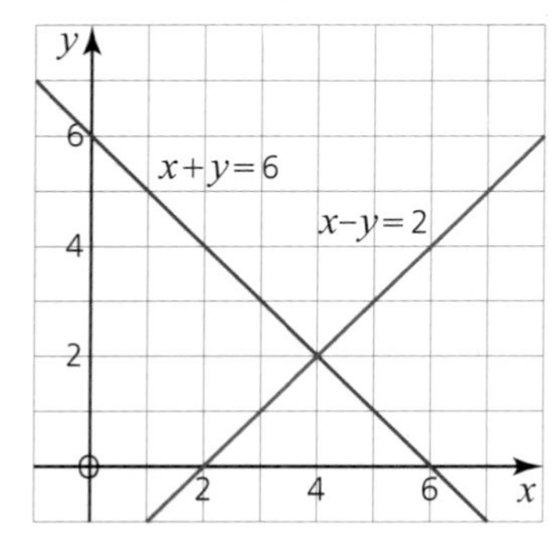

1 Use a graphing package to solve the following simultaneous equations graphically.

a $y = x$ and $x + y = 4$

b $y = 2x$ and $x + y = 3$

c $y = 2x$ and $y = 3$

d $y - x = 2$ and $y + \frac{1}{2}x = 5$

e $y + x = 5$ and $y + \frac{1}{2}x = 3$

2 Check your answers to Q.1 by solving each pair of simultaneous equations algebraically.

ICT activity 2

In this activity you will be using a graphing package or graphical calculator to find the solutions to quadratic and reciprocal functions.

You have seen that if a quadratic equation is plotted, its solution is represented by the points of intersection with the x-axis. For example, when plotting $y = x^2 - 4x + 3$, as shown below, the solution of $x^2 - 4x + 3 = 0$ occurs where the graph crosses the x-axis, i.e. at $x = 1$ and $x = 3$.

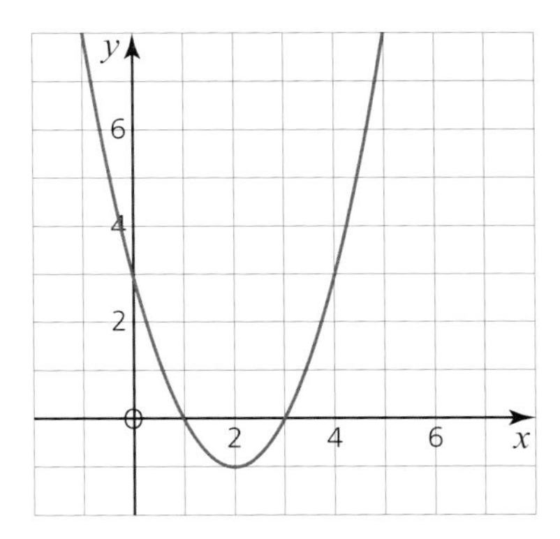

Use a graphing package or a graphical calculator to solve each equation graphically.

a $x^2 + x - 2 = 0$

b $x^2 - 7x + 6 = 0$

c $x^2 + x - 12 = 0$

d $2x^2 + 5x - 3 = 0$

e $\frac{2}{x} - 2 = 0$

f $\frac{2}{x} + 1 = 0$

TOPIC 3

Coordinate geometry

Contents

Course

C4.1
Use and interpret the geometrical terms: point, line, parallel, bearing, right angle, acute, obtuse and reflex angles, perpendicular, similarity and congruence.
Use and interpret vocabulary of triangles, quadrilaterals, circles, polygons and simple solid figures including nets.

C4.2
Measure and draw lines and angles.
Construct a triangle given the three sides using a ruler and a pair of compasses only.

C4.3
Read and make scale drawings.

C4.4
Calculate lengths of similar figures.

C4.5
Recognise congruent shapes.

C4.6
Recognise rotational and line symmetry (including order of rotational symmetry) in two dimensions.

C4.7
Calculate unknown angles using the following geometrical properties:

- angles at a point
- angles at a point on a straight line and intersecting straight lines
- angles formed within parallel lines
- angle properties of triangles and quadrilaterals
- angle properties of regular polygons
- angle in a semi-circle
- angle between tangent and radius of a circle.

The development of geometry

The beginnings of geometry can be traced back to around 2000BCE in ancient Mesopotamia and Egypt. Early geometry was a practical subject concerning lengths, angles, areas and volumes and was used in surveying, construction, astronomy and various crafts.

The earliest known texts on geometry are the Egyptian Rhind Papyrus (c.1650BCE), the Moscow Papyrus (c.1890BCE) and Babylonian clay tablets such as Plimpton 322 (c.1900BCE). For example, the Moscow Papyrus gives a formula for calculating the volume of a truncated pyramid, or frustum.

In the 7th century BCE, the Greek mathematician Thales of Miletus (which is now in Turkey) used geometry to solve problems such as calculating the height of pyramids and the distance of ships from the shore.

Euclid

In around 300BCE, Euclid wrote his book *Elements*, perhaps the most successful textbook of all time. It introduced the concepts of definition, theorem and proof. Its contents are still taught in geometry classes today.

18 Geometrical vocabulary

NB: All diagrams are not drawn to scale.

Angles

Different types of angle have different names:

acute angles lie between 0° and 90°
right angles are exactly 90°
obtuse angles lie between 90° and 180°
reflex angles lie between 180° and 360°.

Two angles which add together to total 180° are called **supplementary angles**.

Two angles which add together to total 90° are called **complementary angles**.

Exercise 18.1

1 Draw and label one example of each of the following types of angle:

right acute obtuse reflex

2 Copy the angles below and write beneath each drawing the type of angle it shows:

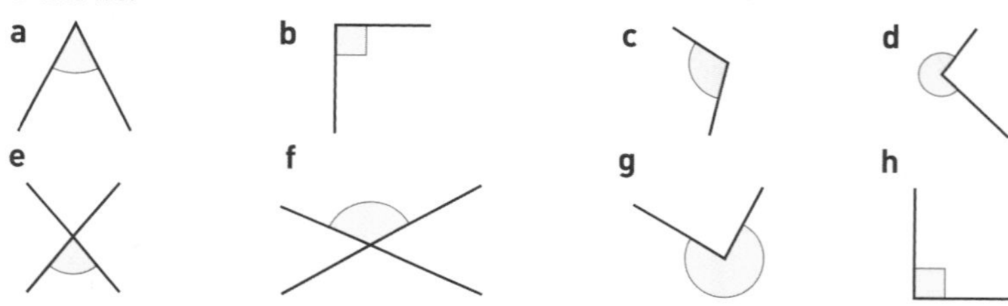

3 State whether the following pairs of angles are supplementary, complementary or neither:

a 70°, 20° **b** 90°, 90° **c** 40°, 50° **d** 80°, 30°
e 15°, 75° **f** 145°, 35° **g** 133°, 57° **h** 33°, 67°
i 45°, 45° **j** 140°, 40°

Angles can be labelled in several ways:

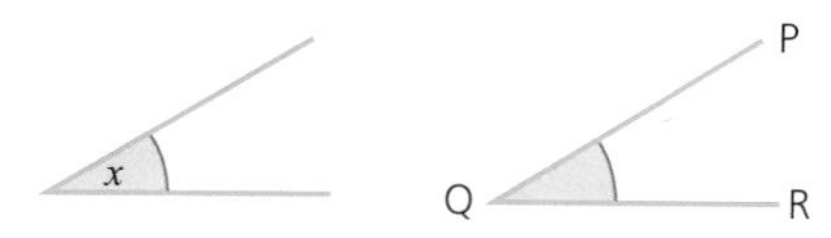

In the first case the angle is labelled directly as x. In the second example the angle can be labelled as angle PQR or angle RQP. In this case three points are chosen, with the angle in question being the middle one.

Exercise 18.2

Sketch the shapes below. Name the angles marked with a single letter, in terms of three vertices.

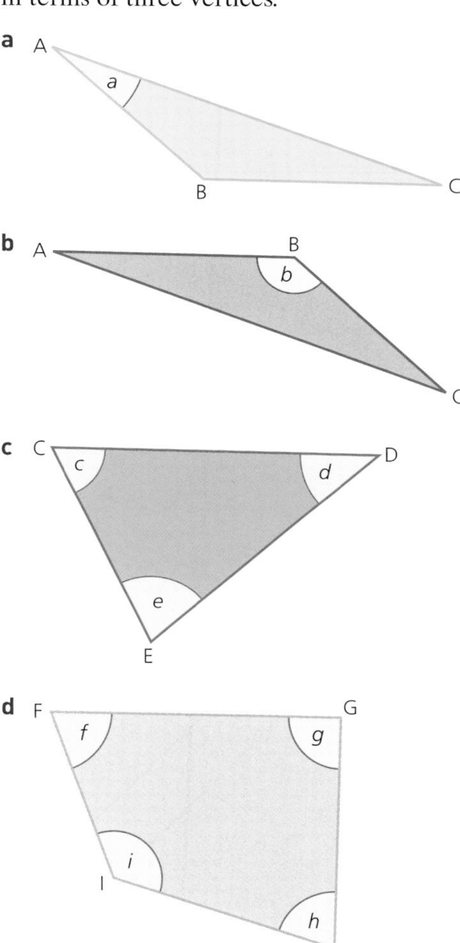

Perpendicular lines

To find the shortest **distance between two points**, you measure the length of the **straight line** which joins them.

Two lines which meet at right angles are **perpendicular** to each other.

So in this diagram CD is perpendicular to AB, and AB is perpendicular to CD.

If the lines AD and BD are drawn to form a triangle, the line CD can be called the **height** or **altitude** of the triangle ABD.

Exercise 18.3 For these diagrams, state which pairs of lines are perpendicular to each other.

a

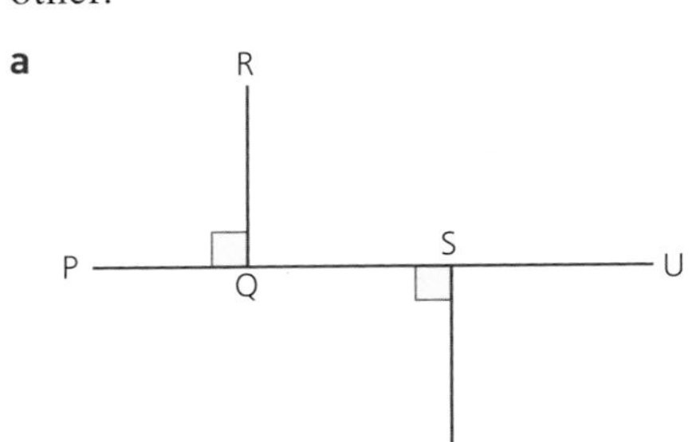

b

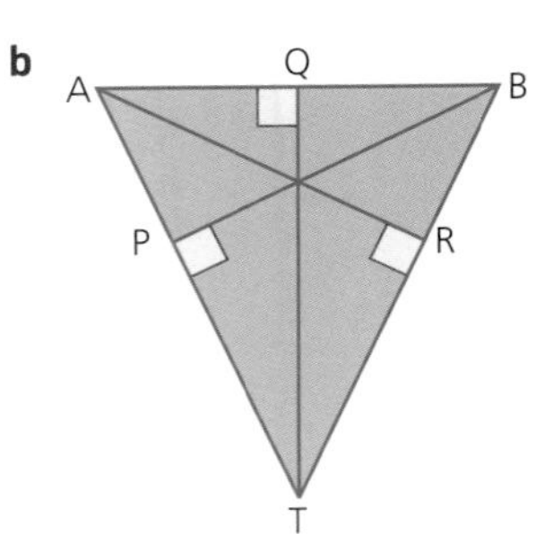

Parallel lines

Parallel lines are straight lines which can be continued to infinity in either direction without meeting.

Railway lines are an example of parallel lines. Parallel lines are marked with arrows as shown:

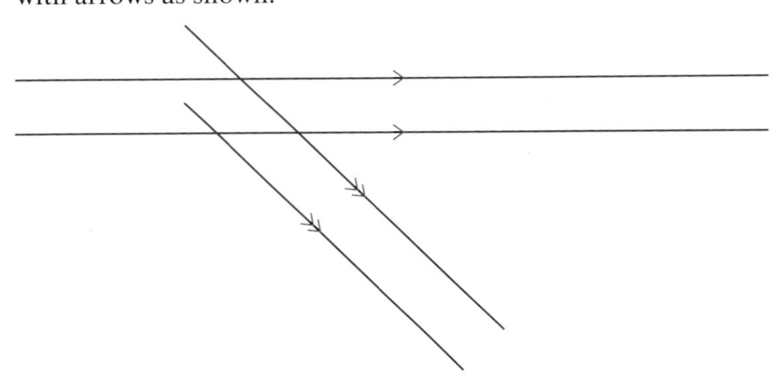

Vocabulary of the circle

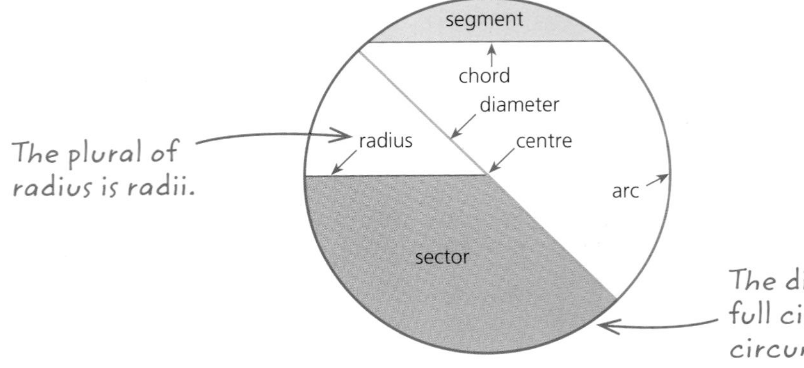

The distance around the full circle is called the circumference.

Polygons

A **polygon** is a closed two-dimensional shape bounded by straight lines. Examples of polygons include triangles, quadrilaterals, pentagons and hexagons. These shapes all belong to the polygon family:

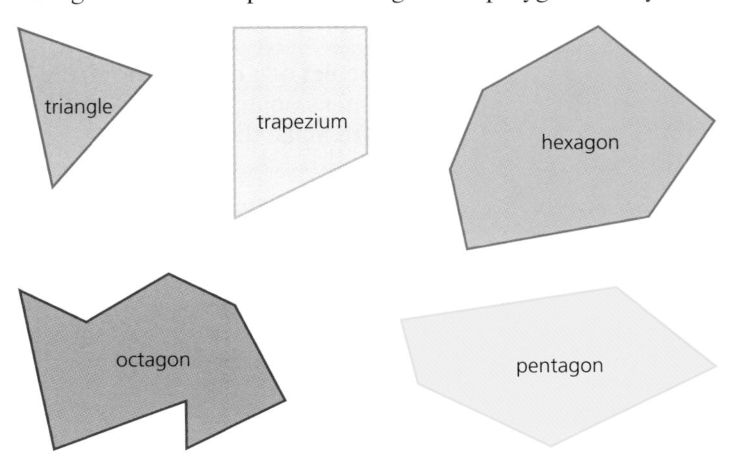

A **regular polygon** is distinctive in that all its sides are of equal length and all its angles are of equal size. These shapes are some examples of regular polygons:

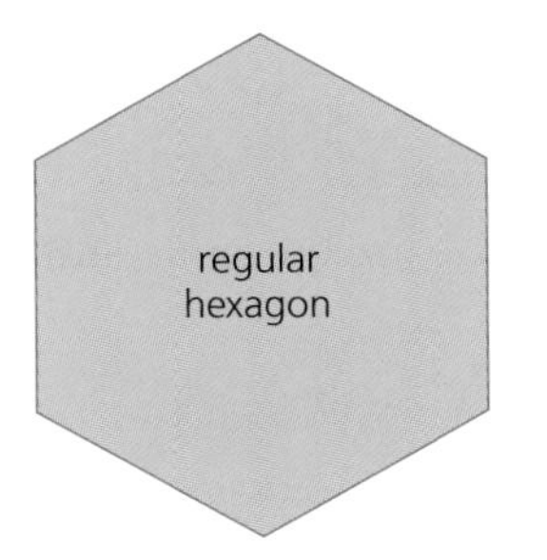

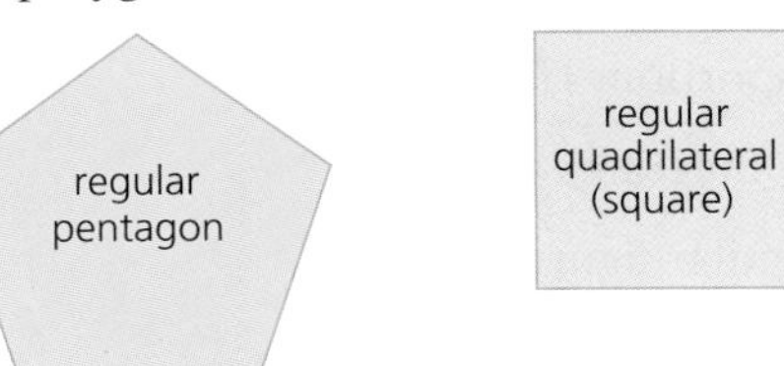

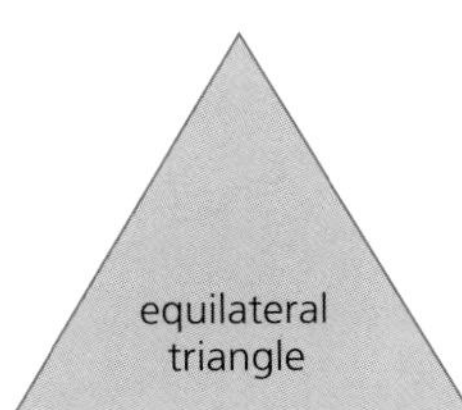

The names of polygons are related to the number of angles they contain:

3 angles = **tri**angle
4 angles = **quad**rilateral (tetragon)
5 angles = **penta**gon
6 angles = **hexa**gon
7 angles = **hepta**gon
8 angles = **octa**gon
9 angles = **nona**gon
10 angles = **deca**gon
12 angles = **dodeca**gon

Exercise 18.4

1 Draw a sketch of each of the shapes listed on the previous page.

2 Draw accurately a regular hexagon, a regular pentagon and a regular octagon.

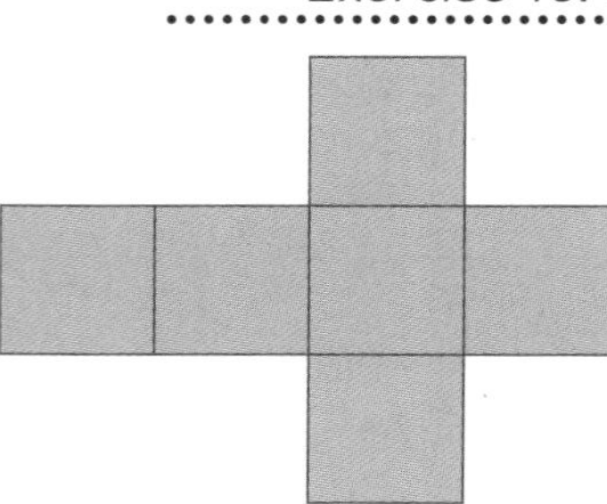

Nets

The diagram is the **net** of a cube. It shows the faces of the cube opened out into a two-dimensional plan. The net of a three-dimensional shape can be folded up to make that shape.

Exercise 18.5

Draw the following on squared paper:

a Two other possible nets of a cube
b The net of a cuboid (rectangular prism)
c The net of a triangular prism
d The net of a cylinder
e The net of a square-based pyramid
f The net of a tetrahedron

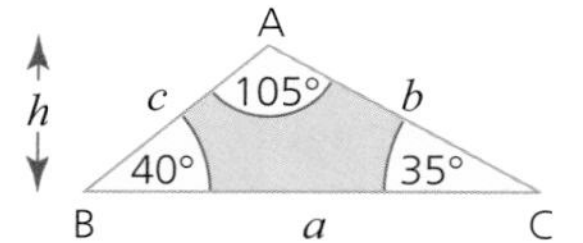

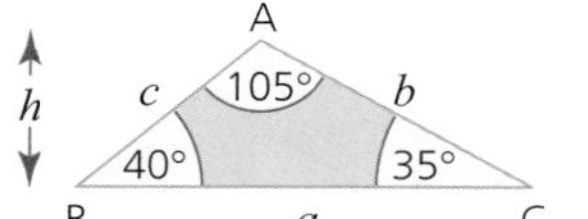

Similar shapes

Two polygons are **similar** if their angles are the same and corresponding sides are in proportion.

For triangles, having equal angles implies that corresponding sides are in proportion. The converse is also true.

In the diagrams (left) triangle ABC and triangle PQR are similar.

For similar figures the ratios of the lengths of the sides are the same and represent the scale factor, i.e.

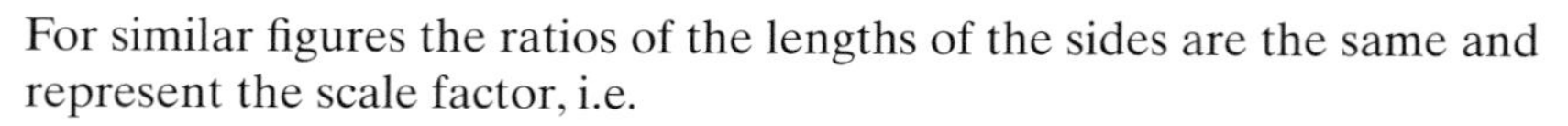

$\frac{p}{a} = \frac{q}{b} = \frac{r}{c} = k$ (where k is the **scale factor of enlargement**)

The heights of similar triangles are proportional also:

$$\frac{H}{h} = \frac{p}{a} = \frac{q}{b} = \frac{r}{c} = k$$

Exercise 18.6

1 a Explain why the two triangles are similar.

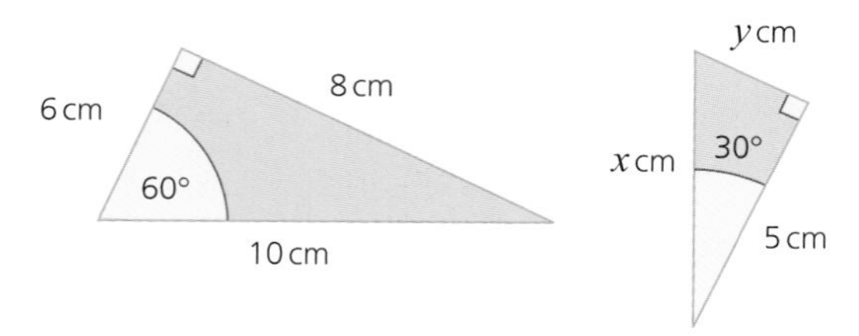

b Calculate the scale factor which reduces the larger triangle to the smaller one.

c Calculate the value of x and the value of y.

2 Which of the triangles below are similar?

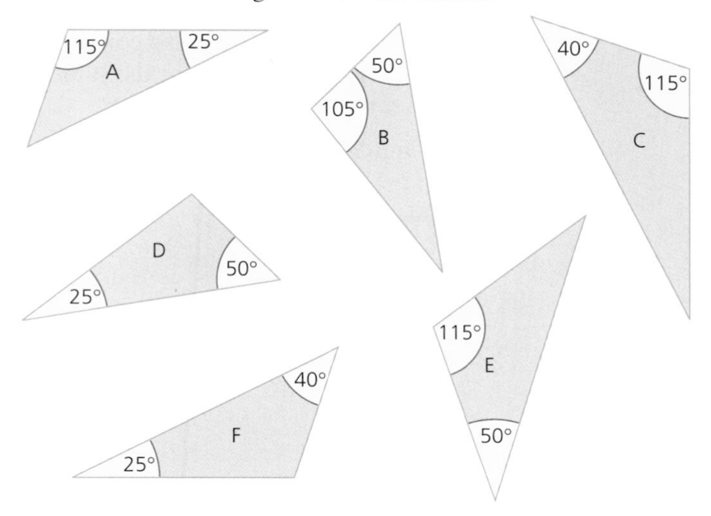

3 The triangles are similar.

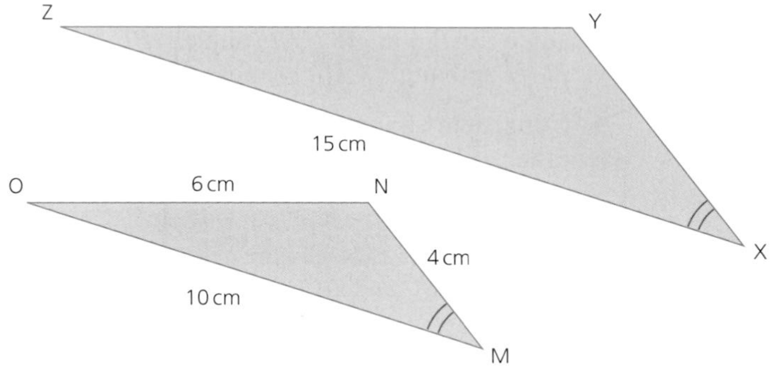

a Calculate the length XY.

b Calculate the length YZ.

4 Calculate the lengths of sides p, q and r in the triangle:

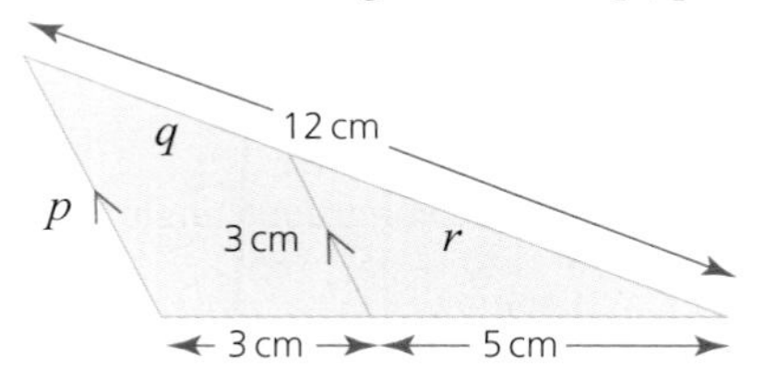

5 Calculate the lengths of sides e and f in the trapezium:

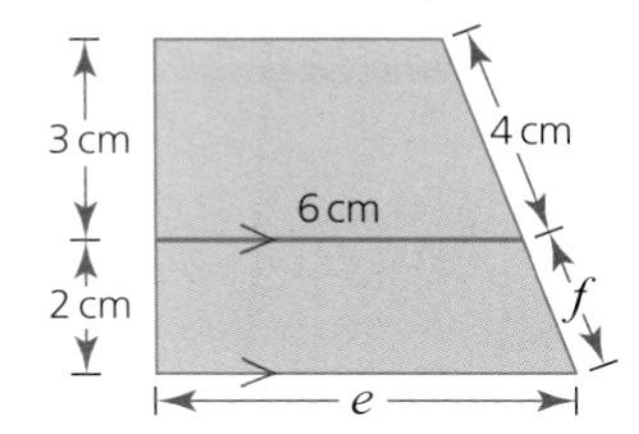

Congruent shapes

Two shapes are **congruent** if their corresponding sides are the same length and corresponding angles the same size, i.e. the shapes are exactly the same size and shape.

Shapes X and Y are congruent:

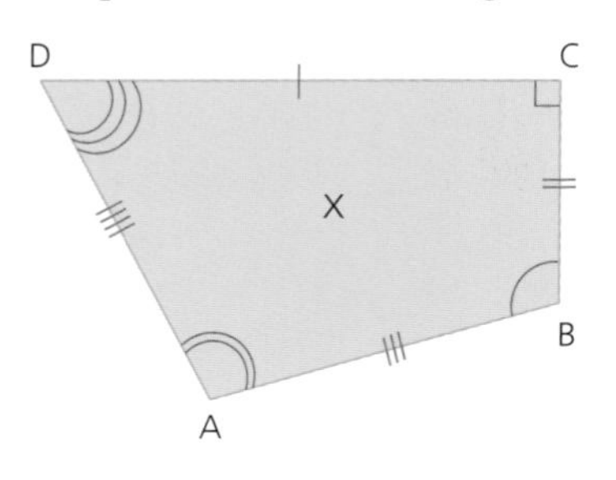

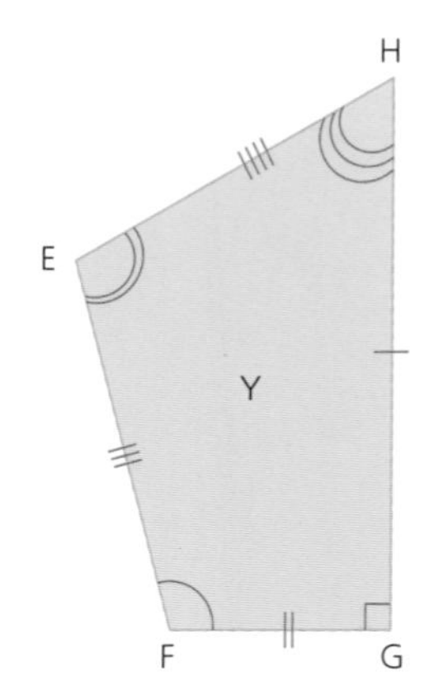

Congruent shapes are by definition also similar, but similar shapes are not necessarily congruent.

They are congruent as AB = EF, BC = FG, CD = GH and DA = HE. Also angle DAB = angle HEF, angle ABC = angle EFG, angle BCD = angle FGH and angle CDA = angle GHE.

Congruent shapes can therefore be reflections and rotations of each other.

Worked example

Triangles ABC and DEF are congruent:

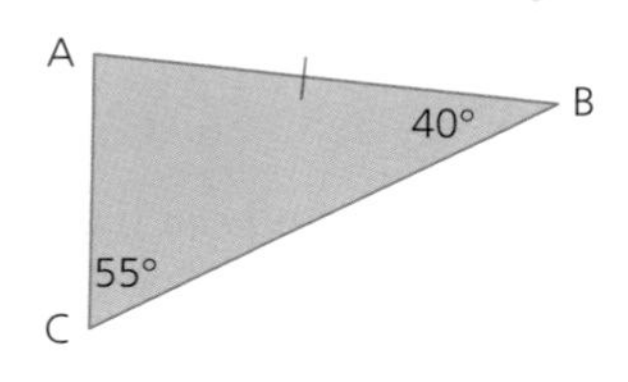

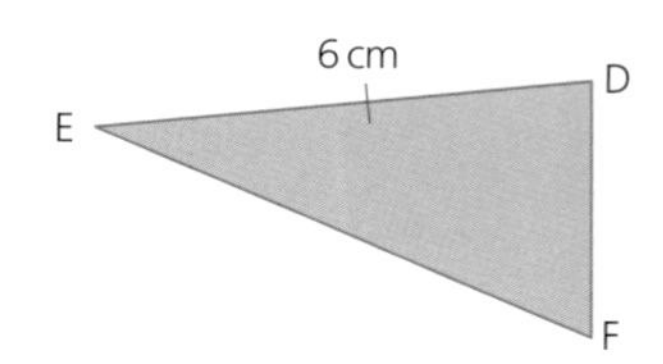

a Calculate the size of angle FDE.
As the two triangles are congruent, angle FDE = angle CAB
Angle $CAB = 180° - 40° - 55° = 85°$
Therefore angle $FDE = 85°$

b Deduce the length of AB.
As AB = DE, AB = 6 cm

Exercise 18.7

1 Look at the shapes on the grid. Which shapes are congruent to shape A?

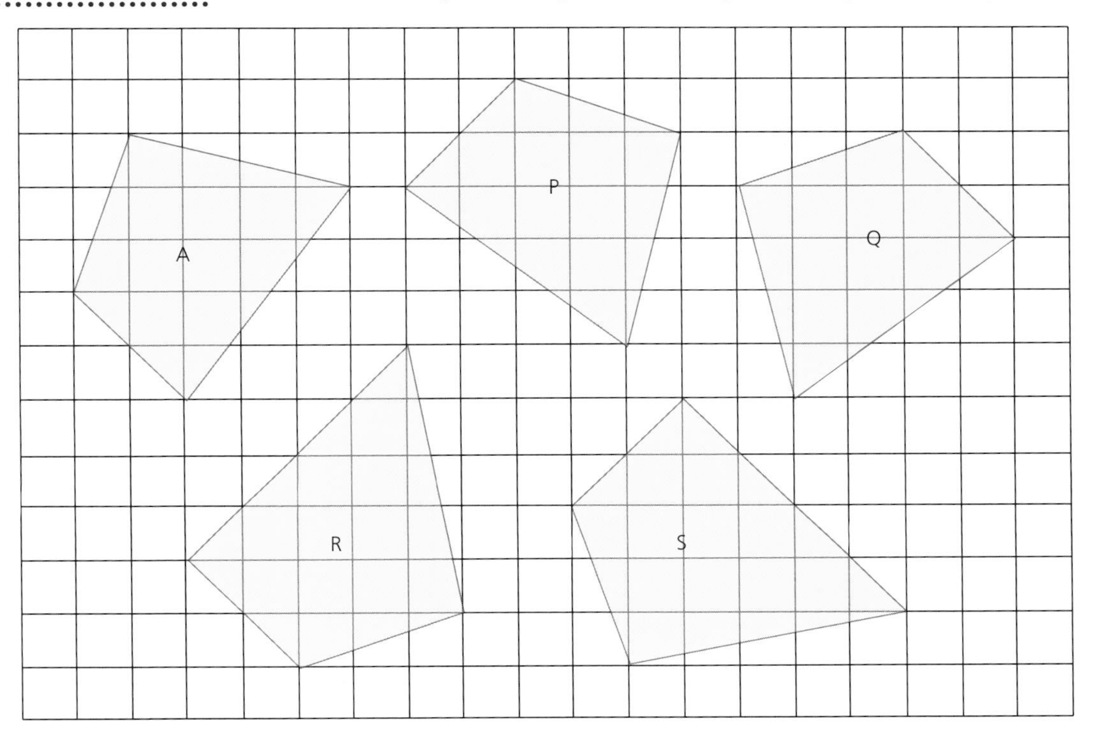

2 Two congruent shapes are shown. Calculate the size of x.

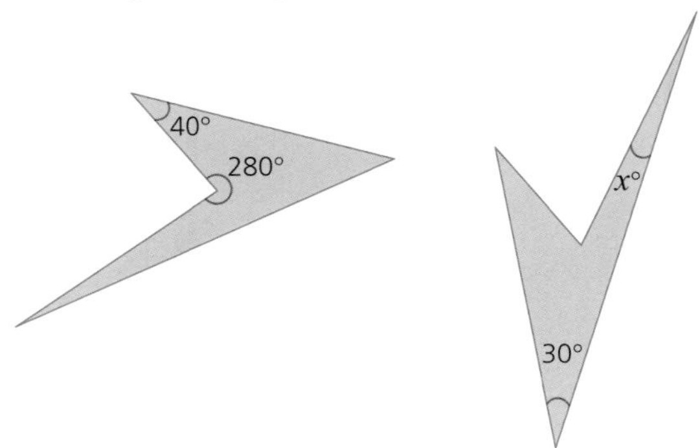

3 A quadrilateral is plotted on a pair of axes. The coordinates of its four vertices are $(0, 1)$, $(0, 5)$, $(3, 4)$ and $(3, 3)$. Another quadrilateral, congruent to the first, is also plotted on the same axes. Three of its vertices have coordinates $(6, 5)$, $(5, 2)$ and $(4, 2)$. Calculate the coordinates of the fourth vertex.

Exercise 18.7 (cont)

4 Triangle P is drawn on a grid. One side of another triangle, Q, is also shown.

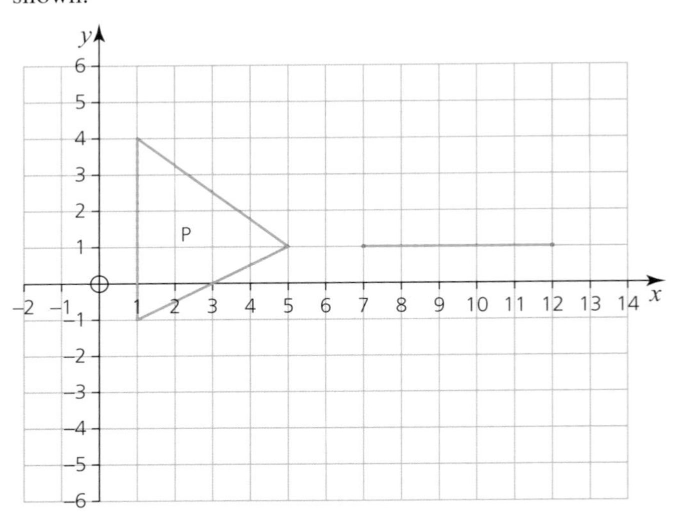

If triangles P and Q are congruent, give all the possible coordinates for the position of the missing vertex.

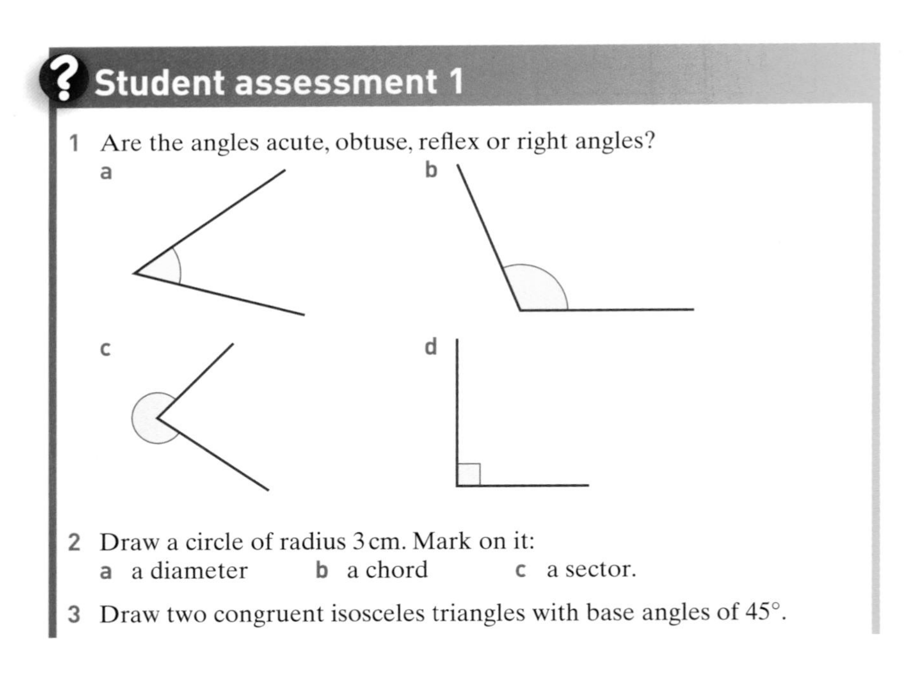

Student assessment 1

1 Are the angles acute, obtuse, reflex or right angles?

a b c d

2 Draw a circle of radius 3 cm. Mark on it:

a a diameter b a chord c a sector.

3 Draw two congruent isosceles triangles with base angles of 45°.

4 Triangle X is drawn on a graph. One side of another triangle congruent to X is also shown:

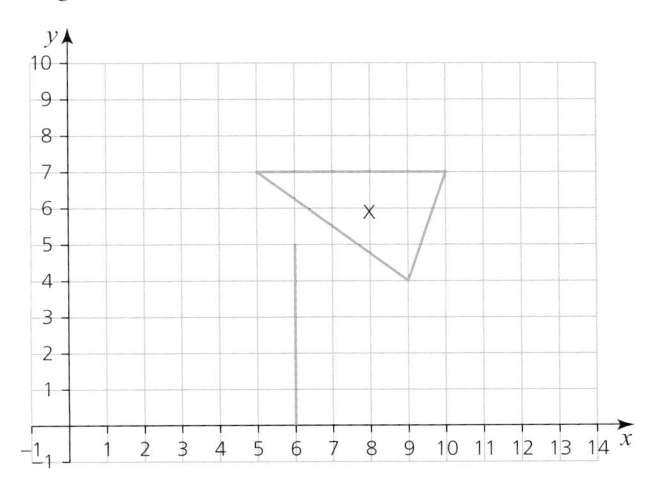

Write the coordinates of all the points where the missing vertex could be.

Student assessment 2

1 Draw and label two pairs of intersecting parallel lines.

2 Make two statements about these two triangles:

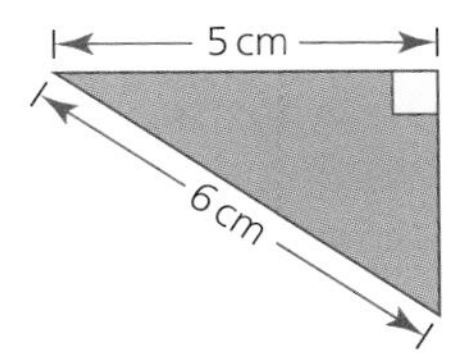

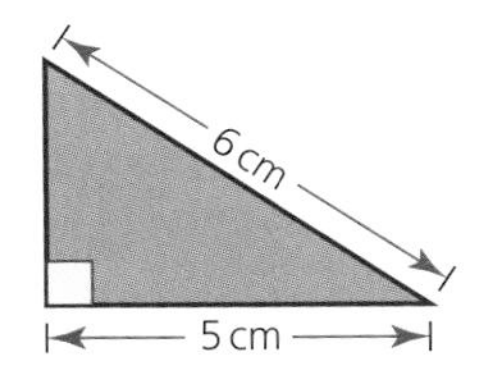

3 On squared paper, draw the net of a triangular prism.

4 The diagram shows an equilateral triangle ABC. The midpoints L, M and N of each side are also joined.
   a Identify a trapezium congruent to trapezium BCLN.
   b Identify a triangle similar to ΔLMN.

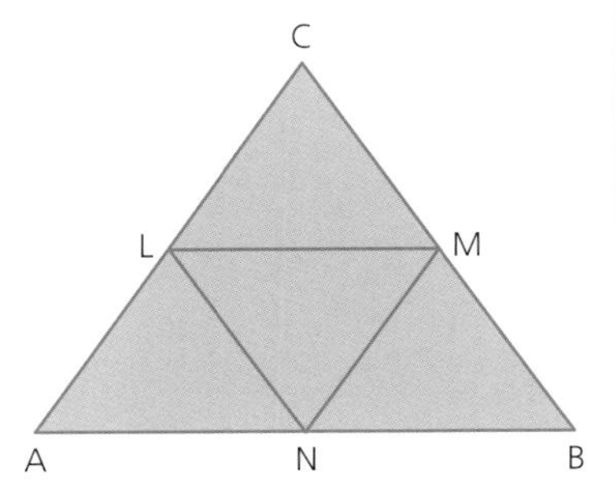

5 Decide whether each of the following statements are true or false.
   a All circles are similar.
   b All squares are similar.
   c All rectangles are similar.
   d All equilateral triangles are congruent.

19 Geometrical constructions and scale drawings

Measuring lines

A straight line can be drawn and measured accurately using a ruler.

Exercise 19.1

1 Using a ruler, measure the length of these lines to the nearest mm:

a

b

c

d

e

f

2 Draw lines of the following lengths using a ruler:

a 3 cm **b** 8 cm **c** 4.6 cm

d 94 mm **e** 38 mm **f** 61 mm

Measuring angles

An angle is a measure of **turn**. When drawn, it can be measured using either a protractor or an angle measurer. The units of turn are degrees (°). Measuring with a protractor needs care, as there are two scales marked on it – an inner one and an outer one.

Worked examples

1 Measure the angle using a protractor:

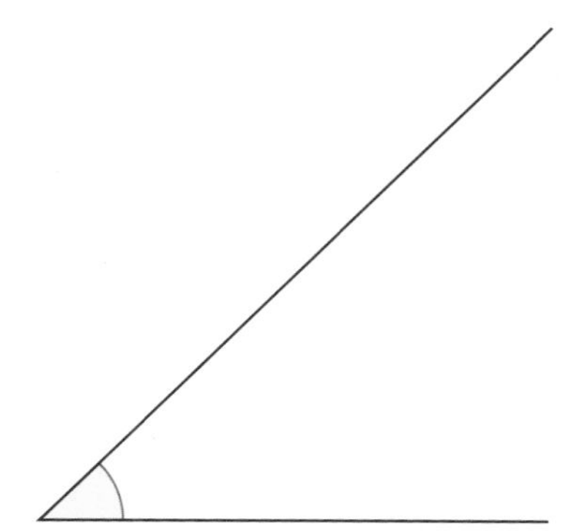

- Place the protractor over the angle so that the cross lies on the point where the two lines meet.
- Align the 0° with one of the lines:

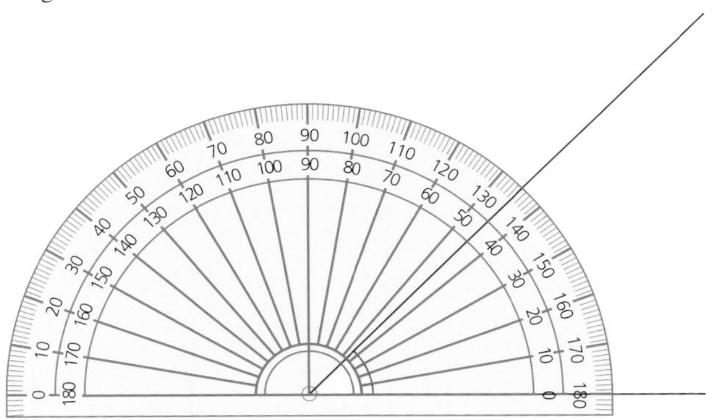

- Decide which scale is appropriate. In this case, it is the inner scale as it starts at 0°.
- Measure the angle using the inner scale.
  The angle is 45°.

2 Draw an angle of 110°.
- Start by drawing a straight line.
- Place the protractor on the line so that the cross is on one of the end points of the line.
  Ensure that the line is aligned with the 0° on the protractor:

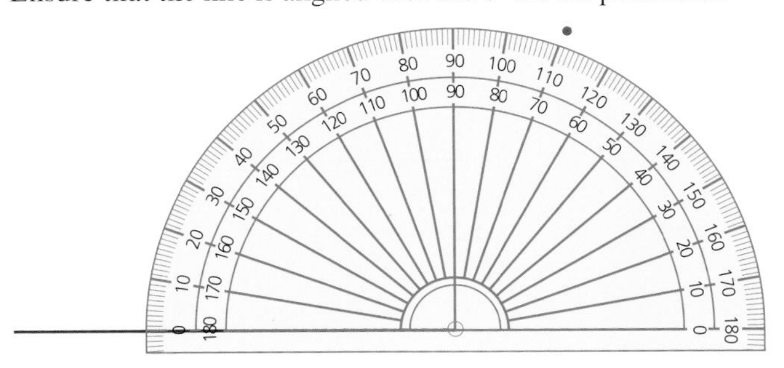

- Decide which scale to use. In this case, it is the outer scale as it starts at 0°.
- Mark where the protractor reads 110°.
- Join the mark made to the end point of the original line.

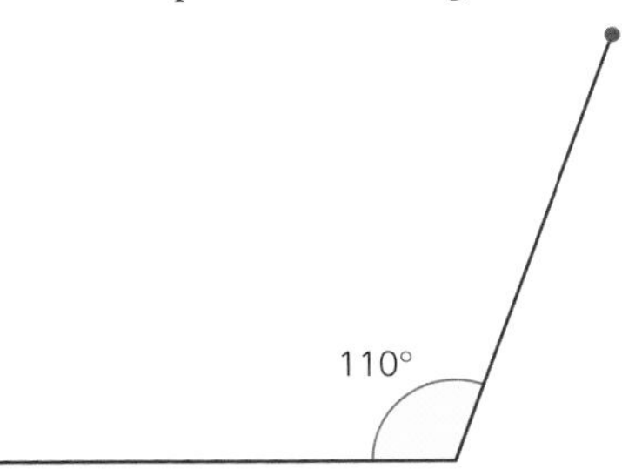

Exercise 19.2

1 Measure each angle:

a

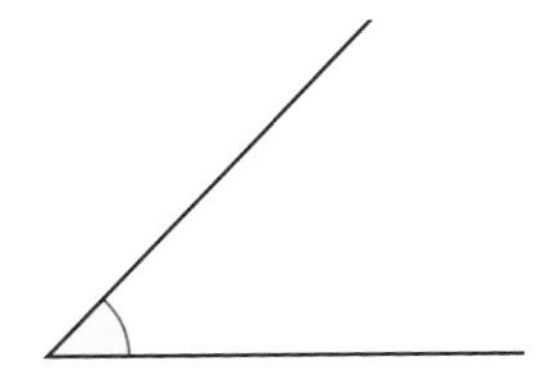

b

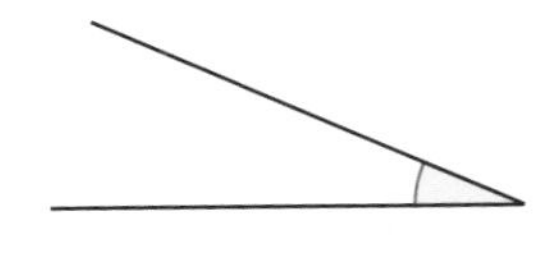

c

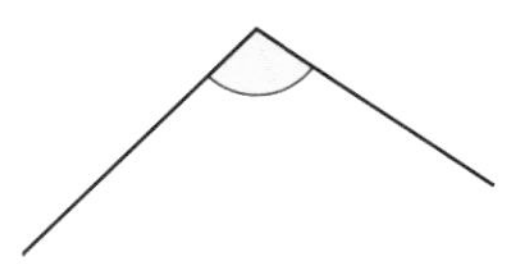

d

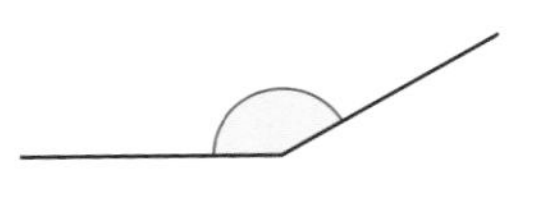

e

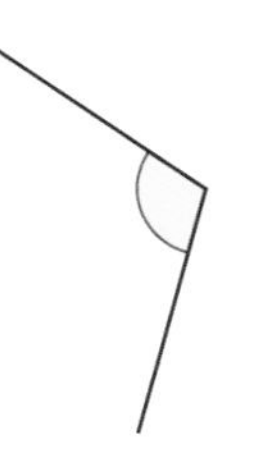

f

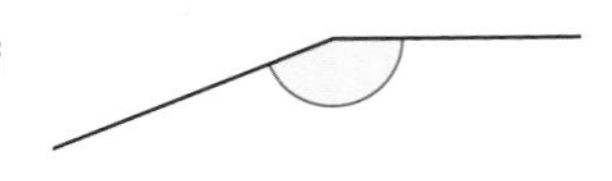

2 Measure each angle:

a

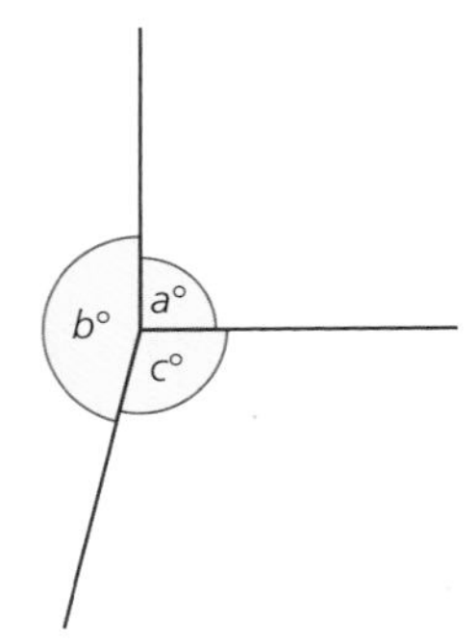

b

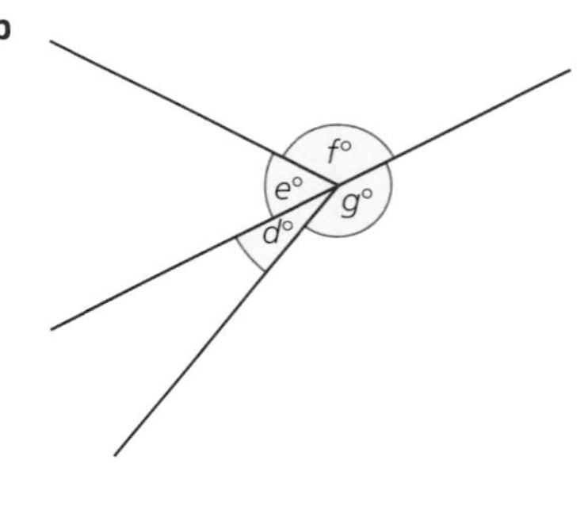

c

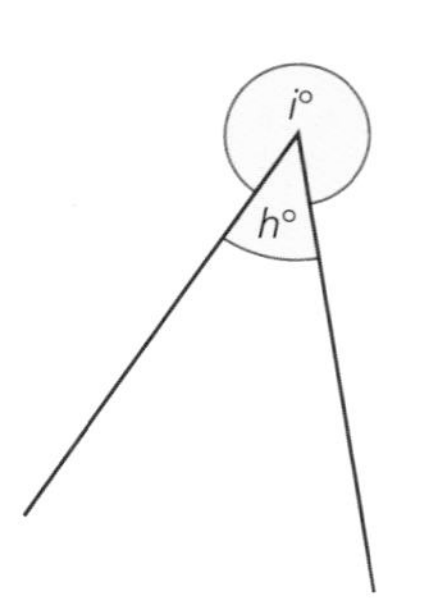

d

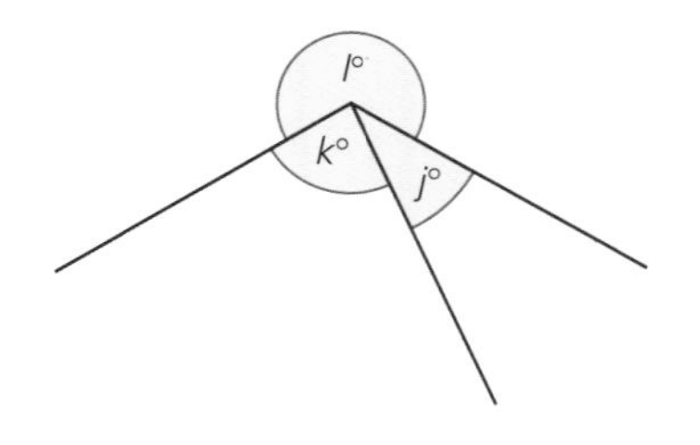

e

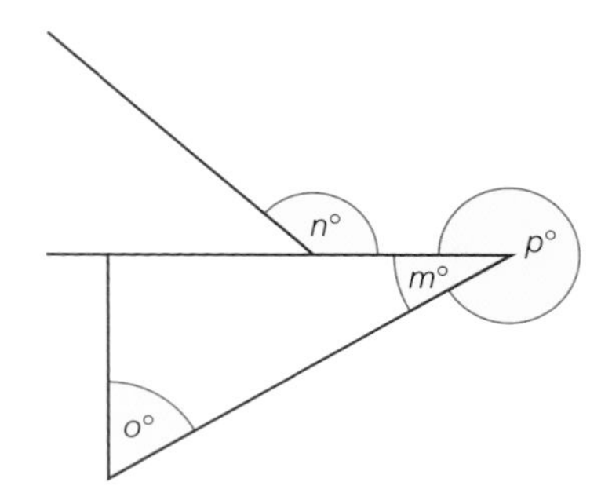

f

s°
q°
t°
r°

3 Draw angles of the following sizes:

a 20° **b** 45° **c** 90° **d** 120°

e 157° **f** 172° **g** 14° **h** 205°

i 311° **j** 283° **k** 198° **l** 352°

Constructing triangles

Triangles can be drawn accurately by using a ruler and a pair of compasses. This is called **constructing** a triangle.

Worked example

Construct the triangle ABC given that:

AB = 8 cm, BC = 6 cm and AC = 7 cm

- Draw the line AB using a ruler:

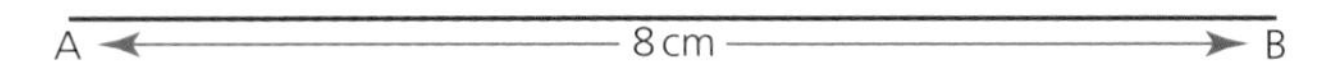

- Open up a pair of compasses to 6 cm. Place the compass point on B and draw an arc:

A 8 cm B

Note that every point on the arc is 6 cm away from B.

- Open up the pair of compasses to 7 cm. Place the compass point on A and draw another arc, with centre A and radius 7 cm, ensuring that it intersects with the first arc. Every point on the second arc is 7 cm from A. Where the two arcs intersect is point C, as it is both 6 cm from B and 7 cm from A.
- Join C to A and C to B:

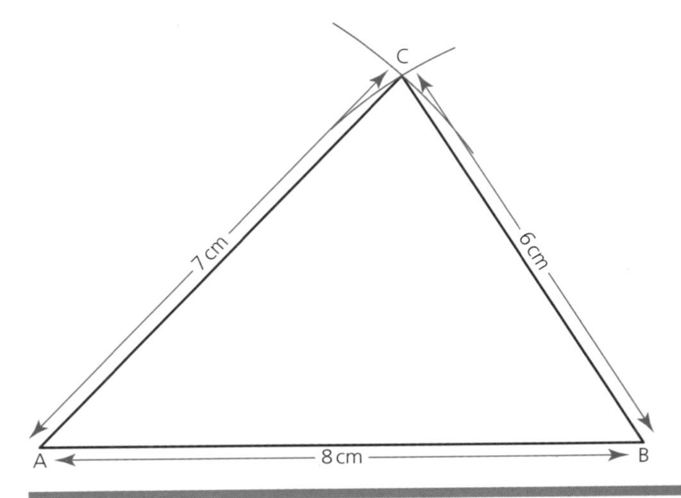

Exercise 19.3

Using only a ruler and a pair of compasses, construct the following triangles:

a ΔABC where $AB = 10$ cm, $AC = 7$ cm and $BC = 9$ cm
b ΔLMN where $LM = 4$ cm, $LN = 8$ cm and $MN = 5$ cm
c ΔPQR, an equilateral triangle of side length 7 cm
d **i** ΔABC where $AB = 8$ cm, $AC = 4$ cm and $BC = 3$ cm
ii Is this triangle possible? Explain your answer.

Scale drawings

Scale drawings are used when an accurate diagram, drawn in proportion, is needed. Common uses of scale drawings include maps and plans. The use of scale drawings involves understanding how to scale measurements.

Worked examples

1 A map is drawn to a scale of 1 : 10000. If two objects are 1 cm apart on the map, how far apart are they in real life? Give your answer in metres.

A scale of 1 : 10000 means that 1 cm on the map represents 10000 cm in real life.

Therefore the distance = 10000 cm
= 100 m

2 A model boat is built to a scale of 1 : 50. If the length of the real boat is 12 m, calculate the length of the model boat in cm.

A scale of 1 : 50 means that 50 cm on the real boat is 1 cm on the model boat.
12 m = 1200 cm
Therefore the length of the model boat = 1200 ÷ 50 cm
= 24 cm

3 a Construct, to a scale of 1 : 1, a triangle ABC such that AB = 6 cm, AC = 5 cm and BC = 4 cm.

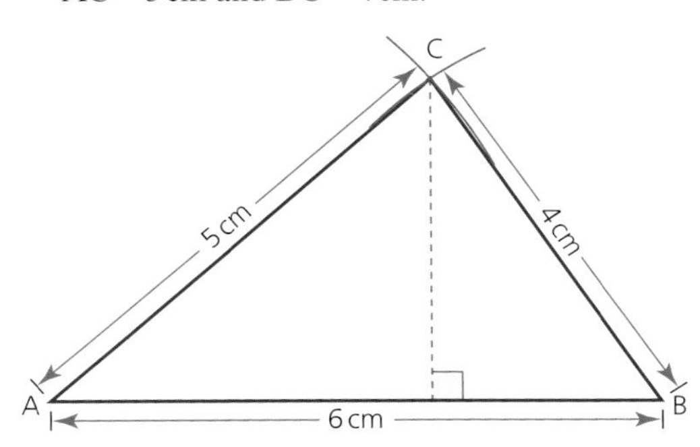

b Measure the perpendicular length of C from AB.
Perpendicular length is 3.3 cm.

c Calculate the area of the triangle.

$$\text{Area} = \frac{\text{base length} \times \text{perpendicular height}}{2}$$

$$\text{Area} = \frac{6 \times 3.3}{2}\text{cm} = 9.9\ \text{cm}^2$$

Exercise 19.4

1 In the following questions, both the scale to which a map is drawn and the distance between two objects on the map are given.

Find the real distance between the two objects, giving your answer in metres.

a 1 : 10 000 3 cm
b 1 : 10 000 2.5 cm
c 1 : 20 000 1.5 cm
d 1 : 8000 5.2 cm

2 In the following questions, both the scale to which a map is drawn and the true distance between two objects are given.

Find the distance between the two objects on the map, giving your answer in cm.

a 1 : 15 000 1.5 km
b 1 : 50 000 4 km
c 1 : 10 000 600 m
d 1 : 25 000 1.7 km

Exercise 19.4 (cont)

3 A rectangular pool measures 20 m by 36 m as shown below:

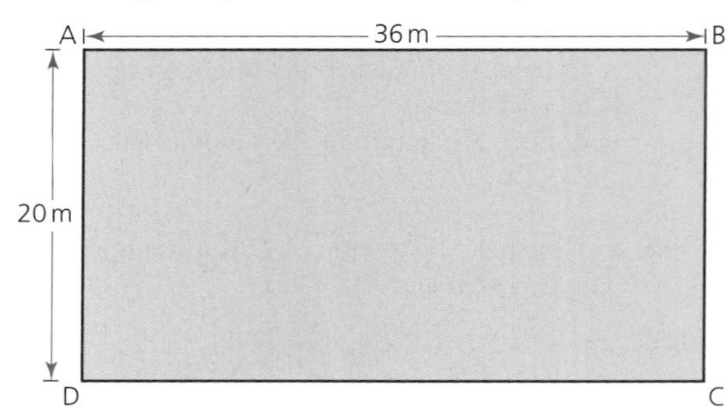

a Construct a scale drawing of the pool, using 1 cm for every 4 m.
b A boy sets off across the pool from D in such a way that his path is in the direction of a point which is 40 m from D and 30 m from C. Work out the distance the boy swam.

4 A triangular enclosure is shown in the diagram below:

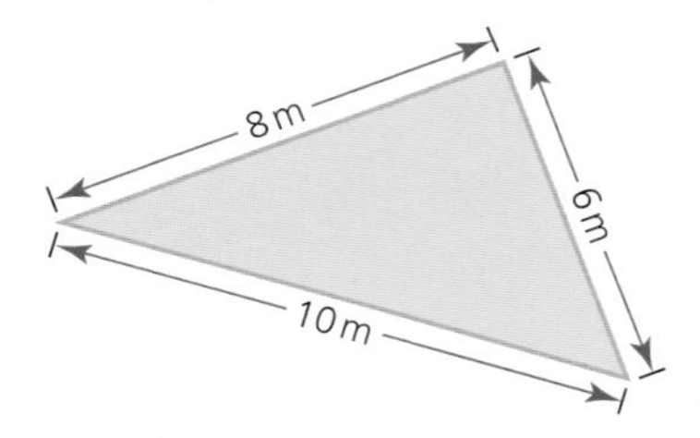

a Using a scale of 1 cm for each metre, construct a scale drawing of the enclosure.
b Calculate the true area of the enclosure.

Student assessment 1

1 a Using a ruler, measure the length of the line:

b Draw a line 4.7 cm long.

2 a Using a protractor, measure the angle shown:

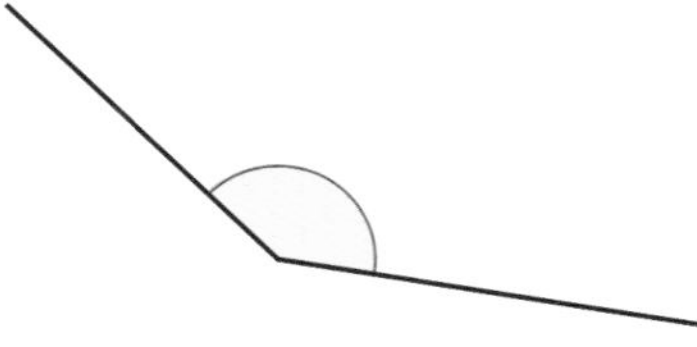

b Draw an angle of 300°.

3 Construct ΔABC such that AB = 8 cm, AC = 6 cm and BC = 12 cm.

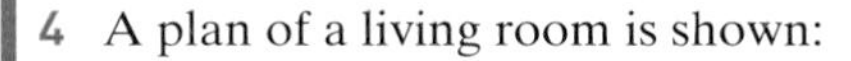

4 A plan of a living room is shown:

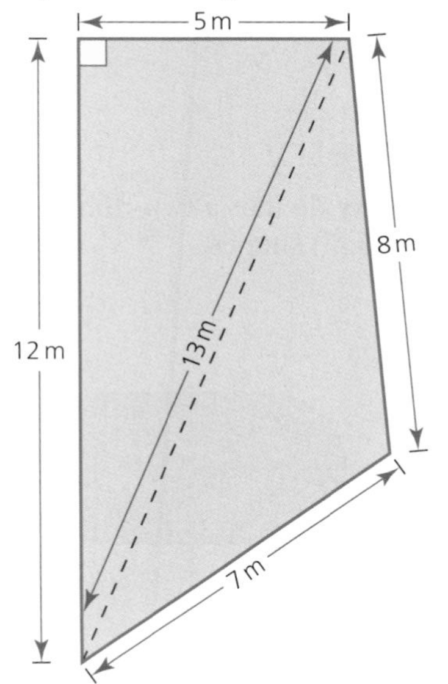

a Using a pair of compasses, construct a scale drawing of the room using 1 cm for every metre.

b Using a set square if necessary, calculate the total area of the actual living room.

5 Measure each of the five angles of the pentagon:

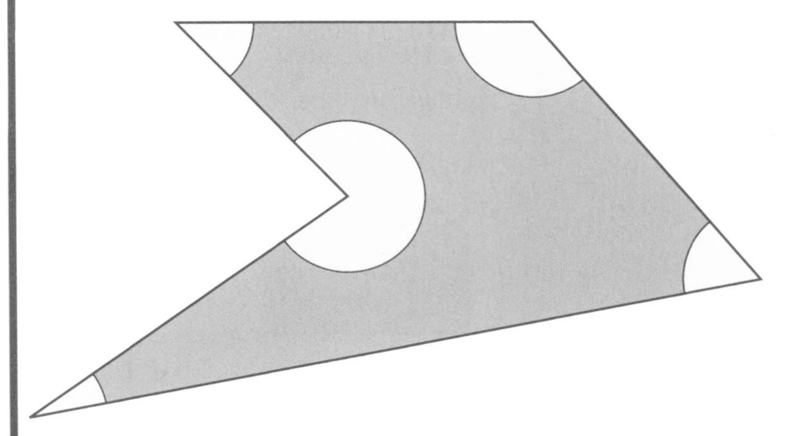

6 Draw, using a ruler and a protractor, a triangle with angles of 40°, 60° and 80°.

7 In the following questions, both the scale to which a map is drawn and the true distance between two objects are given. Find the distance between the two objects on the map, giving your answer in cm.

a 1 : 20000 4.4 km b 1 : 50000 12.2 km

20 Symmetry

Line symmetry

A **line of symmetry** divides a two-dimensional (flat) shape into two congruent (identical) shapes.

e.g.

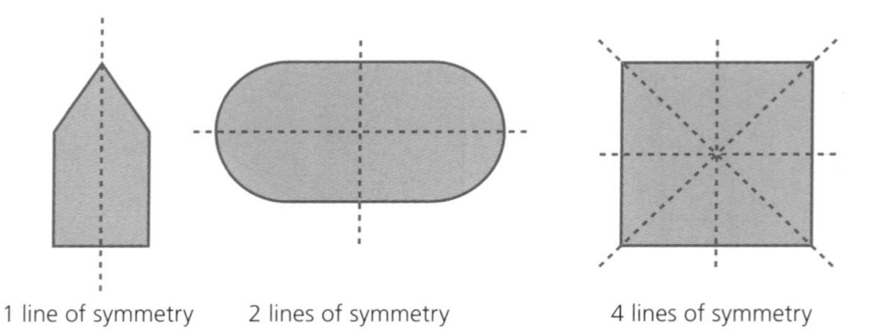

Exercise 20.1

1 Draw the following shapes and, where possible, show all their lines of symmetry:

a square **b** rectangle
c equilateral triangle **d** isosceles triangle
e kite **f** regular hexagon
g regular octagon **h** regular pentagon
i isosceles trapezium **j** circle

2 Copy the shapes and, where possible, show all their lines of symmetry:

a

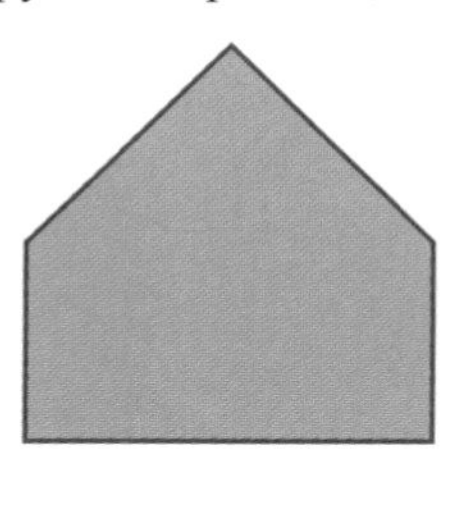

b

c

d

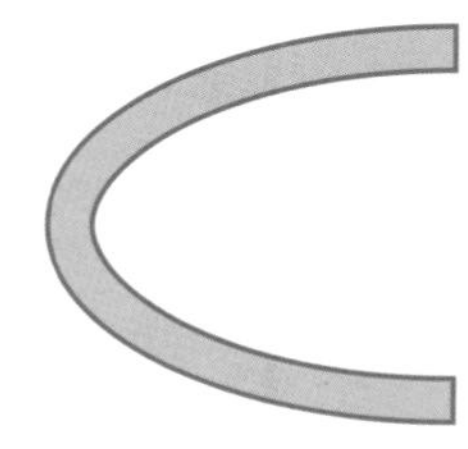

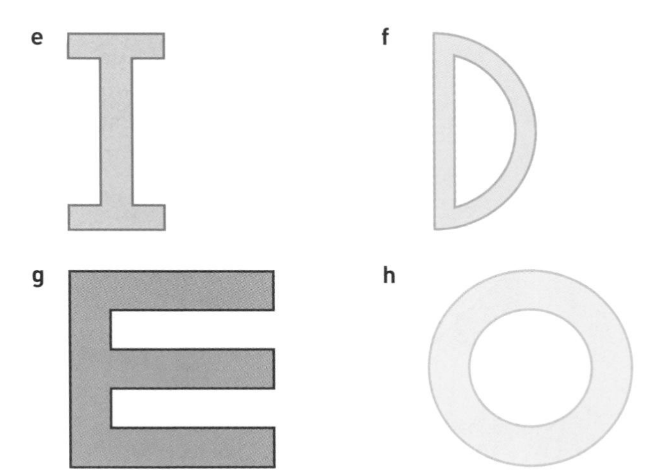

3 Copy the shapes and complete them so that the **bold** line becomes a line of symmetry:

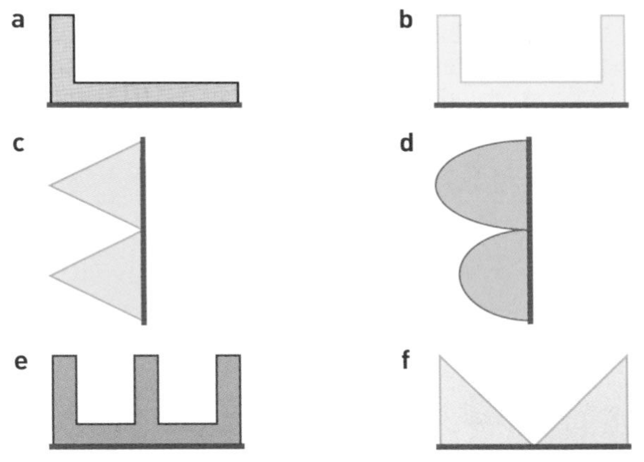

4 Copy the shapes and complete them so that the **bold** lines become lines of symmetry:

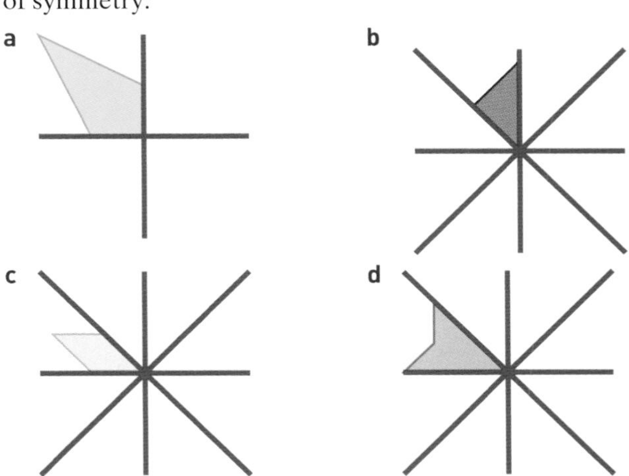

Rotational symmetry

A two-dimensional shape has **rotational symmetry** if, when rotated about a central point, it looks the same as its starting position. The number of times it looks the same during a complete revolution is called the **order of rotational symmetry**.

e.g.

rotational symmetry of order 2

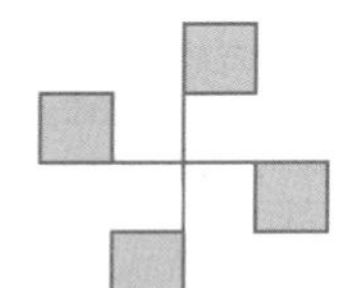

rotational symmetry of order 4

Exercise 20.2

1 Draw the following shapes. Identify the centre of rotation, and state the order of rotational symmetry:

  a square **b** equilateral triangle
  c regular pentagon **d** parallelogram
  e rectangle **f** rhombus
  g regular hexagon **h** regular octagon
  i circle

2 Copy the shapes. Indicate the centre of rotation, and state the order of rotational symmetry:

  a

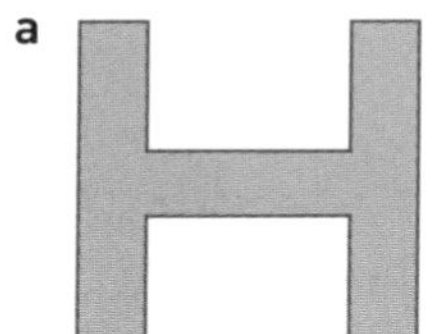

  b

  c

  d

  e

Student assessment 1

1 Draw a shape with exactly:
 a one line of symmetry
 b two lines of symmetry
 c three lines of symmetry.
 Mark the lines of symmetry on each diagram.

2 Draw a shape with:
 a rotational symmetry of order 2
 b rotational symmetry of order 3.
 Mark the position of the centre of rotation on each diagram.

3 Copy and complete the following shapes so that the **bold** lines become lines of symmetry:

a
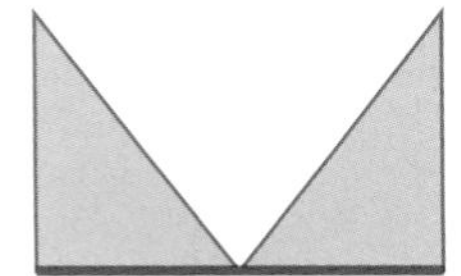

b
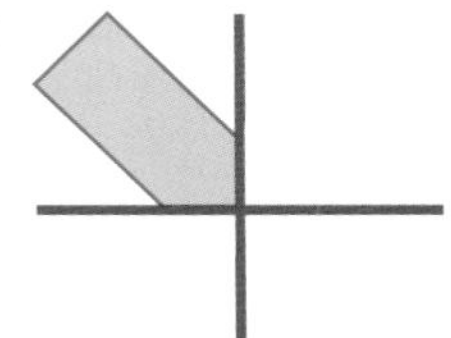

c
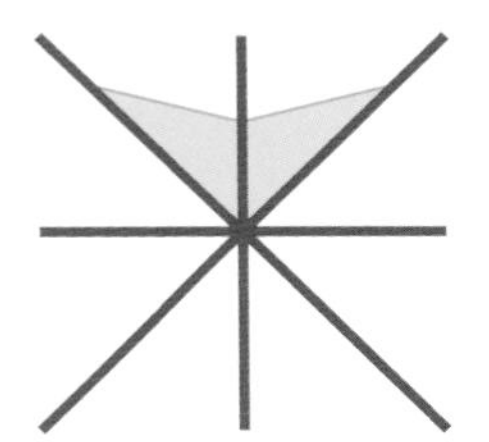

d
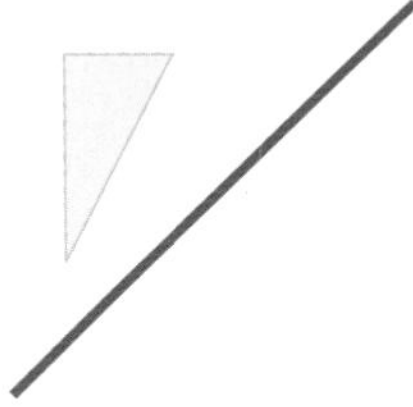

4 State the order of rotational symmetry for the completed drawings in Q.3.

21 Angle properties

Angles at a point and on a line

NB: All diagrams are not drawn to scale.

One complete revolution is equivalent to a rotation of 360° about a point. Similarly, half a complete revolution is equivalent to a rotation of 180° about a point. These facts can be seen clearly by looking at either a circular angle measurer or a semi-circular protractor.

Worked examples

1 Calculate the size of the angle x in the diagram below:

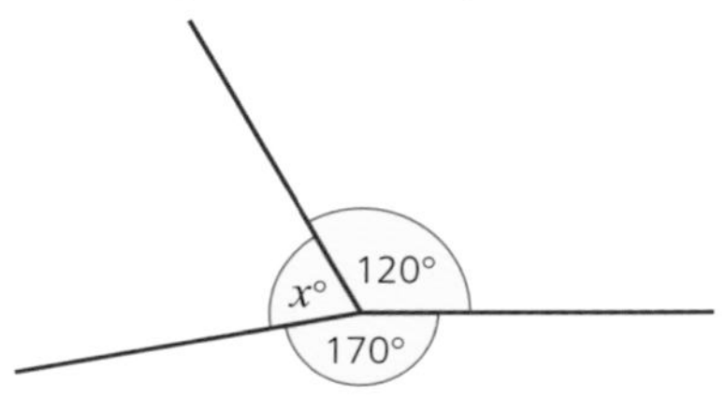

The sum of all the **angles at a point** is 360°. Therefore:

$$120 + 170 + x = 360$$
$$x = 360 - 120 - 170$$
$$x = 70$$

Therefore angle x is 70°.

Note that the size of the angle is **calculated** and *not* **measured**.

2 Calculate the size of angle a in the diagram below:

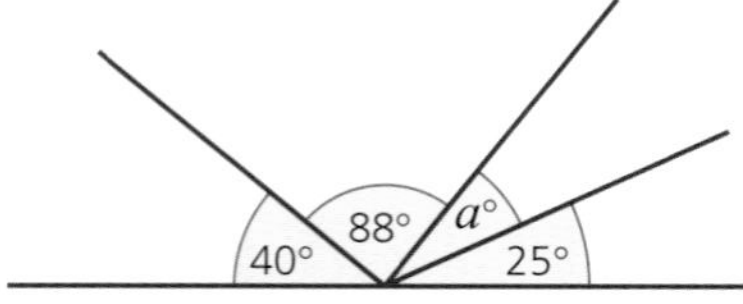

The sum of all the **angles on a straight line** is 180°. Therefore:

$$40 + 88 + a + 25 = 180$$
$$a = 180 - 40 - 88 - 25$$
$$a = 27$$

Therefore angle a is 27°.

Exercise 21.1

1 Calculate the size of angle x:

a

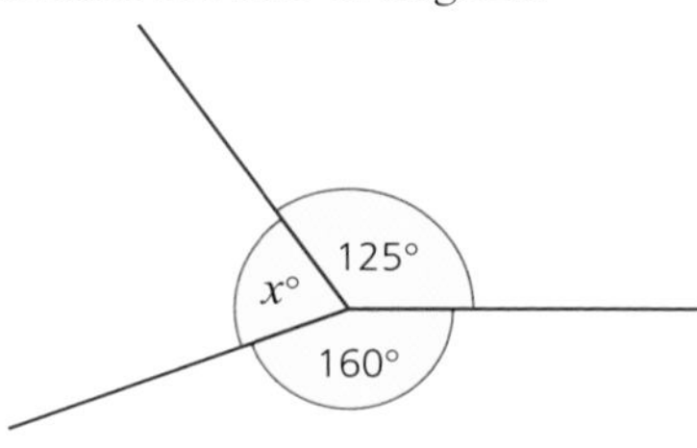

b

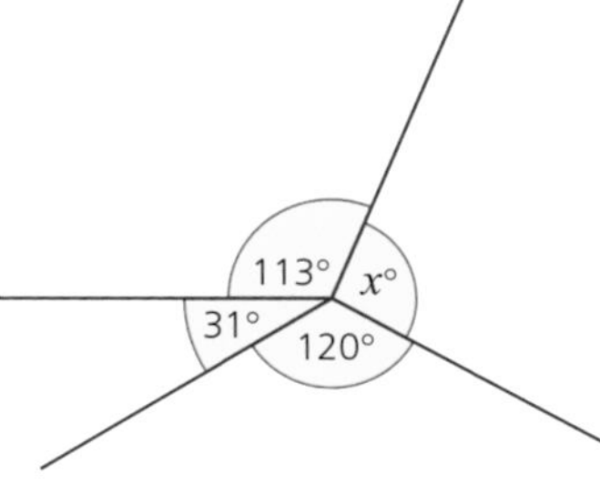

c

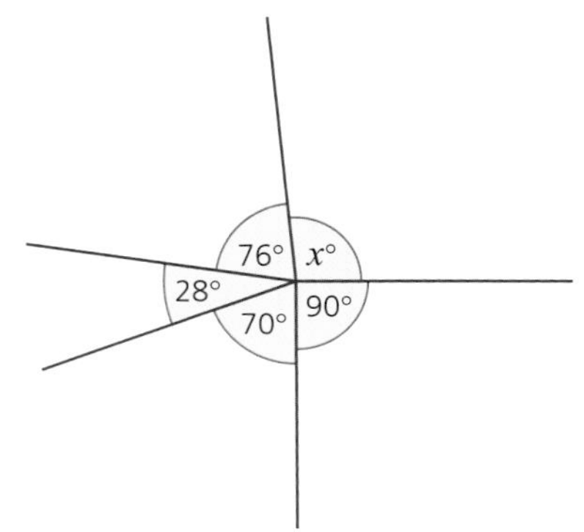

d

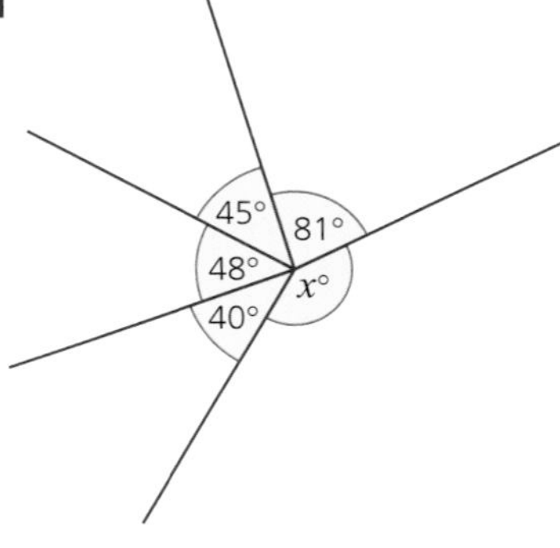

2 In the following questions, the angles lie about a point on a straight line. Find y.

a

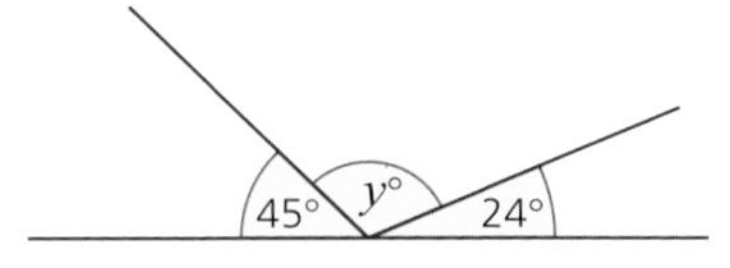

b

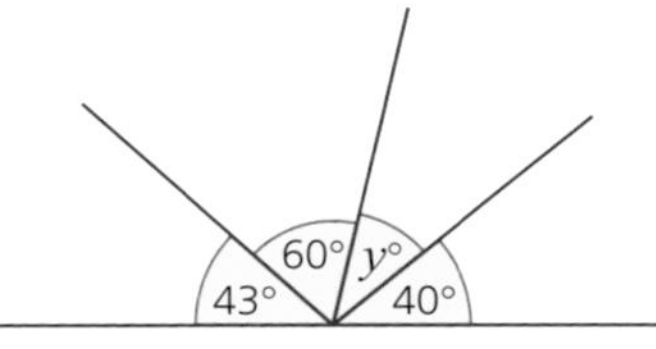

c

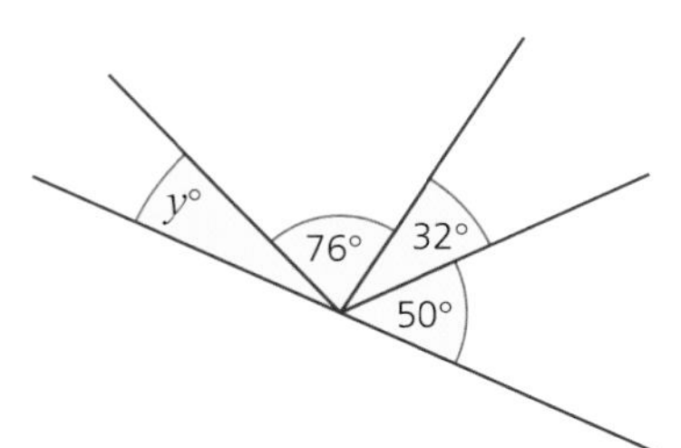

d

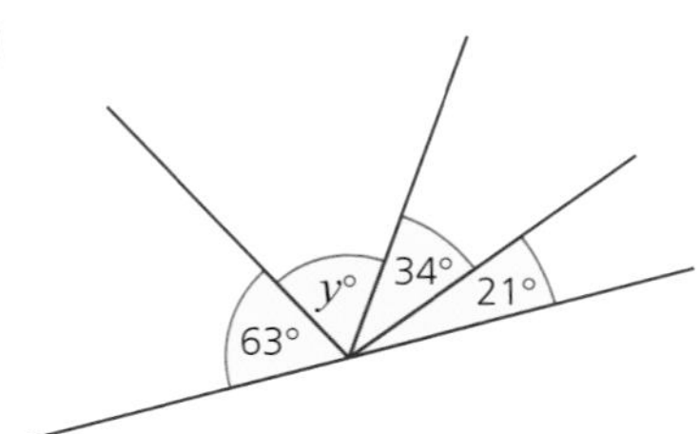

Exercise 21.1 (cont)

3 Calculate the size of angle p:

a

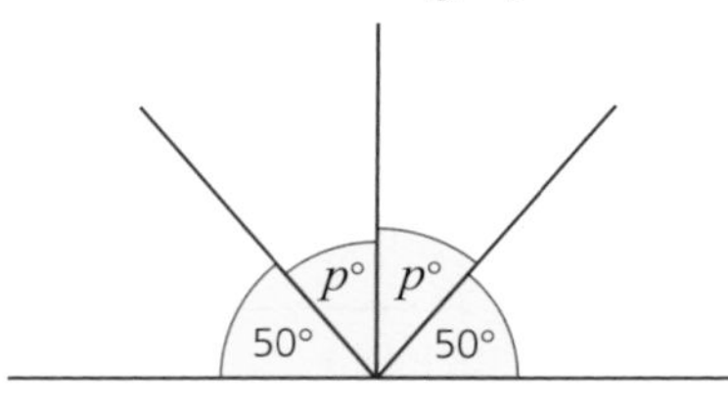

b

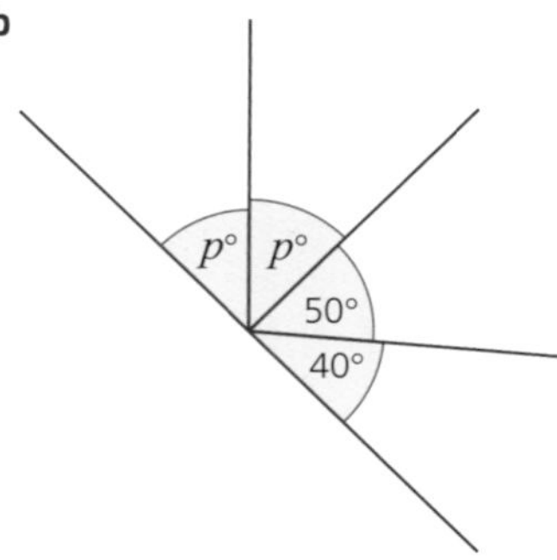

c

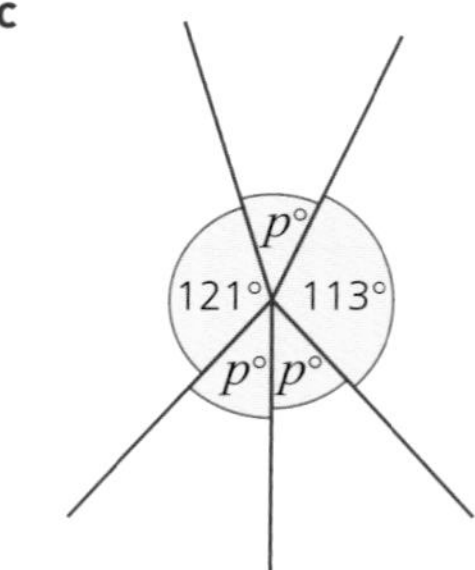

d

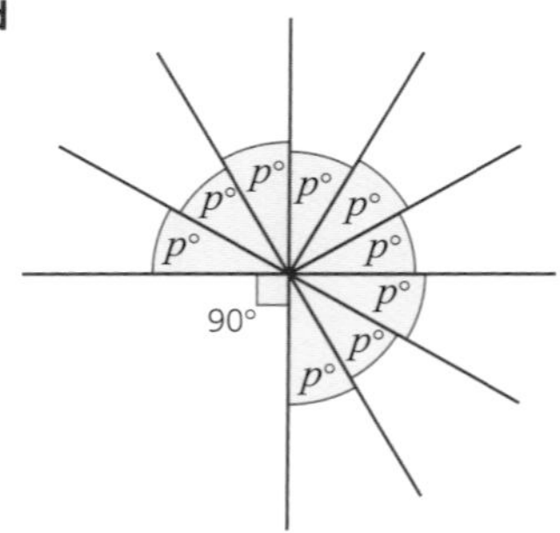

Angles formed by intersecting lines

Exercise 21.2

1 Draw a similar diagram to the one shown. Measure carefully each of the labelled angles and write them down.

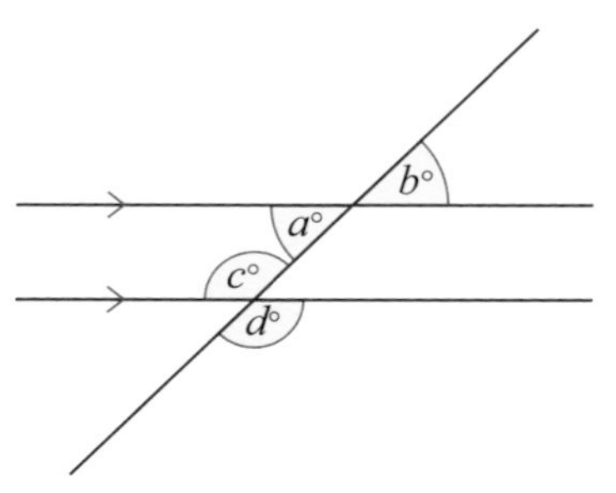

2 Draw a similar diagram to the one shown. Measure carefully each of the labelled angles and write them down.

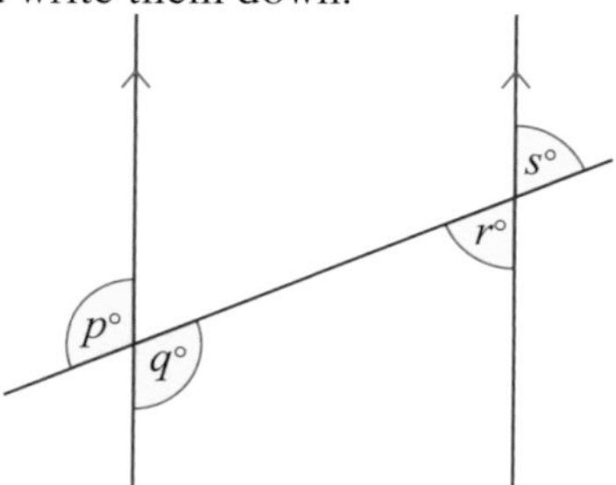

3 Draw a similar diagram to the one shown. Measure carefully each of the labelled angles and write them down.

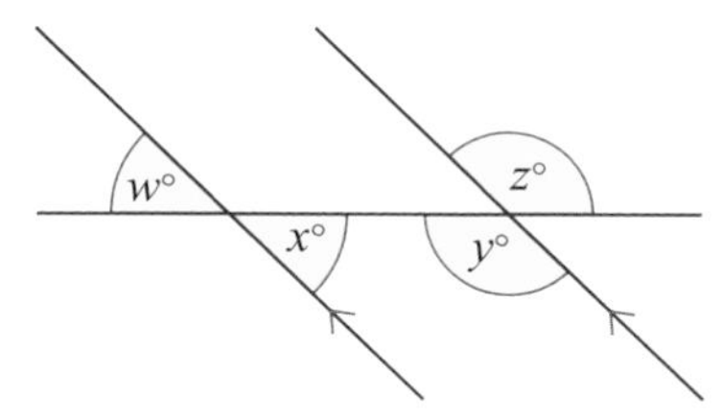

4 Write down what you have noticed about the angles you measured in Q.1–3.

When two straight lines cross, it is found that the angles opposite each other are the same size. They are known as **vertically opposite angles**. By using the fact that angles at a point on a straight line add up to 180°, it can be shown why vertically opposite angles must always be equal in size.

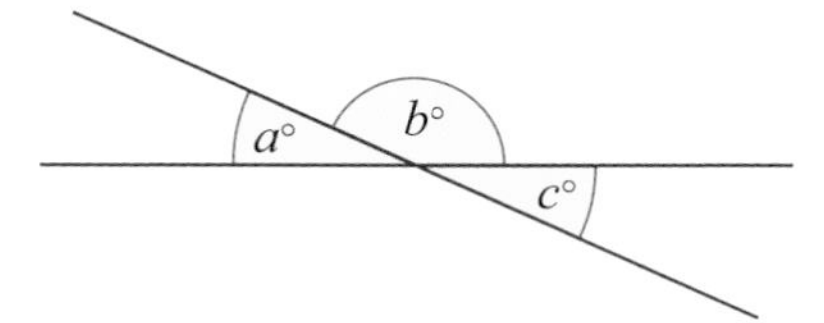

$a + b = 180°$

$c + b = 180°$

Therefore, a is equal to c.

Exercise 21.3

1 Draw a similar diagram to the one shown. Measure carefully each of the labelled angles and write them down.

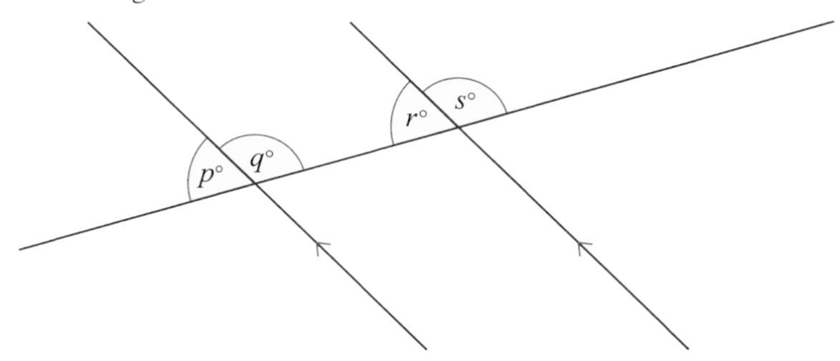

Exercise 21.3 (cont)

2 Draw a similar diagram to the one shown. Measure carefully each of the labelled angles and write them down.

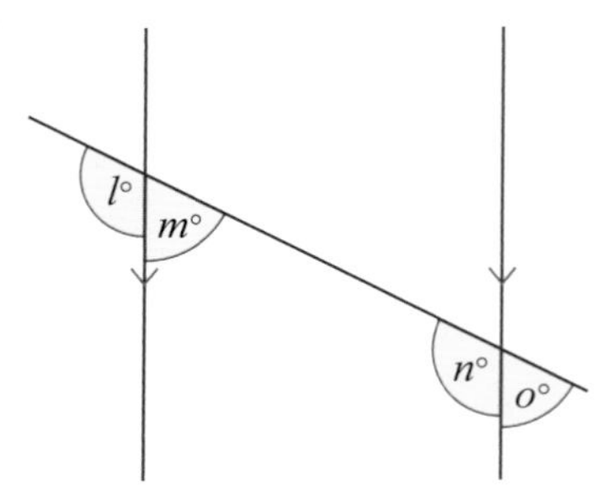

3 Draw a similar diagram to the one shown. Measure carefully each of the labelled angles and write them down.

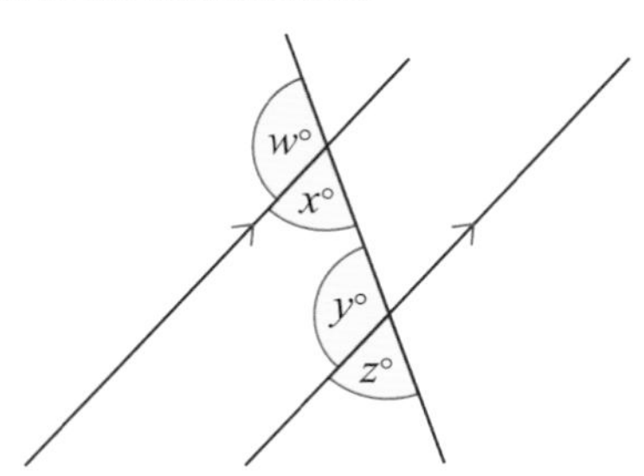

4 Write down what you have noticed about the angles you measured in Q.1–3.

Angles formed within parallel lines

When a line intersects two parallel lines, as in the diagram below, it is found that certain angles are the same size.

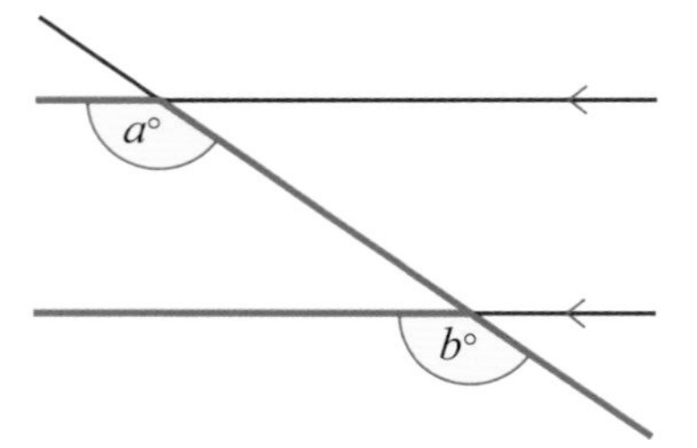

The angles a and b are equal and are known as **corresponding angles**. Corresponding angles can be found by looking for an 'F' formation in a diagram.

A line intersecting two parallel lines also produces another pair of equal angles, known as **alternate angles**. These can be shown to be equal by using the fact that both vertically opposite and corresponding angles are equal.

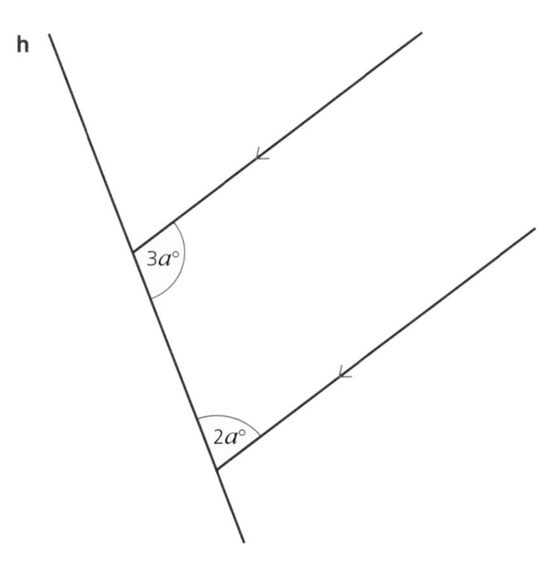

Angle properties of triangles

A **triangle** is a plane (two-dimensional) shape consisting of three angles and three sides. There are six main types of triangle. Their names refer to the sizes of their angles and/or the lengths of their sides, and are as follows:

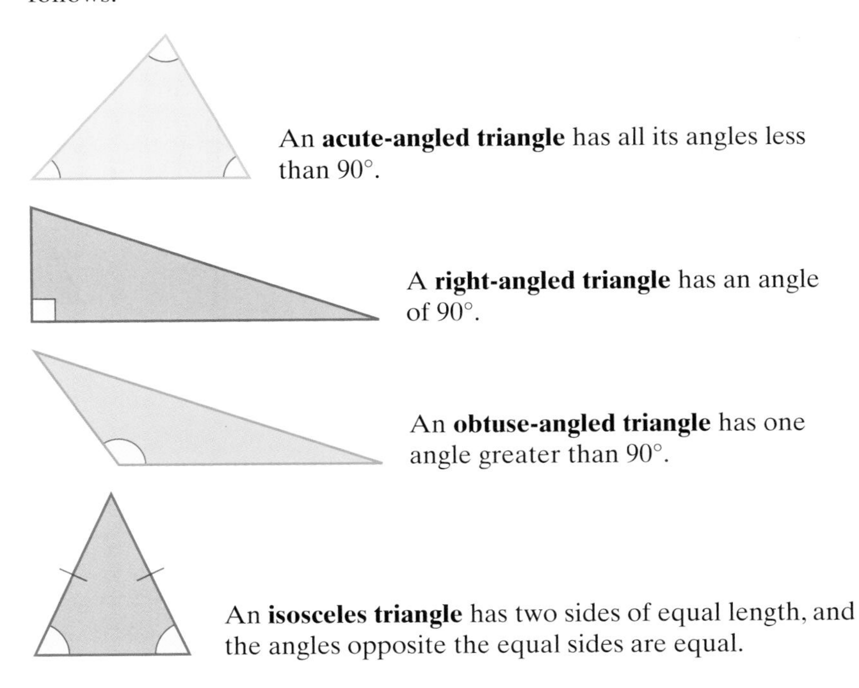

An **acute-angled triangle** has all its angles less than 90°.

A **right-angled triangle** has an angle of 90°.

An **obtuse-angled triangle** has one angle greater than 90°.

An **isosceles triangle** has two sides of equal length, and the angles opposite the equal sides are equal.

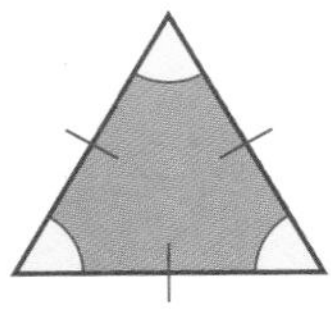

An **equilateral triangle** has three sides of equal length and three equal angles.

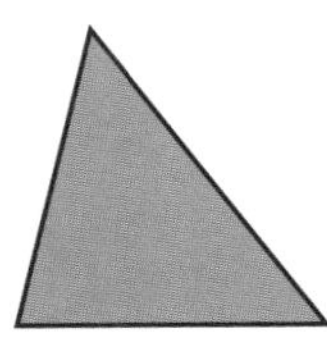

A **scalene triangle** has three sides of different lengths and all three angles are different.

Exercise 21.5

1 Describe the triangles in two ways.

The example shows an acute-angled isosceles triangle.

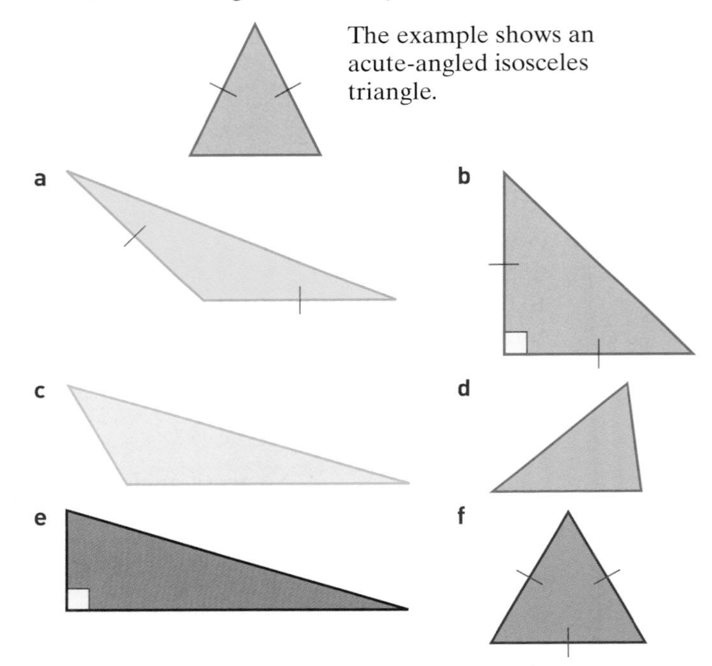

2 Draw the following triangles using a ruler and compasses:

a an acute-angled isosceles triangle of sides 5 cm, 5 cm and 6 cm, and altitude 4 cm

b a right-angled scalene triangle of sides 6 cm, 8 cm and 10 cm

c an equilateral triangle of side 7.5 cm

d an obtuse-angled isosceles triangle of sides 13 cm, 13 cm and 24 cm, and altitude 5 cm.

Exercise 21.6

1 **a** Draw five different triangles. Label their angles x, y and z. As accurately as you can, measure the three angles of each triangle and add them together.

b What do you notice about the sum of the three angles of each of your triangles?

2 **a** Draw a triangle on a piece of paper and label the angles a, b and c. Tear off the corners of the triangle and arrange them as shown below:

b What do you notice about the total angle that a, b and c make?

The sum of the interior angles of a triangle

It can be seen from the previous questions that triangles of any shape have one thing in common. That is, that the sum of their three angles is constant: 180°.

Worked example

Calculate the size of the angle x in the triangle below:

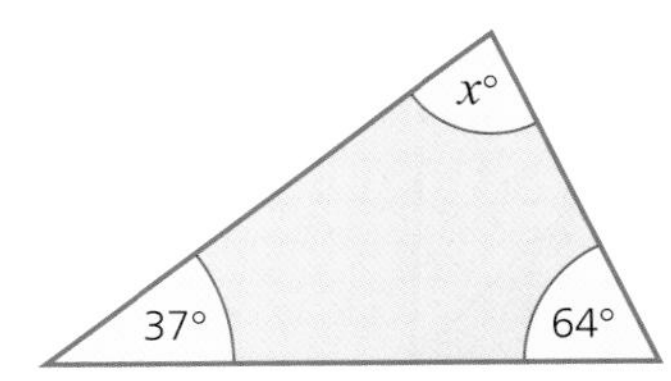

$37 + 64 + x = 180$

$x = 180 - 37 - 64$

Therefore angle x is 79°.

Exercise 21.7

1 For each triangle, use the information given to calculate the size of angle x.

a

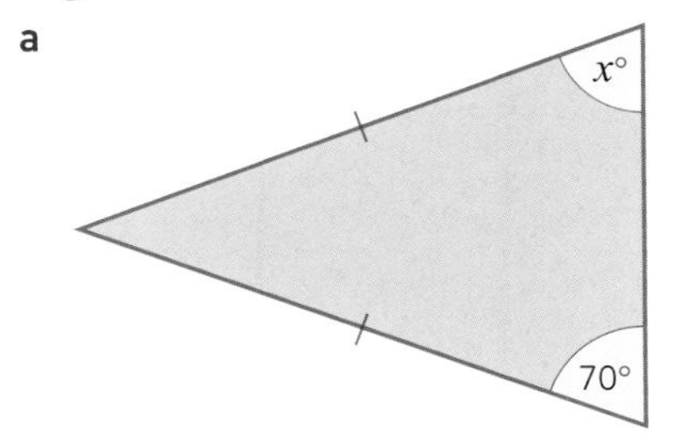

b

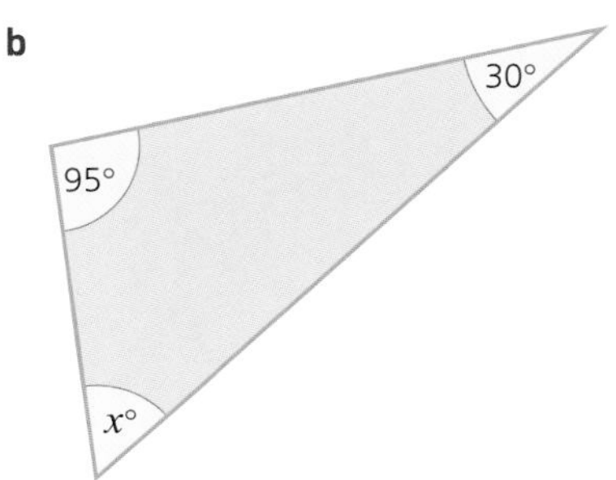

Exercise 21.7 (cont)

c 

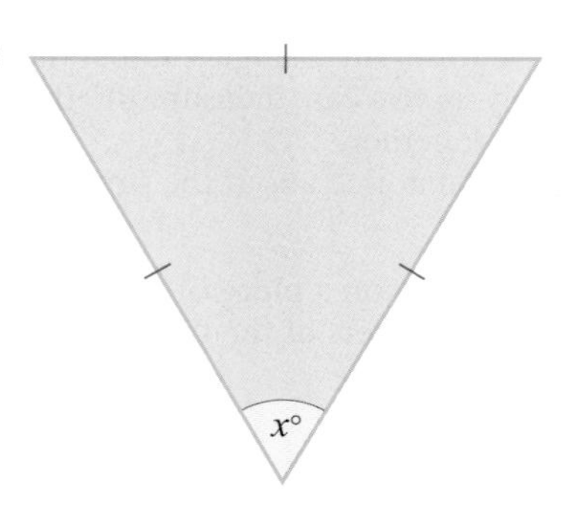

d

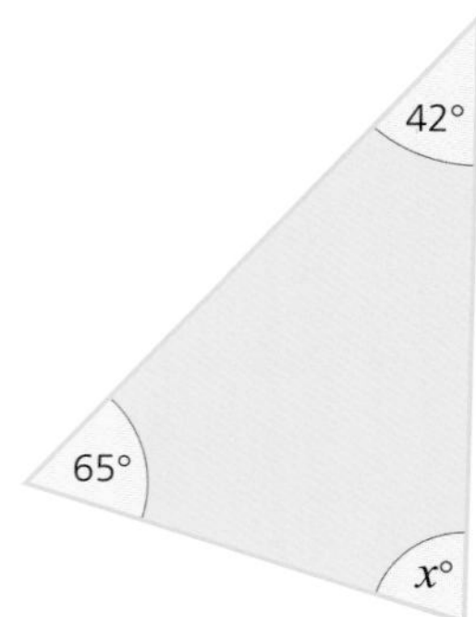

e

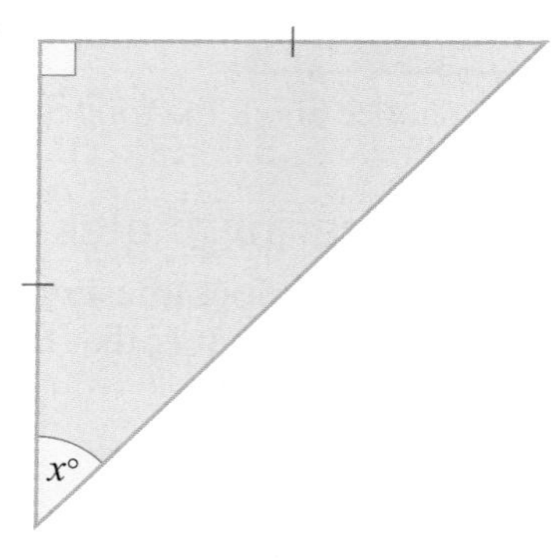

f

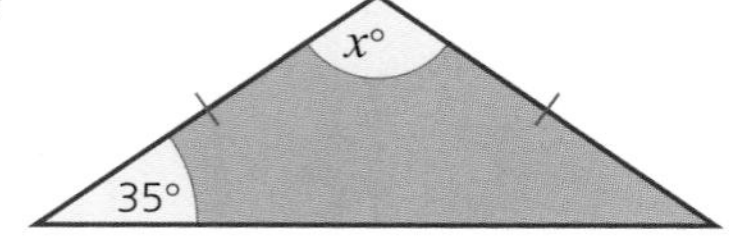

2 In each diagram, calculate the size of the labelled angles.

a

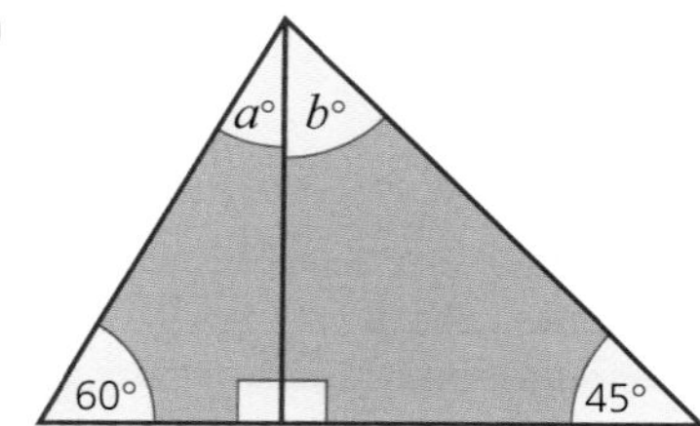

b

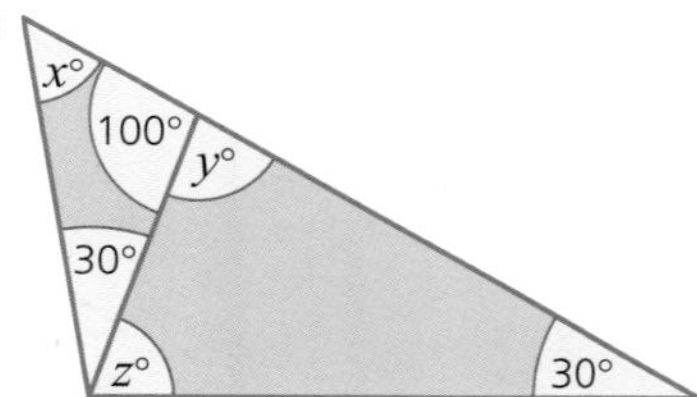

c

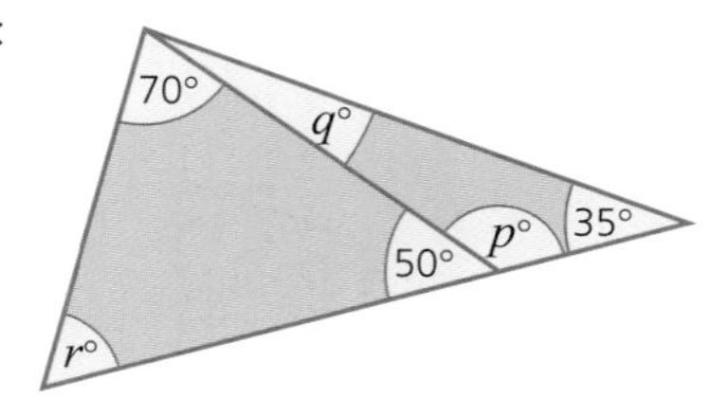

d

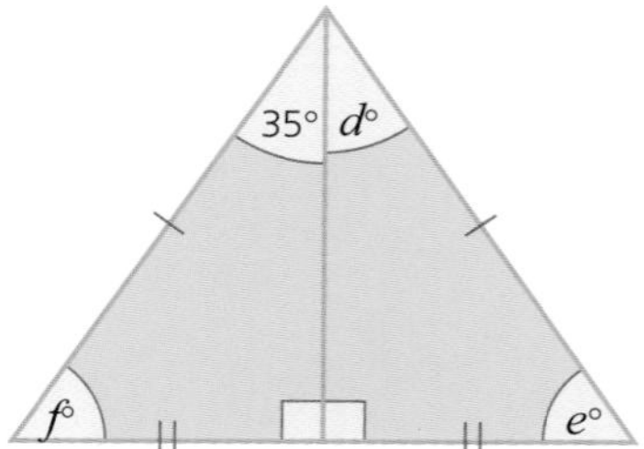

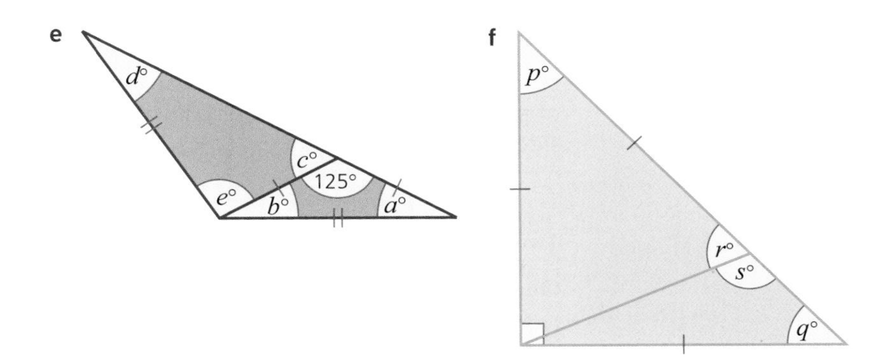

Angle properties of quadrilaterals

A **quadrilateral** is a plane shape consisting of four angles and four sides. There are several types of quadrilateral. The main ones, and their properties, are described below.

Two pairs of parallel sides.

All sides are equal.

All angles are equal.

Diagonals intersect at right angles.

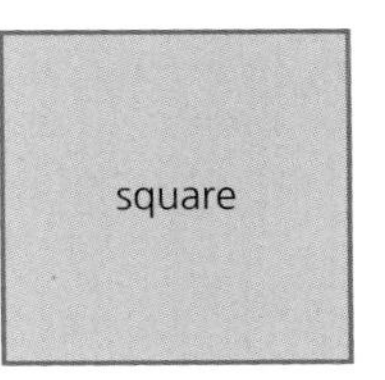

Two pairs of parallel sides.

Opposite sides are equal.

All angles are equal.

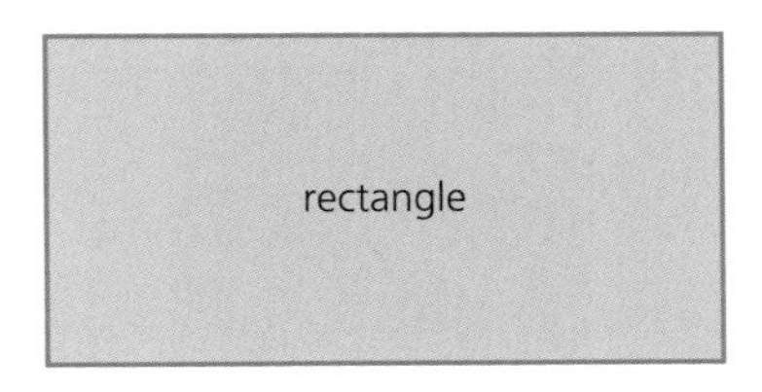

Two pairs of parallel sides.

All sides are equal.

Opposite angles are equal.

Diagonals intersect at right angles.

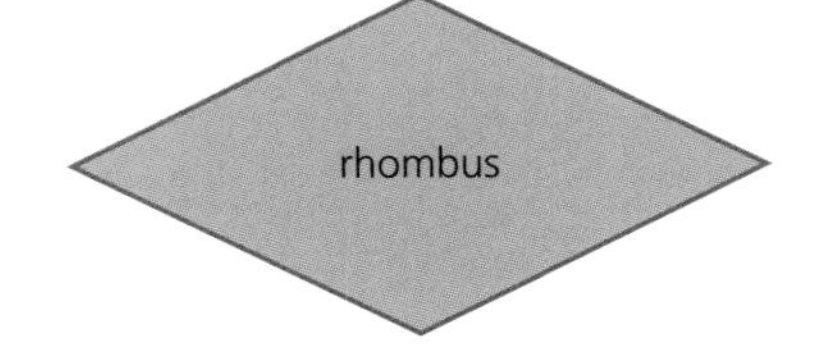

Two pairs of parallel sides.

Opposite sides are equal.

Opposite angles are equal.

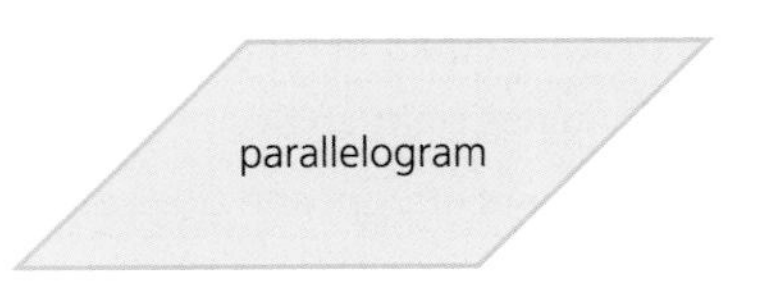

One pair of parallel sides.

An **isosceles trapezium** has one pair of parallel sides and the other pair of sides are equal in length.

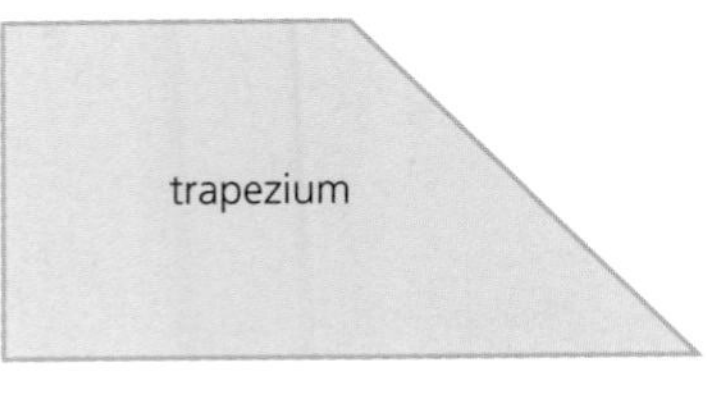

Two pairs of equal sides.

One pair of equal angles.

Diagonals intersect at right angles.

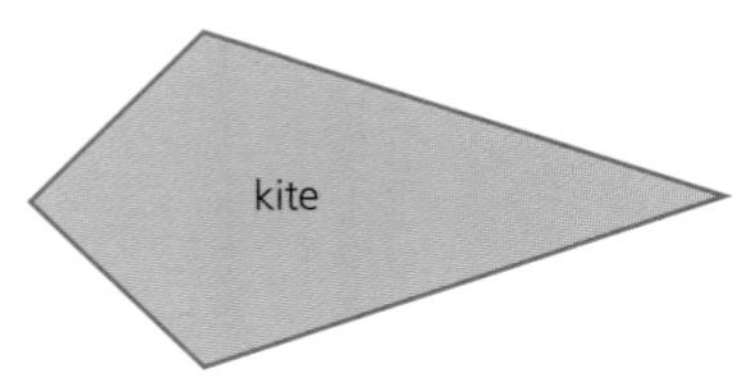

Exercise 21.8

1 Copy the diagrams and name each shape according to the definitions given above.

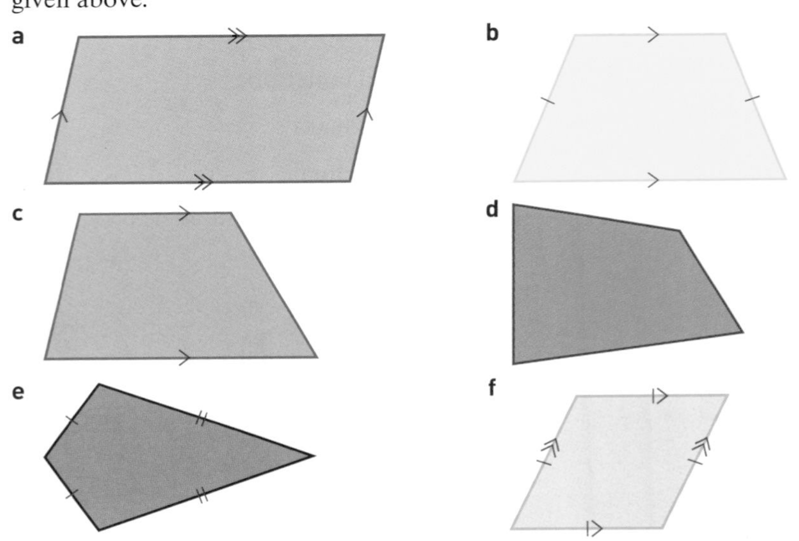

2 Copy and complete the table. The first line has been started for you.

	Rectangle	Square	Parallelogram	Kite	Rhombus	Equilateral triangle
Opposite sides equal in length	Yes		Yes			
All sides equal in length						
All angles right angles						
Both pairs of opposite sides parallel						
Diagonals equal in length						
Diagonals intersect at right angles						
All angles equal						

The sum of the interior angles of a quadrilateral

In the quadrilaterals shown, a straight line is drawn from one of the corners (vertices) to the opposite corner. The result is to split each quadrilateral into two triangles.

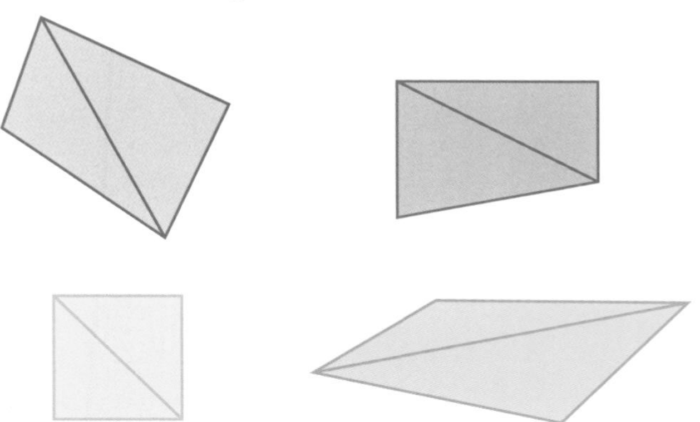

As already shown earlier in the chapter, the sum of the angles of a triangle is 180°. Therefore, as a quadrilateral can be drawn as two triangles, the sum of the four angles of any quadrilateral must be 360°.

Worked example

Calculate the size of angle p in the quadrilateral below:

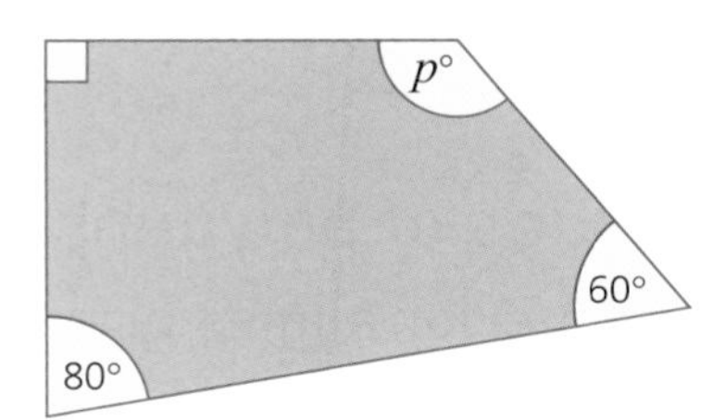

$$90 + 80 + 60 + p = 360$$

$$p = 360 - 90 - 80 - 60$$

Therefore angle p is 130°.

Exercise 21.9 For each diagram, calculate the size of the labelled angles.

a

a°
100°
75°
70°

b

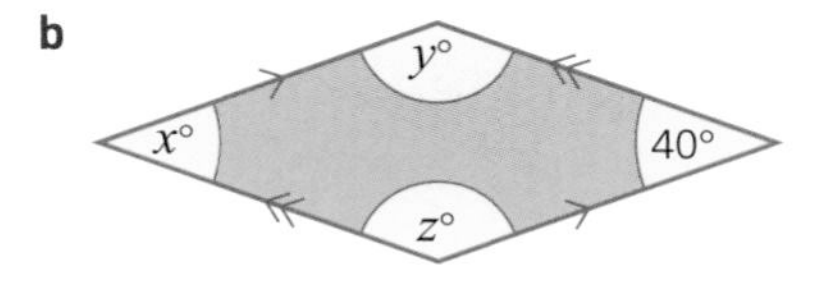

Exercise 21.9 (cont)

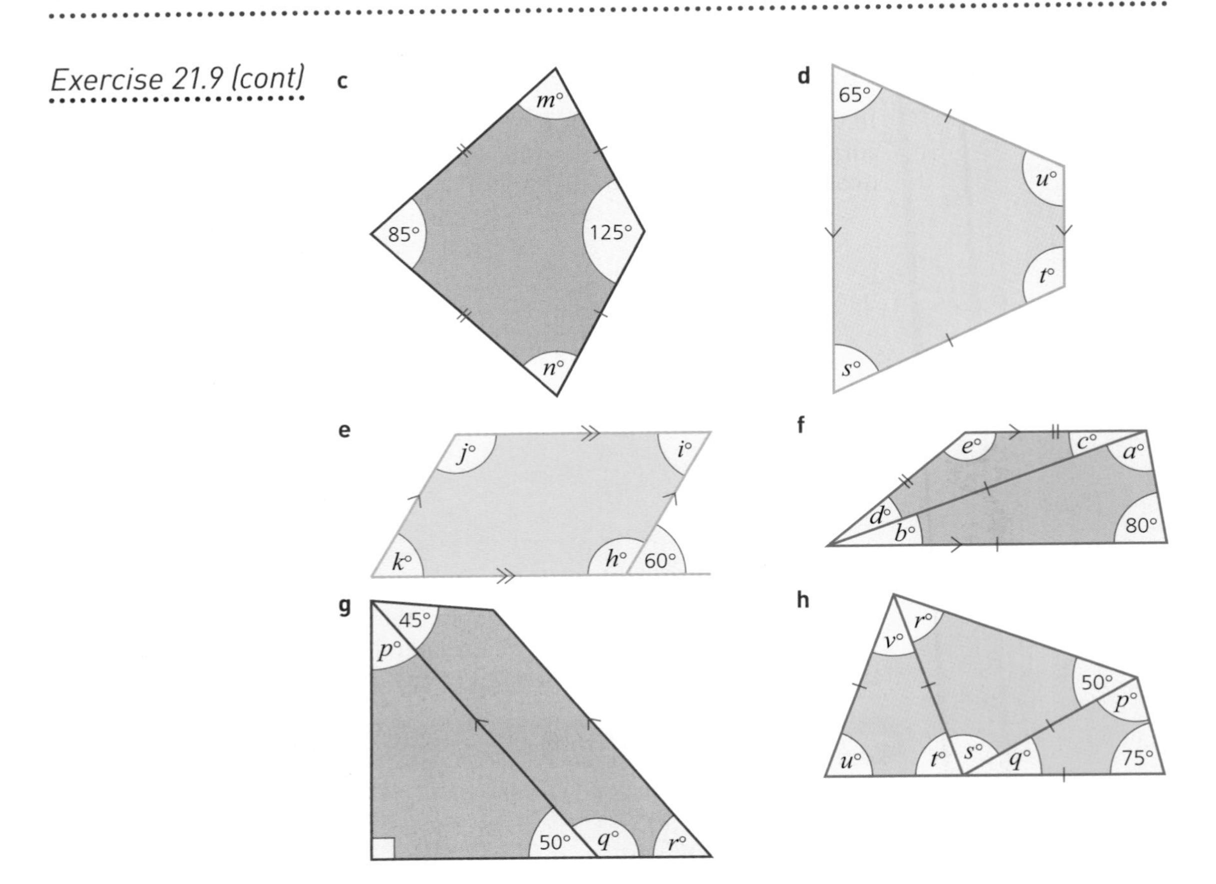

Angle properties of polygons

The sum of the interior angles of a polygon

In the polygons shown, a straight line is drawn from each vertex to vertex A.

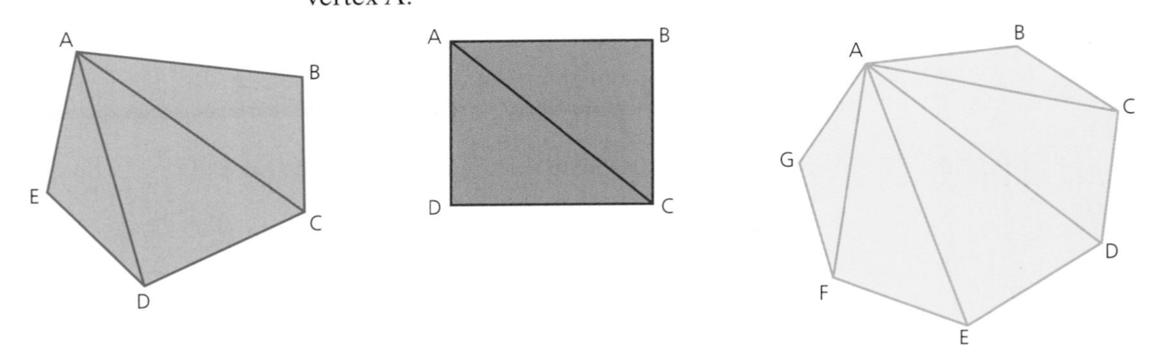

As can be seen, the number of triangles is always two less than the number of sides the polygon has, i.e. if there are n sides, there will be $(n - 2)$ triangles.

Since the angles of a triangle add up to 180°, the sum of the **interior angles** of a polygon is therefore $180(n-2)°$.

Worked example

Find the sum of the interior angles of a regular pentagon and hence the size of each interior angle.

For a pentagon, $n = 5$.

Therefore the sum of the interior angles $= 180(5 - 2)°$
$= 180 \times 3°$
$= 540°$

For a regular pentagon the interior angles are of equal size.

Therefore each angle $= \frac{540°}{5} = 108°$.

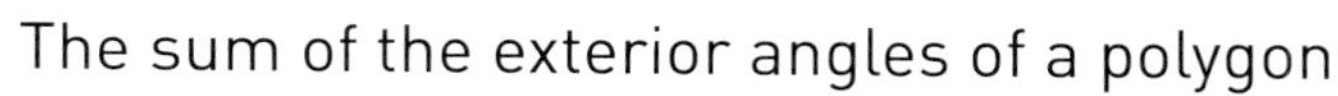

The sum of the exterior angles of a polygon

The angles marked a, b, c, d, e and f represent the **exterior angles** of the regular hexagon drawn.

For any convex polygon the sum of the exterior angles is 360°.

If the polygon is regular and has n sides, then each exterior angle $= \frac{360°}{n}$.

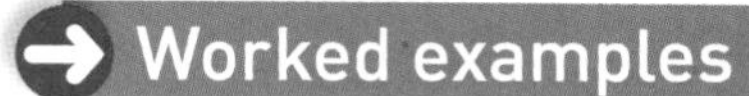

Worked examples

1 Find the size of an exterior angle of a regular nonagon.

$\frac{360°}{9} = 40°$

2 Calculate the number of sides a regular polygon has if each exterior angle is 15°.

$n = \frac{360°}{15}$
$= 24$

The polygon has 24 sides.

Exercise 21.10

1 Find the sum of the interior angles of the following polygons:
a a hexagon **b** a nonagon **c** a heptagon

2 Find the value of each interior angle of the following regular polygons:
a an octagon **b** a square **c** a decagon **d** a dodecagon

3 Find the size of each exterior angle of the following regular polygons:
a a pentagon **b** a dodecagon **c** a heptagon

4 The exterior angles of regular polygons are given. In each case calculate the number of sides the polygon has.
a 20° **b** 36° **c** 10°
d 45° **e** 18° **f** 3°

Exercise 21.10 (cont)

5 The interior angles of regular polygons are given. In each case calculate the number of sides the polygon has.

a 108° b 150° c 162°
d 156° e 171° f 179°

6 Calculate the number of sides a regular polygon has if an interior angle is five times the size of an exterior angle.

7 Copy and complete the table for regular polygons:

Number of sides	Name	Sum of exterior angles	Size of an exterior angle	Sum of interior angles	Size of an interior angle
3					
4					
5					
6					
7					
8					
9					
10					
12					

The angle in a semi-circle

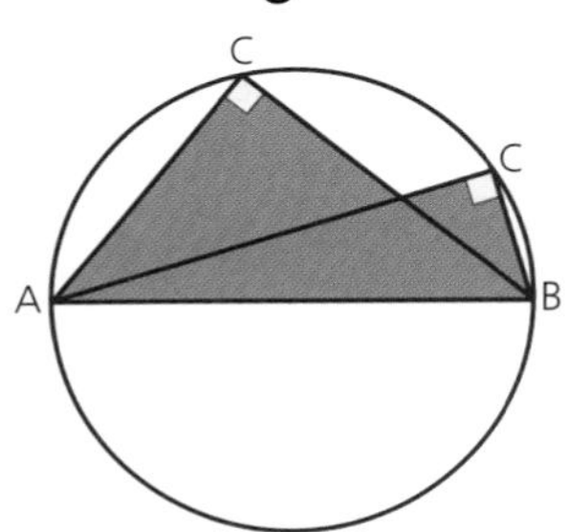

If AB represents the diameter of the circle, then the angle at C is 90°.

Exercise 21.11

In each of the following diagrams, O marks the centre of the circle. Calculate the value of x in each case.

a

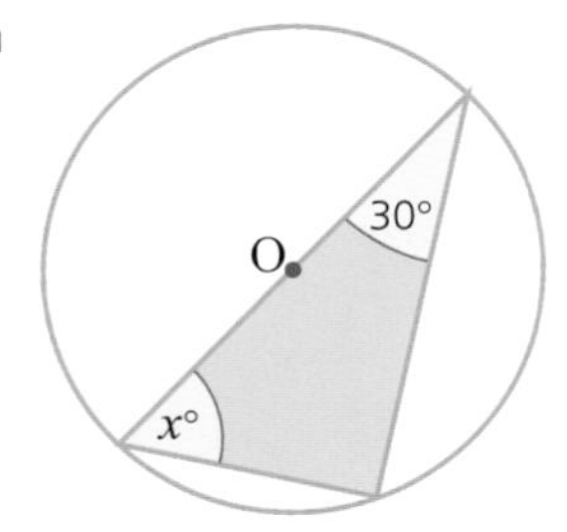

b

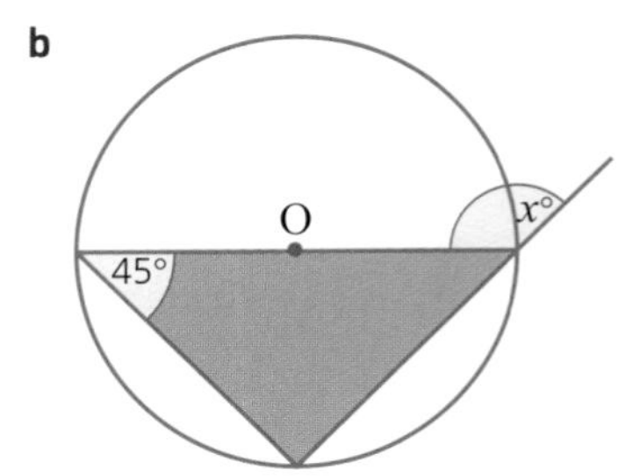

c

d

e

f

The angle between a tangent and a radius of a circle

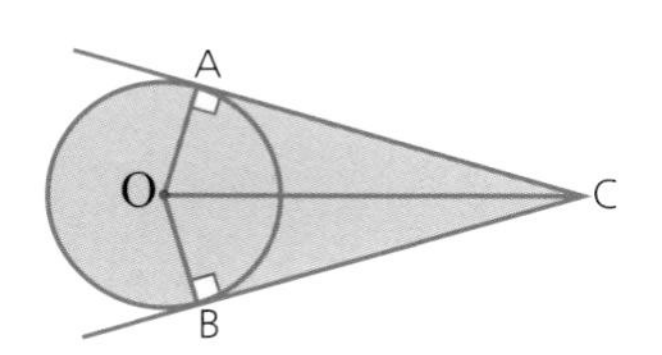

The angle between a tangent at a point and the radius to the same point on the circle is a right angle.

Triangles OAC and OBC (left) are congruent as angle *OAC* and angle *OBC* are right angles, OA = OB because they are both radii and OC is common to both triangles.

Exercise 21.12

1 In each of the following diagrams, O marks the centre of the circle. Calculate the value of x in each case.

a

b

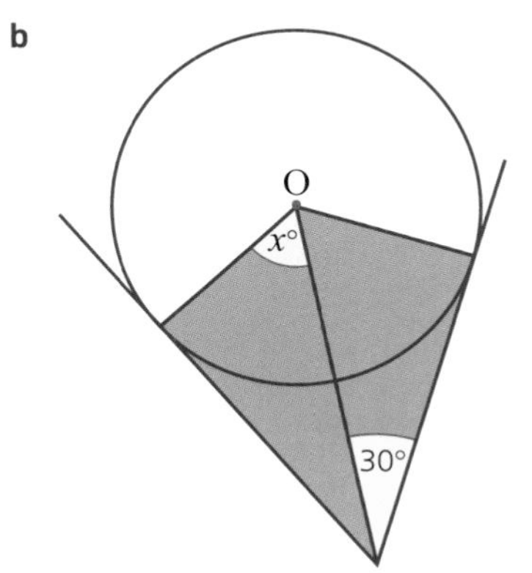

Exercise 21.12 (cont)

c

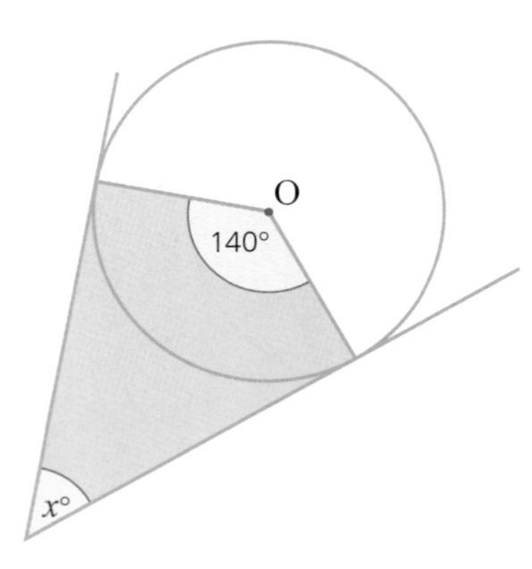

d

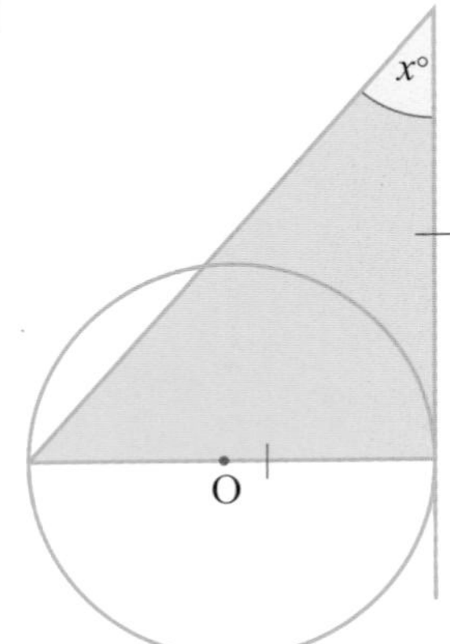

e

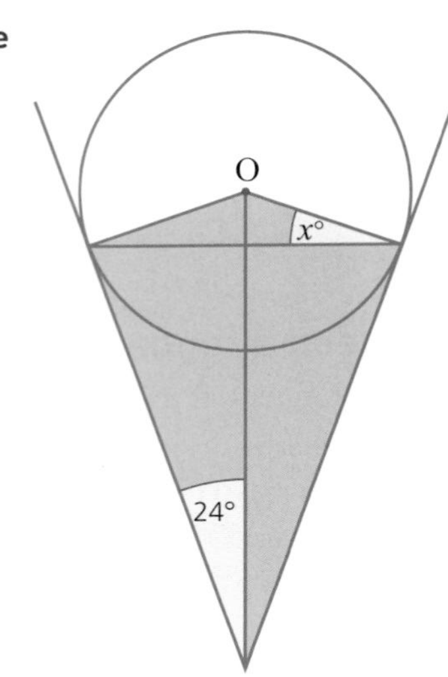

f

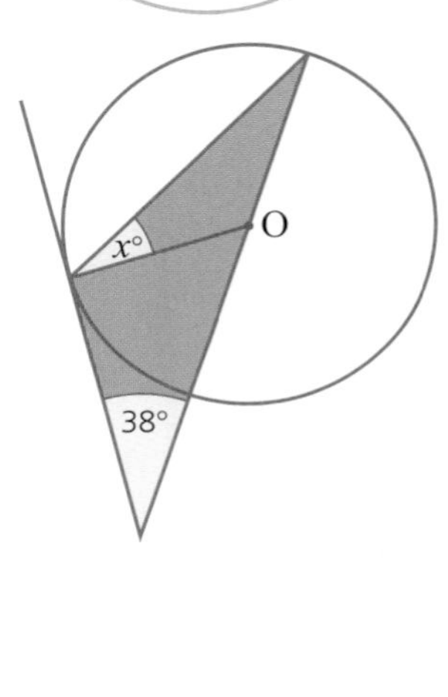

2 In the following diagrams, calculate the value of x.

a

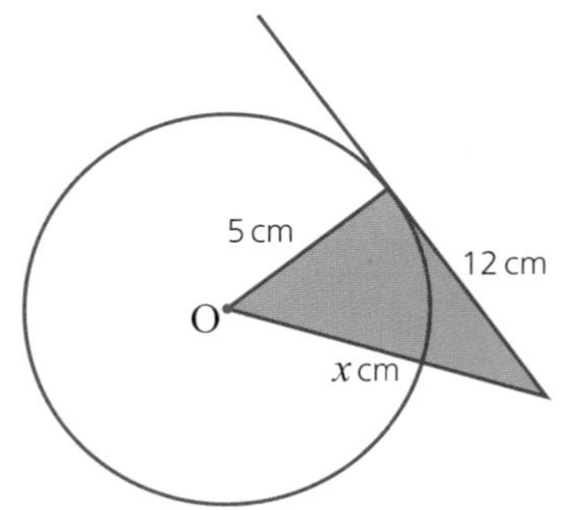

b

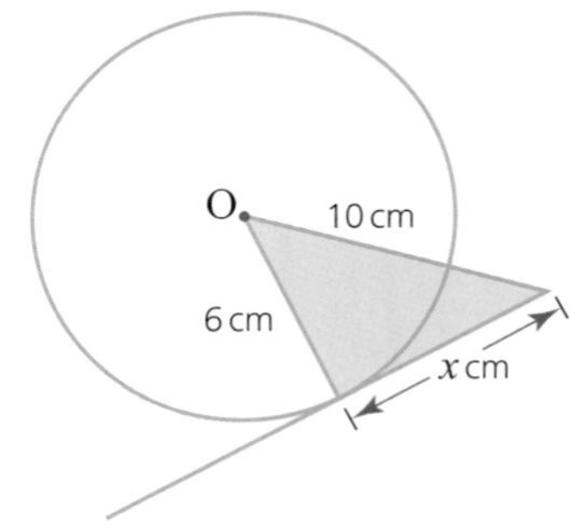

c

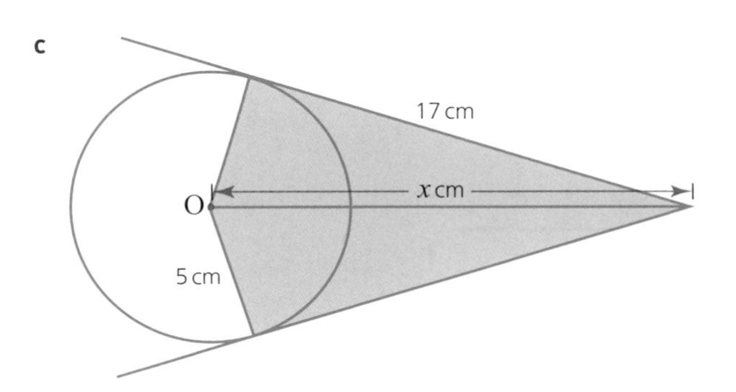

Student assessment 1

1 For each diagram, calculate the size of the labelled angles.

a

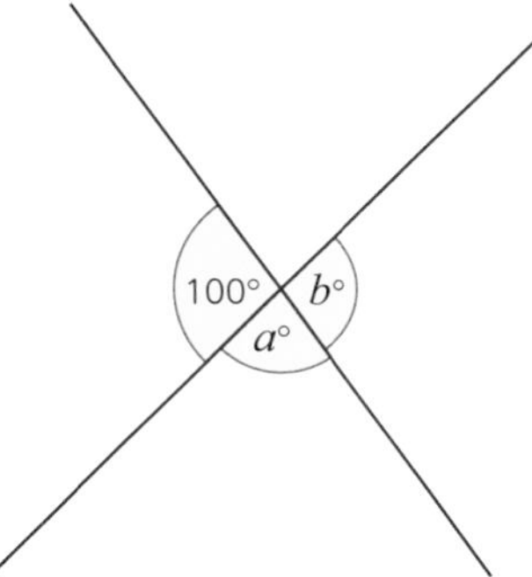

b

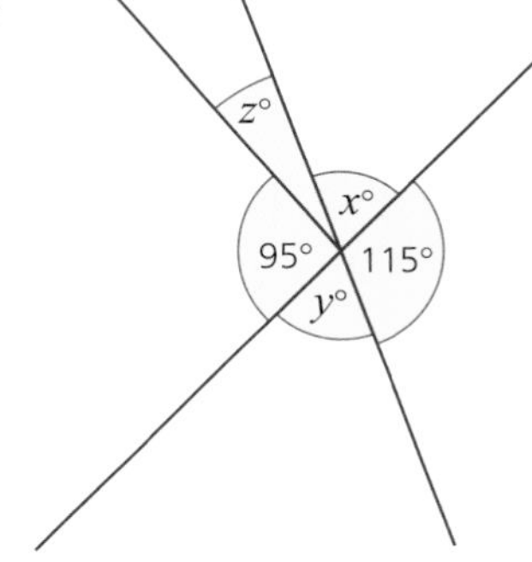

c

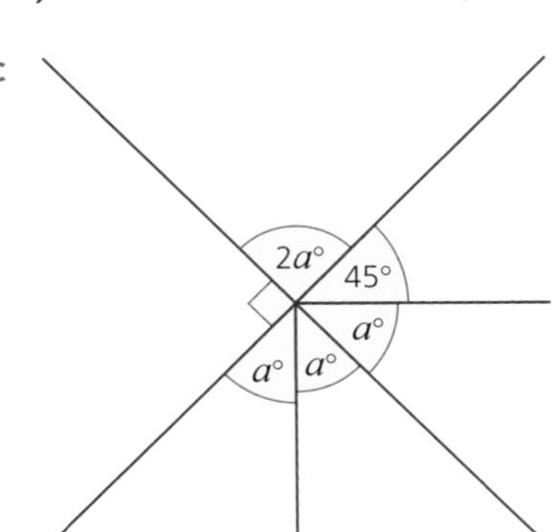

d

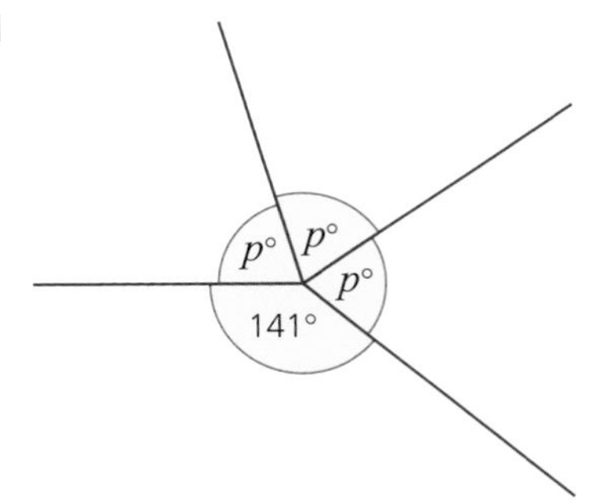

2 For each diagram, calculate the size of the labelled angles.

a

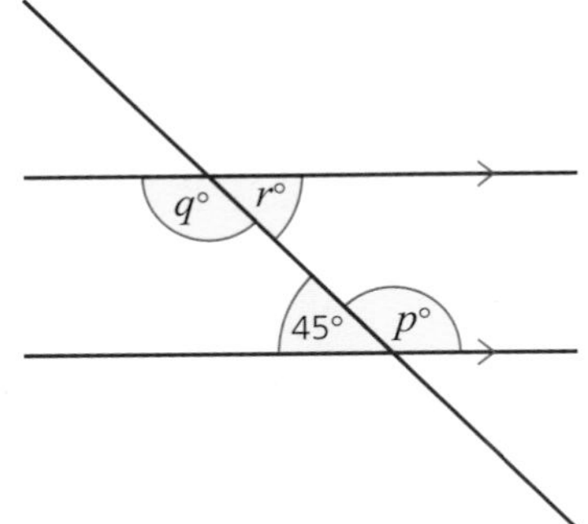

b

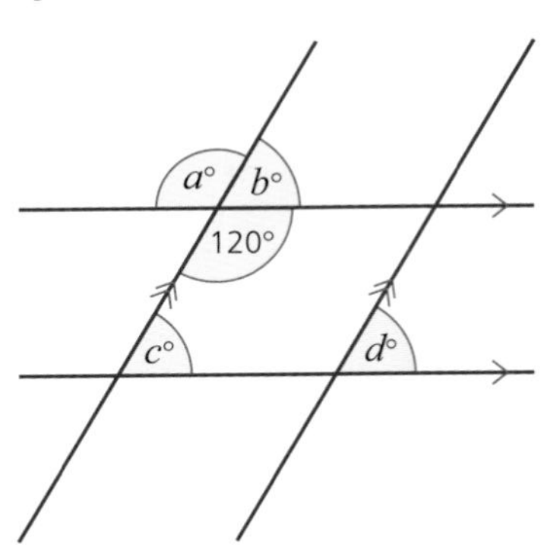

3 For each diagram, calculate the size of the labelled angles.

a

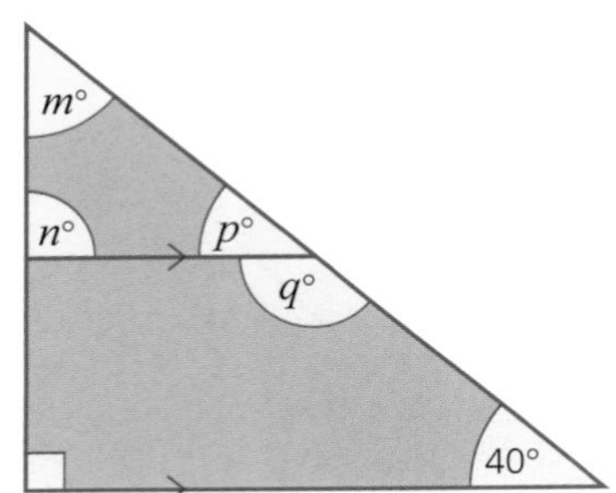

b

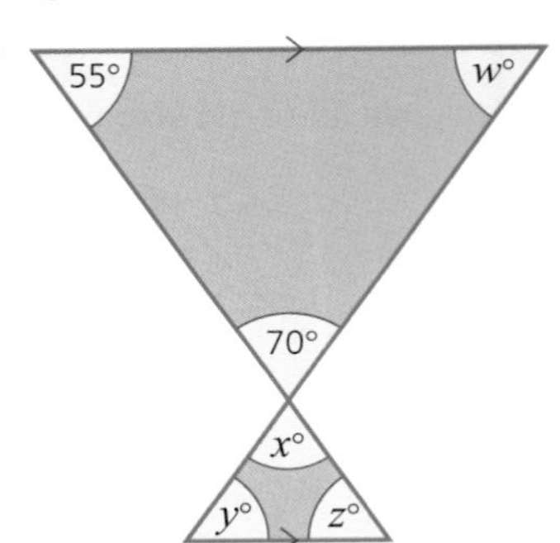

c

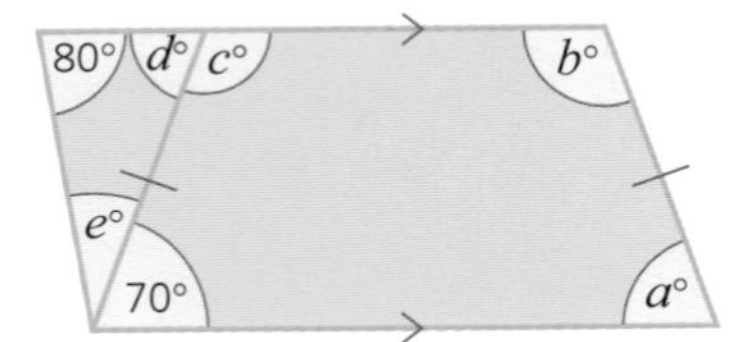

d

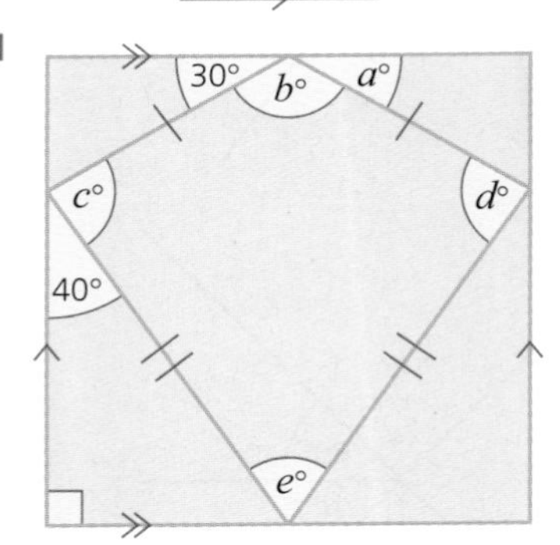

Student assessment 2

1 Draw a diagram of an octagon to help illustrate the fact that the sum of the internal angles of an octagon is given by $180 \times (8 - 2)°$.

2 Find the size of each interior angle of a 20-sided regular polygon.

3 What is the sum of the interior angles of a nonagon?

4 What is the sum of the exterior angles of a polygon?

5 What is the size of the exterior angle of a regular pentagon?

6 If AB is the diameter of the circle and AC = 5 cm and BC = 12 cm, calculate:

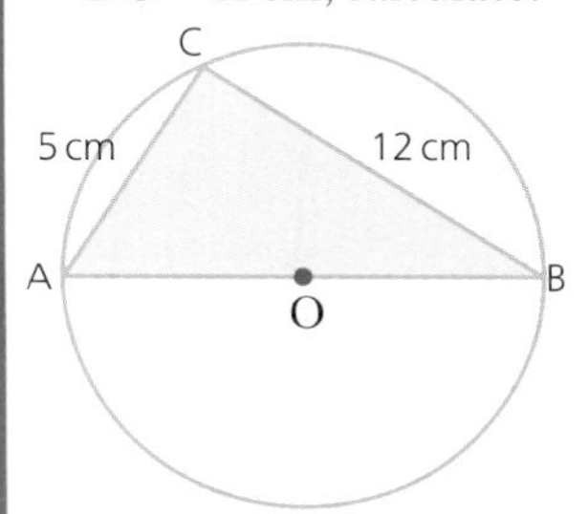

a the size of angle ACB

b the length of the radius of the circle.

In Q.7–10, O marks the centre of the circle. Calculate the size of the angle marked x in each case.

7

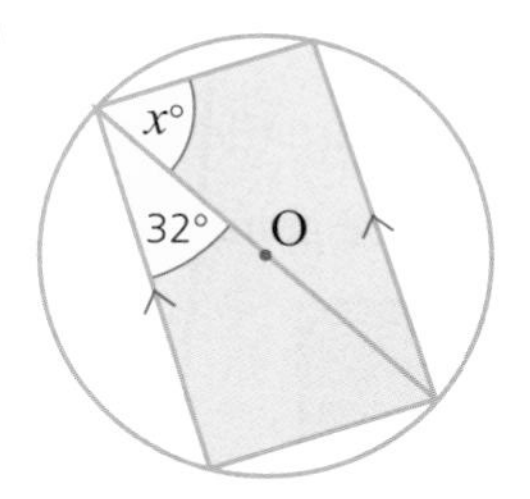

8

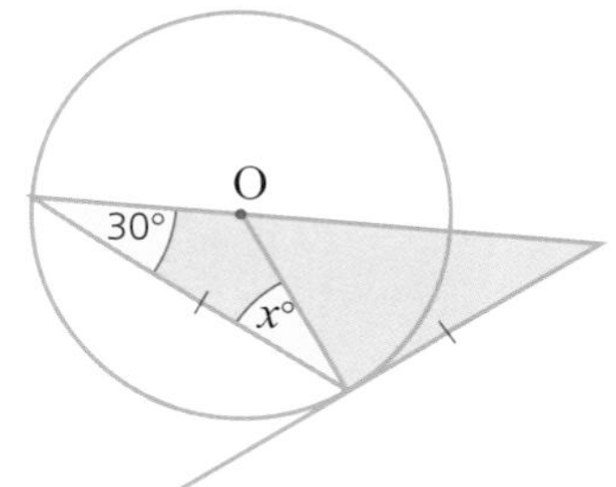

9

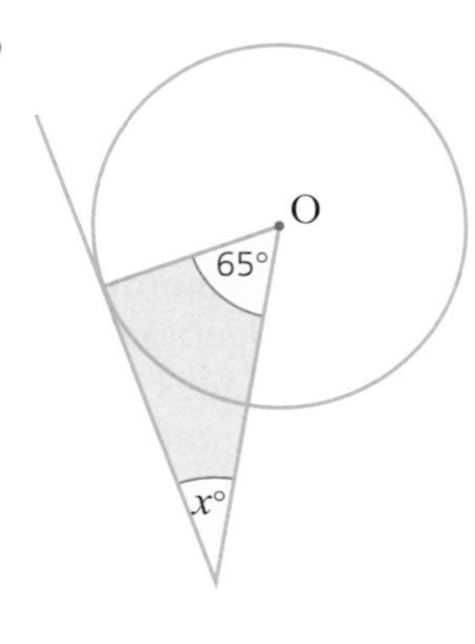

10

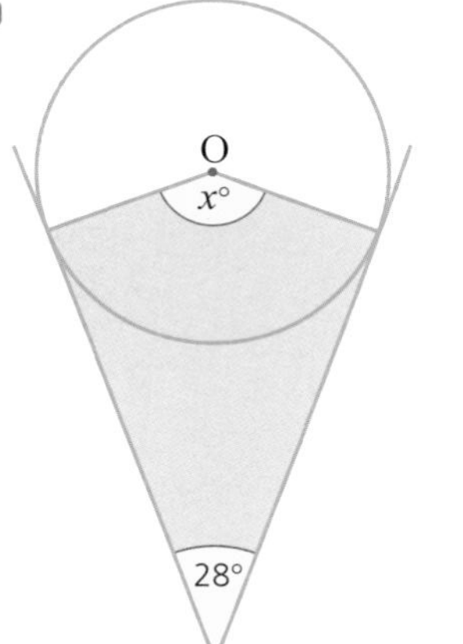

TOPIC 4

Mathematical investigations and ICT

Fountain borders

The Alhambra Palace in Granada, Spain, has many fountains which pour water into pools. Many of the pools are surrounded by beautiful ceramic tiles. This investigation looks at the number of square tiles needed to surround a particular shape of pool.

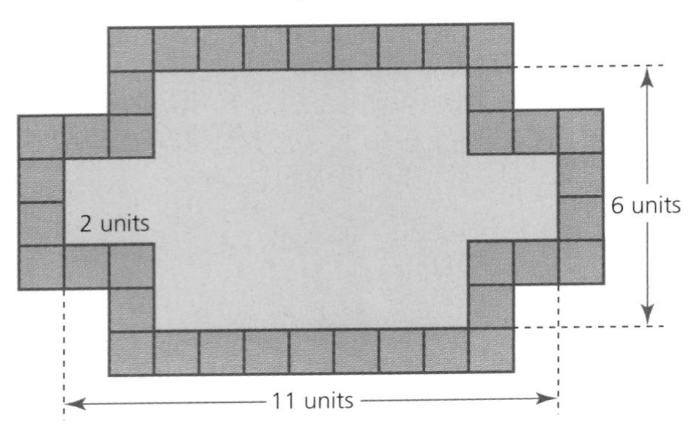

The diagram above shows a rectangular pool 11×6 units, in which a square of dimension 2×2 units is taken from each corner.

The total number of unit square tiles needed to surround the pool is 38.

The shape of the pools can be generalised as shown below:

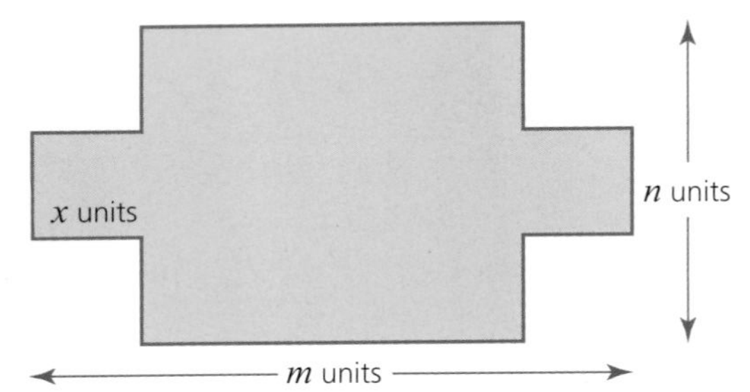

1 Investigate the number of unit square tiles needed for different sized pools. Record your results in an ordered table.
2 From your results write an algebraic rule in terms of m, n and x (if necessary) for the number of tiles T needed to surround a pool.
3 Justify, in words and using diagrams, why your rule works.

Tiled walls

Many cultures have used tiles to decorate buildings. Putting tiles on a wall takes skill. These days, to make sure that each tile is in the correct position, 'spacers' are used between the tiles.

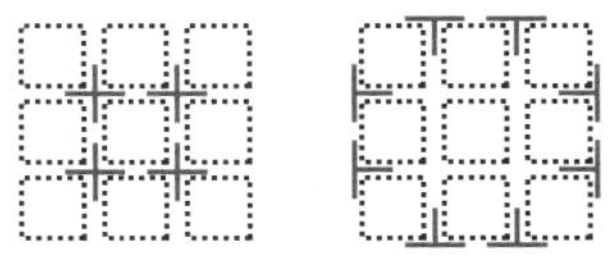

You can see from the diagram that there are + shaped and T shaped spacers.

1 Draw other sized squares and rectangles, and investigate the relationship between the dimensions of each shape (length and width) and the number of + shaped and T shaped spacers.
2 Record your results in an ordered table.
3 Write an algebraic rule for the number of + shaped spacers c in a rectangle l tiles long by w tiles wide.
4 Write an algebraic rule for the number of T shaped spacers t in a rectangle l tiles long by w tiles wide.

ICT activity

In this activity, you will use a spreadsheet to calculate the sizes of interior and exterior angles of regular polygons.

Set up a spreadsheet as shown below:

	A	B	C	D	E	F
1			Regular Polygons			
2	Number of sides	Name	Sum of exterior angles	Size of an exterior angle	Size of an interior angle	Sum of interior angles
3	3					
4	4					
5	5					
6	6					
7	7					
8	8					
9	9					
10	10					
11	12					
12	20					
13						
14						
15						
16						

Use formulae to generate the results in these columns

1 By using formulae, use the spreadsheet to generate the results for the sizes of the interior and exterior angles.
2 Write down the general formulae you would use to calculate the sizes of the interior and exterior angles of an n-sided regular polygon.

TOPIC 5

Mensuration

Contents

Course

C5.1
Use current units of mass, length, area, volume and capacity in practical situations and express quantities in terms of larger or smaller units.

C5.2
Carry out calculations involving the perimeter and area of a rectangle, triangle, parallelogram and trapezium and compound shapes derived from these.

C5.3
Carry out calculations involving the circumference and area of a circle.

Solve simple problems involving the arc length and sector area as fractions of the circumference and area of a circle.

C5.4
Carry out calculations involving the surface area and volume of a cuboid, prism and cylinder.

Carry out calculations involving the surface area and volume of a sphere, pyramid and cone.

C5.5
Carry out calculations involving the areas and volumes of compound shapes.

Measurement

A measurement is the ratio of a physical quantity, such as a length, time or temperature, to a unit of measurement, such as the metre, the second or the degree Celsius. So if someone is 1.68 m tall they are 1.68 times bigger than the standard measure called a metre.

The International System of Units (or SI units from the French language name *Système International d'Unités*) is the world's most widely used system of units. The SI units for the seven basic physical quantities are:

- the metre (m) – the SI unit of length
- the kilogram (kg) – the SI unit of mass
- the second (s) – the SI unit of time
- the ampere (A) – the SI unit of electric current
- the kelvin (K) – the SI unit of temperature
- the mole (mol) – the SI unit of amount of substance
- the candela (cd) – the SI unit of luminous intensity.

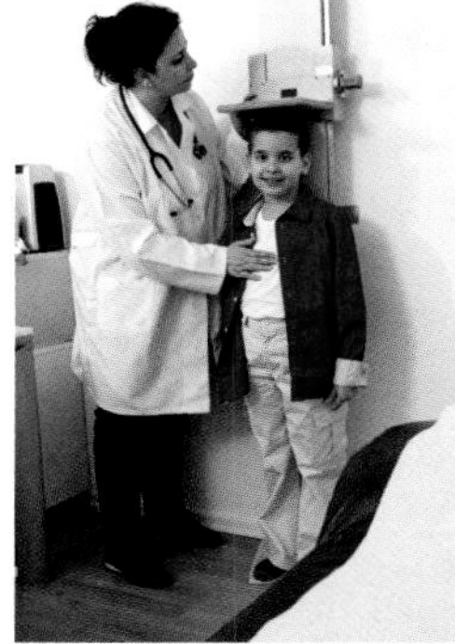

This system was a development of the metric system which was first used in the 1790s during the French Revolution. This early system used just the metre and the kilogram and was intended to give fair and consistent measures in France.

22 Measures

Metric units

The metric system uses a variety of units for length, mass and capacity.

- The common units of length are: kilometre (km), metre (m), centimetre (cm) and millimetre (mm).
- The common units of mass are: tonne (t), kilogram (kg), gram (g) and milligram (mg).
- The common units of capacity are: litre (L or l) and millilitre (ml).

Note

'centi' comes from the Latin *centum* meaning hundred (a centimetre is one hundredth of a metre).
'milli' comes from the Latin *mille* meaning thousand (a millimetre is one thousandth of a metre).
'kilo' comes from the Greek *khilloi* meaning thousand (a kilometre is one thousand metres).
It may be useful to have some practical experience of estimating lengths, volumes and capacities before starting the following exercises.

Exercise 22.1

1 Copy and complete the sentences:
 a There are ... centimetres in one metre.
 b There are ... millimetres in one metre.
 c One metre is one ... of a kilometre.
 d There are ... kilograms in one tonne.
 e There are ... grams in one kilogram.
 f One milligram is one ... of a gram.
 g One thousand kilograms is one ... .
 h One thousandth of a gram is one ... .
 i One thousand millilitres is one ... .
 j One thousandth of a litre is one ... .

2 Which of the units given would be used to measure the following?
 mm, cm, m, km, mg, g, kg, tonnes, ml, litres
 a your height
 b the length of your finger
 c the mass of a shoe
 d the amount of liquid in a cup
 e the height of a van
 f the mass of a ship
 g the capacity of a swimming pool
 h the length of a highway
 i the mass of an elephant
 j the capacity of the petrol tank of a car.

3 Use a ruler to draw lines of these lengths:
a 6 cm
b 18 cm
c 41 mm
d 8.7 cm
e 67 mm

4 Draw four lines and label them A, B, C and D.
a Estimate their lengths in mm.
b Measure them to the nearest mm.

5 Copy the sentences and put in the correct unit:
a A tree in the school grounds is 28 ... tall.
b The distance to the nearest big city is 45 ... .
c The depth of a lake is 18 ... .
d A woman's mass is about 60 ... .
e The capacity of a bowl is 5 ... .
f The distance Ahmet can run in 10 seconds is about 70 ... .
g The mass of my car is about 1.2 ... .
h Ayse walks about 1700 ... to school.
i A melon has a mass of 650 ... .
j The amount of blood in your body is 5 ... .

Converting from one unit to another

Length

1 km = 1000 m

Therefore 1 m = $\frac{1}{1000}$ km

1 m = 100 cm

Therefore 1 cm = $\frac{1}{100}$ m

1 m = 1000 mm

Therefore 1 mm = $\frac{1}{1000}$ m

1 cm = 10 mm

Therefore 1 mm = $\frac{1}{10}$ cm

Worked examples

1 Change 5.8 km into m.

Since 1 km = 1000 m,

5.8 km is 5.8 × 1000 m

5.8 km = 5800 m

2 Change 4700 mm to m.

Since 1 m is 1000 mm,

4700 mm is 4700 ÷ 1000 m

4700 mm = 4.7 m

3 Convert 2.3 km into cm.

2.3 km is 2.3 × 1000 m = 2300 m

2300 m is 2300 × 100 cm

2.3 km = 230 000 cm

Exercise 22.2

1 Put in the missing unit to make the statements correct:

a 3 cm = 30 ...
b 25 ... = 2.5 cm
c 3200 m = 3.2 ...
d 7.5 km = 7500 ...
e 300 ... = 30 cm
f 6000 mm = 6 ...
g 3.2 m = 3200 ...
h 4.2 ... = 4200 mm
i 1 million mm = 1 ...
j 2.5 km = 2500 ...

2 Convert to millimetres:

a 2 cm
b 8.5 cm
c 23 cm
d 1.2 m
e 0.83 m
f 0.05 m
g 62.5 cm
h 0.087 m
i 0.004 m

3 Convert to metres:

a 3 km
b 4700 mm
c 560 cm
d 6.4 km
e 0.8 km
f 96 cm
g 62.5 cm
h 0.087 km
i 0.004 km

4 Convert to kilometres:

a 5000 m
b 6300 m
c 1150 m
d 2535 m
e 250 000 m
f 500 m
g 70 m
h 8 m
i 1 million m
j 700 million m

Mass

1 tonne is 1000 kg

Therefore 1 kg = $\frac{1}{1000}$ tonne

1 g is 1000 mg

Therefore 1 mg = $\frac{1}{1000}$ g

1 kilogram is 1000 g

Therefore 1 g = $\frac{1}{1000}$ kg

Worked examples

1 Convert 8300 kg to tonnes.

Since 1000 kg = 1 tonne, 8300 kg is 8300 ÷ 1000 tonnes

8300 kg = 8.3 tonnes

2 Convert 2.5 g to mg.

Since 1 g is 1000 mg, 2.5 g is 2.5 × 1000 mg

2.5 g = 2500 mg

Exercise 22.3

1 Convert:
- a 3.8 g to mg
- b 28 500 kg to tonnes
- c 4.28 tonnes to kg
- d 320 mg to g
- e 0.5 tonnes to kg

2 An item has a mass of 630 g, another item has a mass of 720 g. Express the total mass in kg.

3 a Express the total weight in kg:
1.2 tonne, 760 kg, 0.93 tonne and 640 kg
- b Express the total weight in g:
460 mg, 1.3 g, 1260 mg and 0.75 g
- c A cat weighs 2800 g and a dog weighs 6.5 kg. What is the total weight in kg of the two animals?
- d In one bag of shopping, Imran has items of total mass 1350 g. In another bag there are items of total mass 3.8 kg. What is the mass in kg of both bags of shopping?
- e Find the total mass in kg of the fruit listed:
apples 3.8 kg, peaches 1400 g, bananas 0.5 kg, oranges 7500 g, grapes 0.8 kg

Capacity

1 litre is 1000 millilitres

Therefore 1 ml = $\frac{1}{1000}$ litre

Exercise 22.4

1 Convert to litres:
- a 8400 ml
- b 650 ml
- c 87 500 ml
- d 50 ml
- e 2500 ml

2 Convert to millilitres:
- a 3.2 litres
- b 0.75 litre
- c 0.087 litre
- d 8 litres
- e 0.008 litre
- f 0.3 litre

3 Calculate the following and give the totals in millilitres:
- a 3 litres + 1500 ml
- b 0.88 litre + 650 ml
- c 0.75 litre + 6300 ml
- d 450 ml + 0.55 litre

4 Calculate the following and give the totals in litres:
- a 0.75 litre + 450 ml
- b 850 ml + 490 ml
- c 0.6 litre + 0.8 litre
- d 80 ml + 620 ml + 0.7 litre

Student assessment 1

1 Convert the lengths to the units indicated:
- a 2.6 cm to mm
- b 62.5 cm to mm
- c 0.88 m to cm
- d 0.007 m to mm
- e 4800 mm to m
- f 7.81 km to m
- g 6800 m to km
- h 0.875 km to m
- i 2 m to mm
- j 0.085 m to mm

2 Convert the masses to the units indicated:
- a 4.2 g to mg
- b 750 mg to g
- c 3940 g to kg
- d 4.1 kg to g
- e 0.72 tonnes to kg
- f 4100 kg to tonnes
- g 6 280 000 mg to kg
- h 0.83 tonnes to g
- i 47 million kg to tonnes
- j 1 kg to mg

3 Add the masses, giving your answer in kg:

3.1 tonnes, 4860 kg and 0.37 tonnes

4 Convert the liquid measures to the units indicated:
- a 1800 ml to litres
- b 3.2 litres to ml
- c 0.083 litre to ml
- d 250 000 ml to litres

Student assessment 2

1 Convert the lengths to the units indicated:
- a 4.7 cm to mm
- b 0.003 m to mm
- c 3100 mm to cm
- d 6.4 km to m
- e 49 000 m to km
- f 4 m to mm
- g 0.4 cm to mm
- h 0.034 m to mm
- i 460 mm to cm
- j 50 000 m to km

2 Convert the masses to the units indicated:
- a 3.6 mg to g
- b 550 mg to g
- c 6500 g to kg
- d 6.7 kg to g
- e 0.37 tonnes to kg
- f 1510 kg to tonnes
- g 380 000 kg to tonnes
- h 0.077 kg to g
- i 6 million mg to kg
- j 2 kg to mg

3 Subtract 1570 kg from 2 tonnes.

4 Convert the measures of capacity to the units indicated:
- a 3400 ml to litres
- b 6.7 litres to ml
- c 0.73 litre to ml
- d 300 000 ml to litres

23 Perimeter, area and volume

All diagrams are not drawn to scale.

The perimeter and area of a rectangle

The **perimeter** of a shape is the distance around the outside of the shape. Perimeter can be measured in mm, cm, m, km, etc.

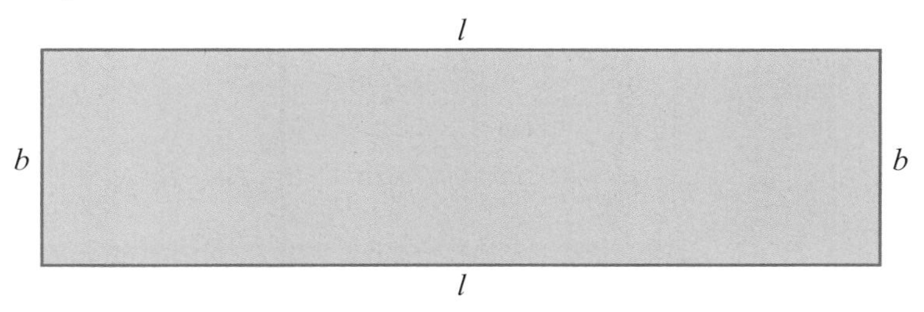

The perimeter of the rectangle above of **length** l and **breadth** b is:

$$\text{Perimeter} = l + b + l + b$$

This can be rearranged to give:

$$\text{Perimeter} = 2l + 2b$$

This can be factorised to give:

$$\text{Perimeter} = 2(l + b)$$

The **area** of a shape is the amount of surface that it covers. Area is measured in mm^2, cm^2, m^2, km^2, etc.

The area A of the rectangle above is given by the formula:

$$A = lb$$

Worked example

Find the area of the rectangle.

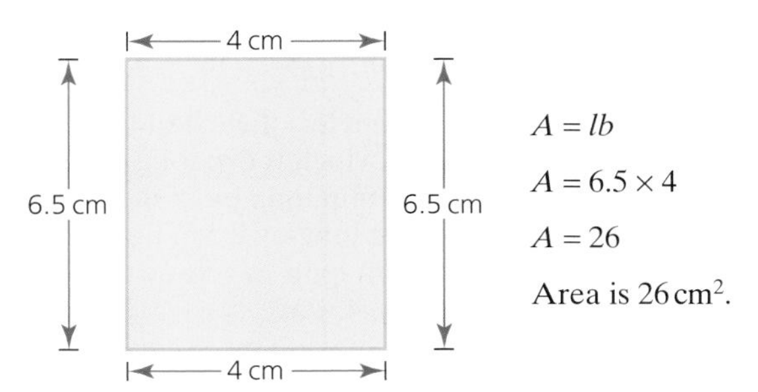

$A = lb$

$A = 6.5 \times 4$

$A = 26$

Area is $26\,cm^2$.

Exercise 23.1

Calculate the area and perimeter of the rectangles described in the table.

	Length	Breadth	Area	Perimeter
a	6 cm	4 cm		
b	5 cm	9 cm		
c	4.5 cm	6 cm		
d	3.8 m	10 m		
e	5 m	4.2 m		
f	3.75 cm	6 cm		
g	3.2 cm	4.7 cm		
h	18.7 m	5.5 cm		
i	85 cm	1.2 m		
j	3.3 m	75 cm		

Worked example

Calculate the breadth of a rectangle with an area of 200 cm^2 and length 25 cm.

$A = lb$
$200 = 25b$
$b = 8$
So the breadth is 8 cm.

Exercise 23.2

1 Use the formula for the area of a rectangle to find the value of A, l or b as indicated in the table.

	Length	Breadth	Area
a	8.5 cm	7.2 cm	A cm^2
b	25 cm	b cm	250 cm^2
c	l cm	25 cm	400 cm^2
d	7.8 m	b m	78 m^2
e	l cm	8.5 cm	102 cm^2
f	22 cm	b cm	330 cm^2
g	l cm	7.5 cm	187.5 cm^2

2 Find the area and perimeter of each of these squares or rectangles:
- **a** the floor of a room which is 8 m long by 7.5 m wide
- **b** a stamp which is 35 mm long by 25 mm wide
- **c** a wall which is 8.2 m long by 2.5 m high
- **d** a field which is 130 m long by 85 m wide
- **e** a chessboard of side 45 cm
- **f** a book which is 25 cm wide by 35 cm long
- **g** an airport runway which is 3.5 km long by 800 m wide
- **h** a street which is 1.2 km long by 25 m wide
- **i** a sports hall which is 65 m long by 45 m wide
- **j** a tile which is a square of side 125 mm

The area of a triangle

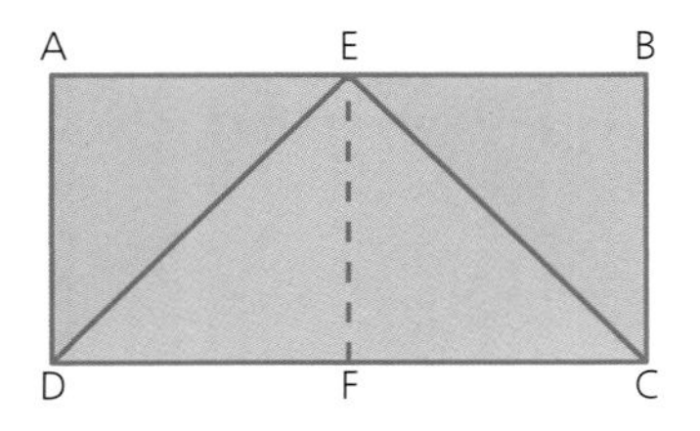

Rectangle ABCD has triangle CDE drawn inside it.

Point E is a **vertex** of the triangle.

EF is the **height** or **altitude** of the triangle.

CD is the length of the rectangle, but is called the **base** of the triangle.

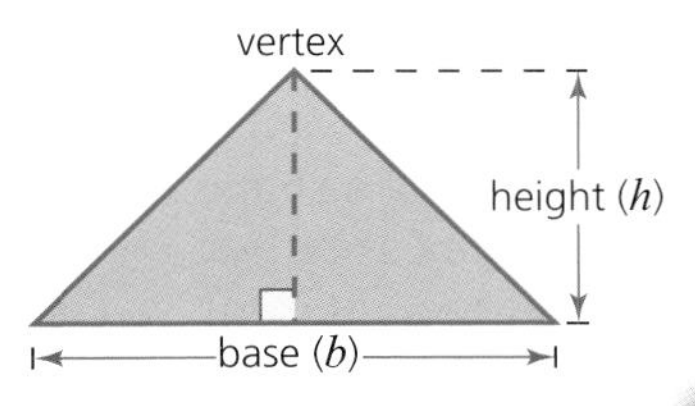

It can be seen from the diagram that triangle DEF is half the area of the rectangle AEFD.

Also, triangle CFE is half the area of rectangle EBCF.

It follows that triangle CDE is half the area of rectangle ABCD.

Area of a triangle is $A = \frac{1}{2}bh$, where b is the base and h is the height.

Note

It does not matter which side is called the base, but the height **must** be measured at right angles from the base to the opposite vertex.

Exercise 23.3

Calculate the areas of the triangles:

a

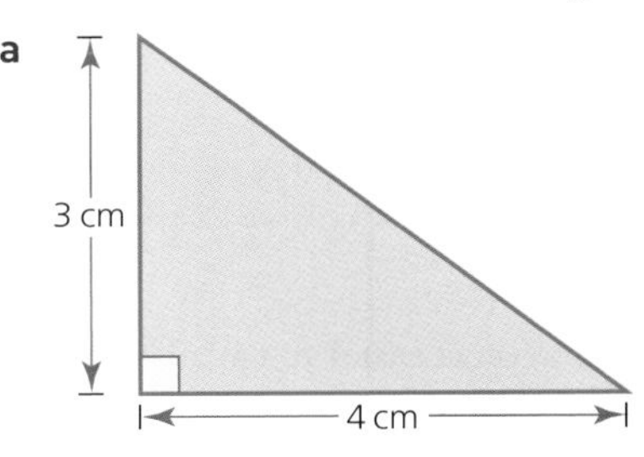

b

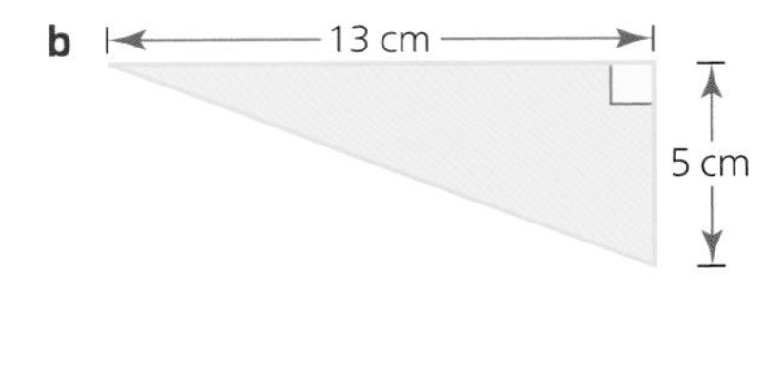

c

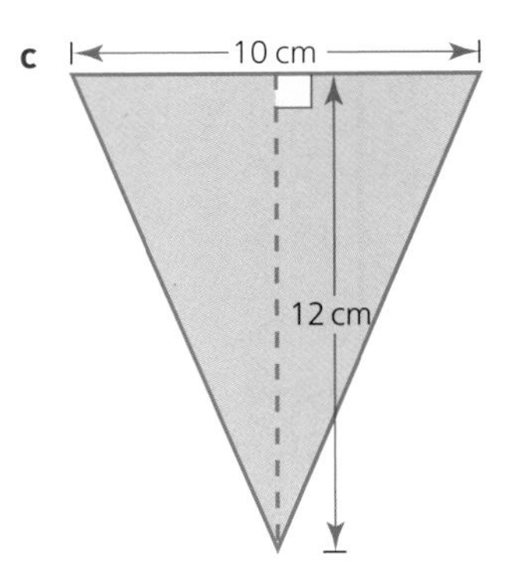

d

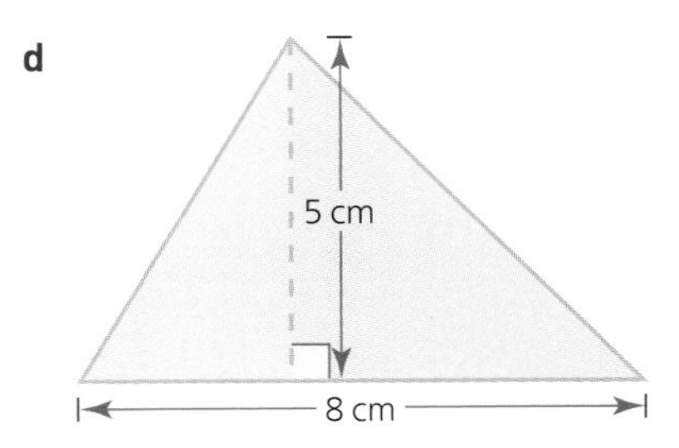

Exercise 23.3 (cont)

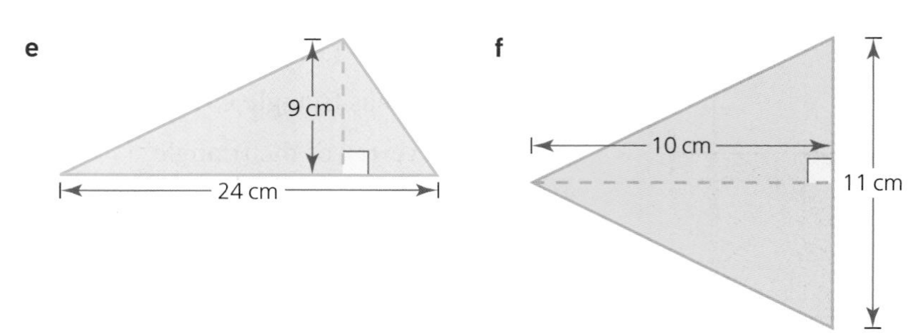

Compound shapes

Sometimes, being asked to work out the perimeter and area of a shape can seem difficult. However, calculations can often be made easier by splitting a shape up into simpler shapes. A shape that can be split into simpler ones is known as a **compound shape**.

Worked example

The diagram shows a pentagon and its dimensions. Calculate the area of the shape.

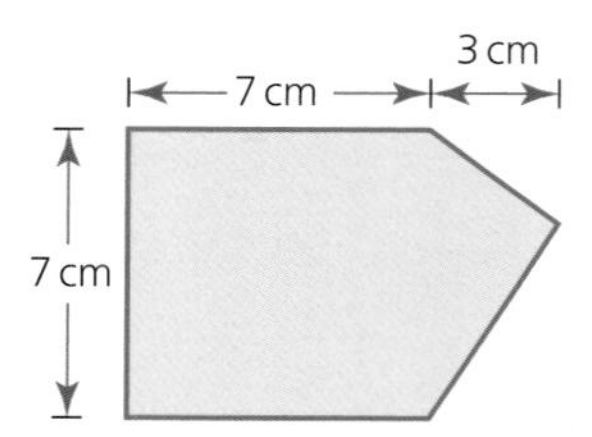

The area of the pentagon is easier to calculate if it is split into two simpler shapes; a square and a triangle:

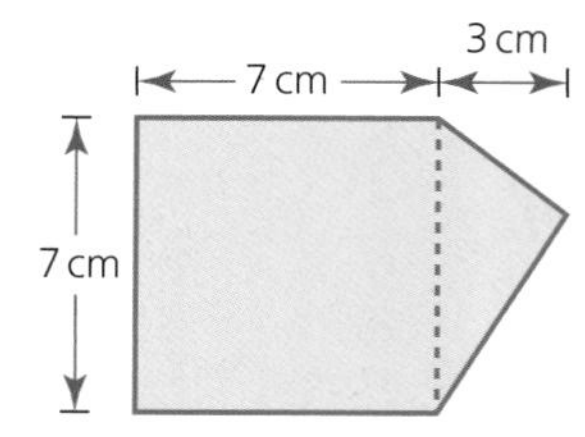

The area of the square is $7 \times 7 = 49\,\text{cm}^2$.

The area of the triangle is $\frac{1}{2} \times 7 \times 3 = 10.5\,\text{cm}^2$.

Therefore the total area of the pentagon is $49 + 10.5 = 59.5\,\text{cm}^2$.

Exercise 23.4

Calculate the areas of the compound shapes:

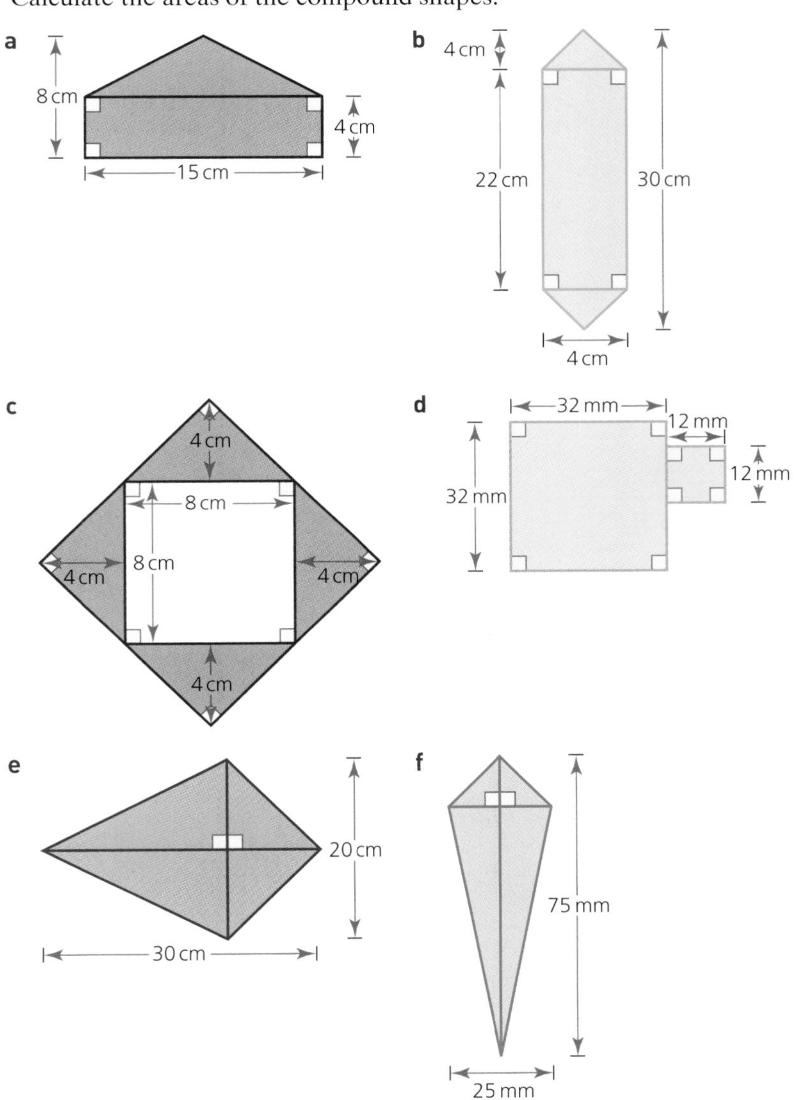

The area of a parallelogram and a trapezium

A **parallelogram** can be rearranged to form a rectangle in the following way:

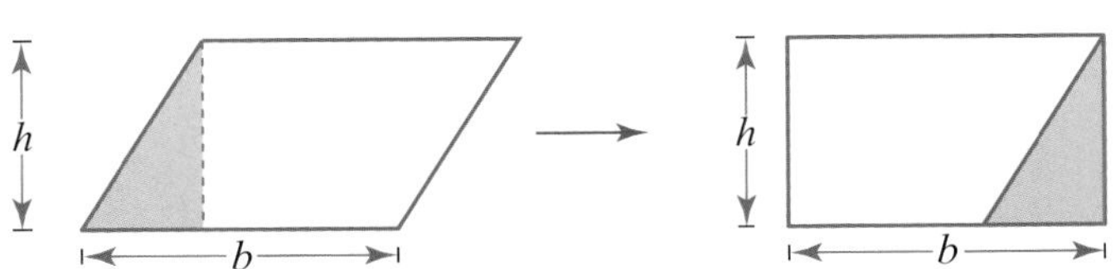

Therefore:

area of parallelogram = base length × perpendicular height.

A **trapezium** can be split into two triangles:

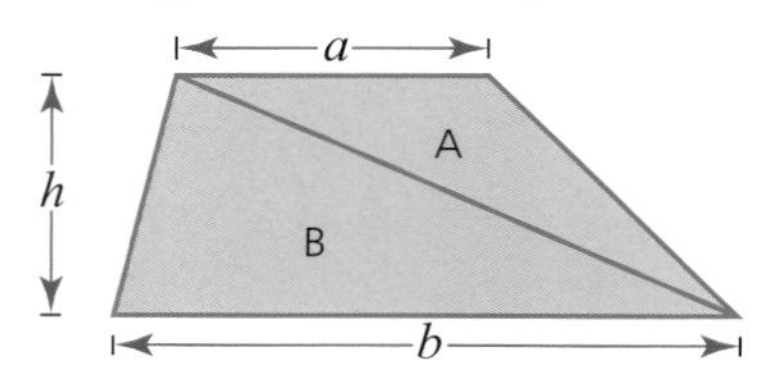

Area of triangle A $= \frac{1}{2} \times a \times h$

Area of triangle B $= \frac{1}{2} \times b \times h$

Area of trapezium

$= \text{area of triangle A} + \text{area of triangle B}$

$= \frac{1}{2}ah + \frac{1}{2}bh$

$= \frac{1}{2}h(a + b)$

Worked examples

1 Calculate the area of the parallelogram:

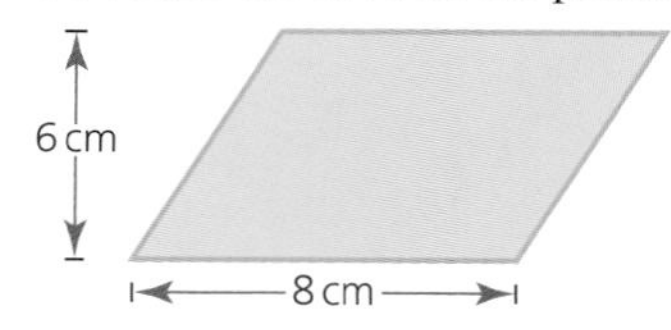

Area = base length × perpendicular height

$= 8 \times 6$

$= 48\,\text{cm}^2$

2 Calculate the shaded area in the shape:

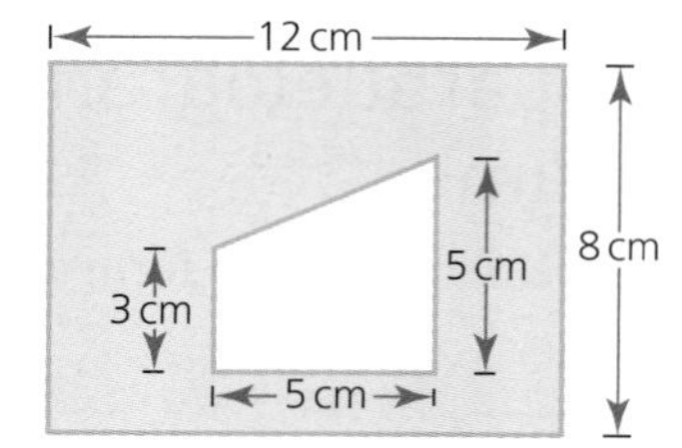

$$\text{Area of rectangle} = 12 \times 8$$
$$= 96\,\text{cm}^2$$
$$\text{Area of trapezium} = \tfrac{1}{2} \times 5(3 + 5)$$
$$= 2.5 \times 8$$
$$= 20\,\text{cm}^2$$
$$\text{Shaded area} = 96 - 20$$
$$= 76\,\text{cm}^2$$

Exercise 23.5

Find the area of these shapes:

a

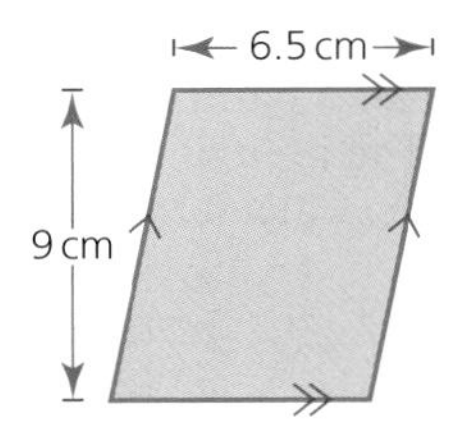

b

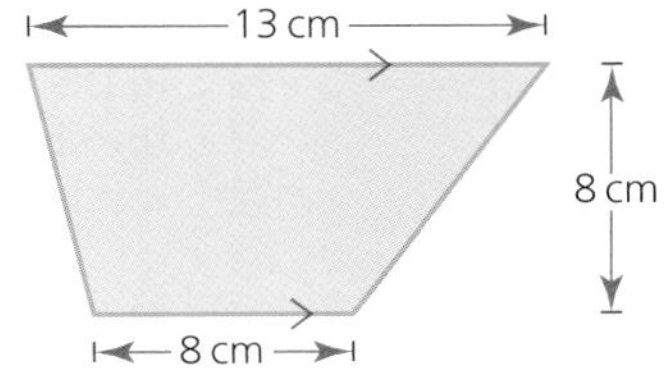

c

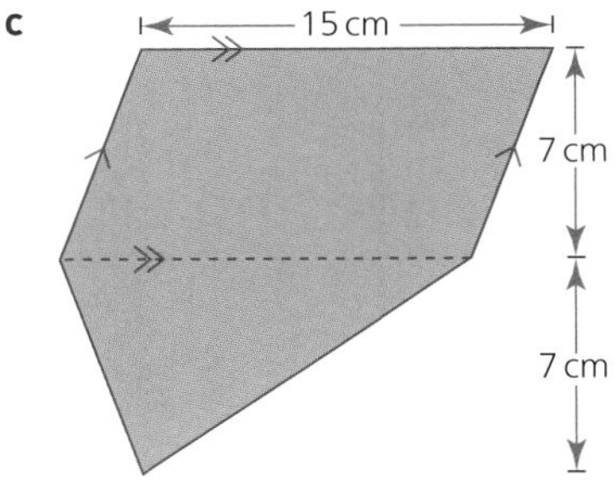

Exercise 23.6

1 Calculate the value of a.

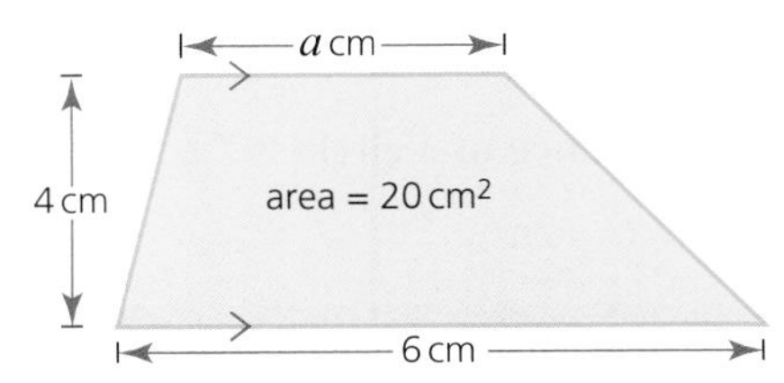

2 If the areas of the trapezium and parallelogram in the diagram are equal, calculate the value of x.

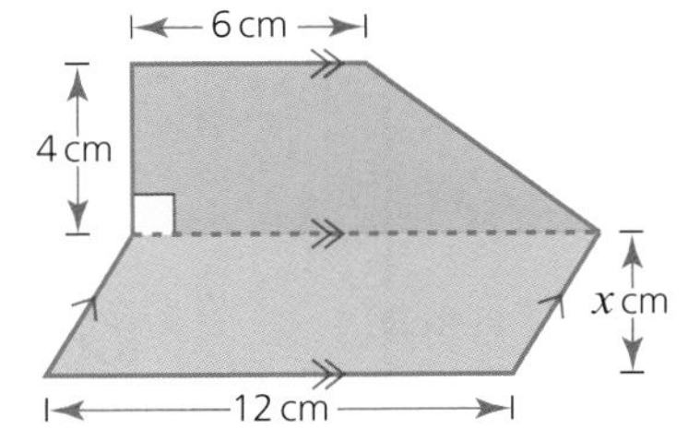

Exercise 23.6 (cont)

3 The end view of a house is as shown in the diagram.

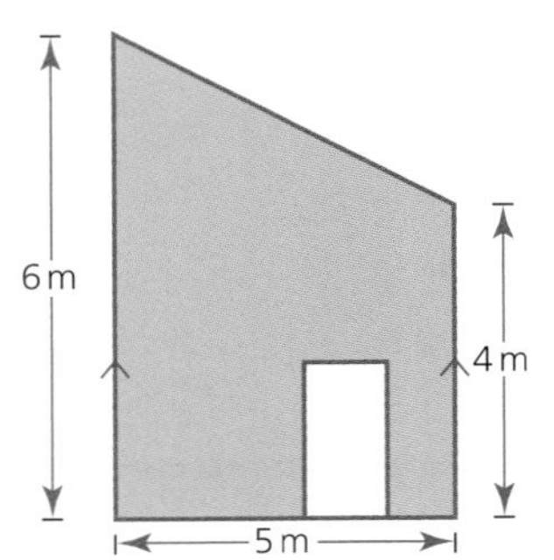

If the door has a width and height of 0.75 m and 2 m, respectively, calculate the area of brickwork.

4 A garden in the shape of a trapezium is split into three parts: flower beds in the shape of a triangle and a parallelogram, and a section of grass in the shape of a trapezium. The area of the grass is two and a half times the total area of the flower beds. Calculate:

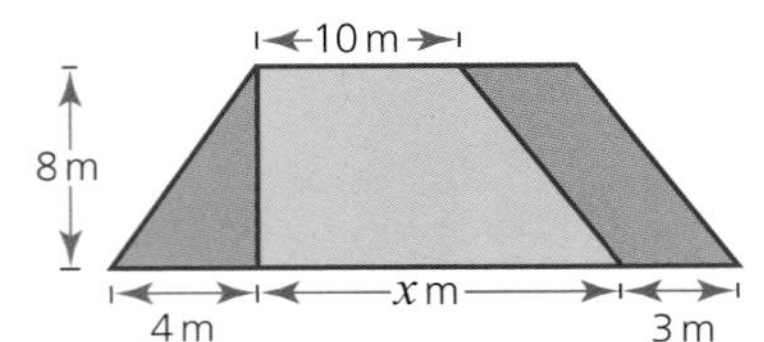

a the area of each flower bed
b the area of grass
c the value of x.

The circumference and area of a circle

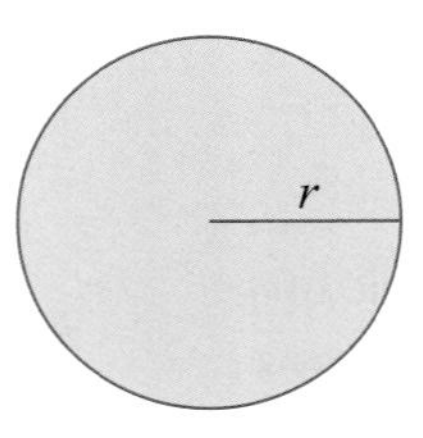

The **circumference of a circle** is $2\pi r$.

$$C = 2\pi r$$

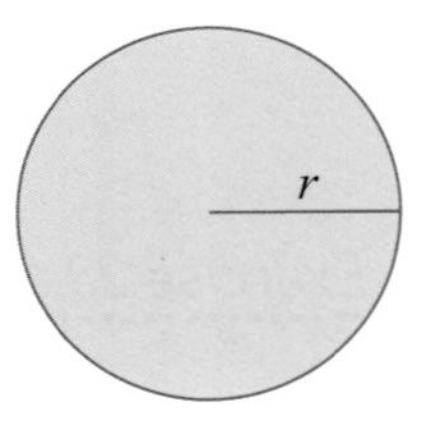

The **area of a circle** is πr^2.

$$A = \pi r^2$$

Worked examples

1 Calculate the circumference of this circle, giving your answer to 3 s.f.

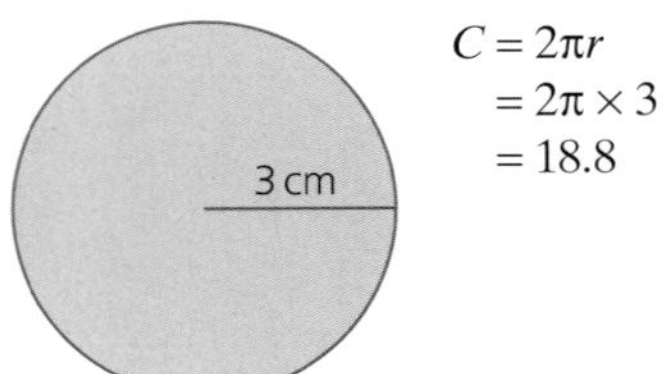

$$C = 2\pi r$$
$$= 2\pi \times 3$$
$$= 18.8$$

The circumference is 18.8 cm.

The answer 18.8 cm is only correct to 3 s.f. and as such is only an approximation. An **exact** answer involves leaving the answer in terms of π, i.e.

$$C = 2\pi r$$
$$= 2\pi \times 3$$
$$= 6\pi \text{ cm}$$

2 If the circumference of this circle is 12 cm, calculate the radius, giving your answer:

a to 3 s.f.

b in terms of π.

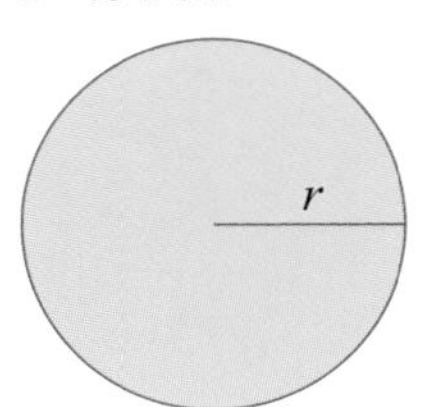

a $C = 2\pi r$

$r = \frac{C}{2\pi}$

$r = \frac{12}{2\pi}$

$= 1.91$

The radius is 1.91 cm.

b $r = \frac{C}{2\pi} = \frac{12}{2\pi}$

$= \frac{6}{\pi}$ cm

3 Calculate the area of this circle, giving your answer:

a to 3 s.f.

b in exact form.

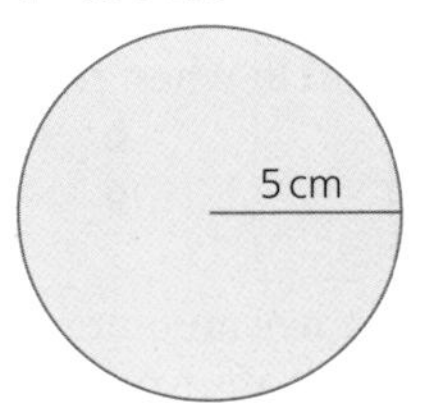

a $A = \pi r^2$

$= \pi \times 5^2 = 78.5$

The area is 78.5 cm^2.

b $A = \pi r^2$

$= \pi \times 5^2$

$= 25\pi \text{ cm}^2$

4 The area of a circle is 34 cm^2, calculate the radius, giving your answer:

a to 3 s.f.

b in terms of π.

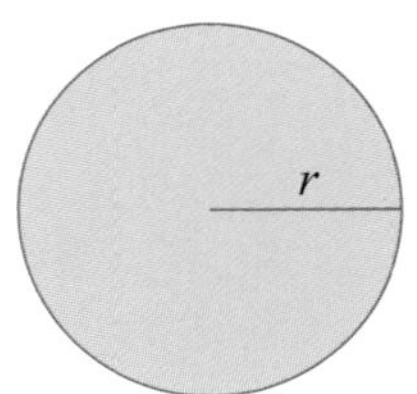

a $A = \pi r^2$

$r = \sqrt{\frac{A}{\pi}}$

$r = \sqrt{\frac{34}{\pi}} = 3.29$

The radius is 3.29 cm.

b $r = \sqrt{\frac{A}{\pi}} = \sqrt{\frac{34}{\pi}}$ cm

Exercise 23.7

1 Calculate the circumference of each circle, giving your answers to 2 d.p.

a
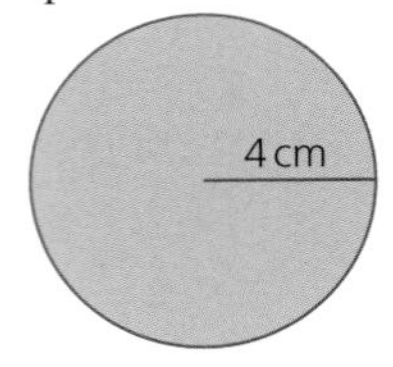

b
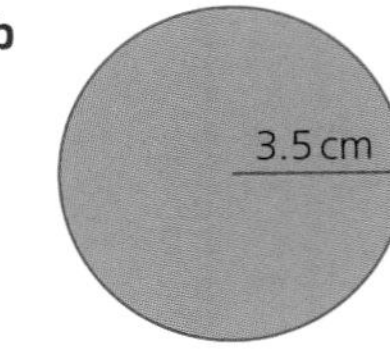

c
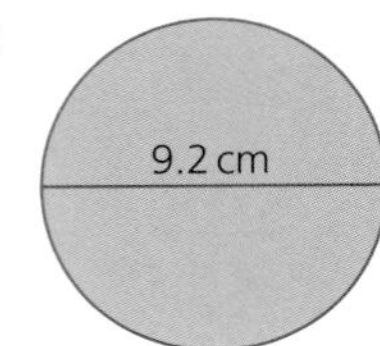

d
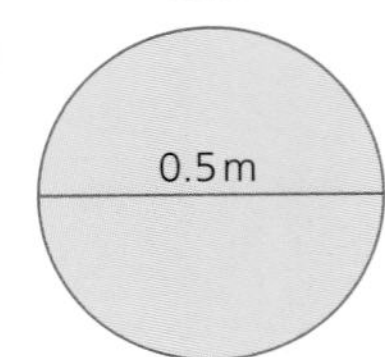

2 Calculate the area of each circle in Q.1. Give your answers to 2 d.p.

3 Calculate the radius of a circle when the circumference is:

a 15 cm
b π cm
c 4 m
d 8 mm

4 Calculate the diameter of a circle when the area is:

a $16\,\text{cm}^2$
b $9\pi\,\text{cm}^2$
c $8.2\,\text{m}^2$
d $14.6\,\text{mm}^2$

Exercise 23.8

1 The wheel of a child's toy car has an outer radius of 25 cm. Calculate:

a how far the car has travelled after one complete turn of the wheel
b how many times the wheel turns for a journey of 1 km.

2 If the wheel of a bicycle has a diameter of 60 cm, calculate how far a cyclist will have travelled after the wheel has rotated 100 times.

3 A circular ring has a cross-section. If the outer radius is 22 mm and the inner radius 20 mm, calculate the cross-sectional area of the ring. Give your answer in terms of π.

4 Four circles are drawn in a line and enclosed by a rectangle, as shown. If the radius of each circle is 3 cm, calculate the unshaded area within the rectangle giving your answer in exact form.

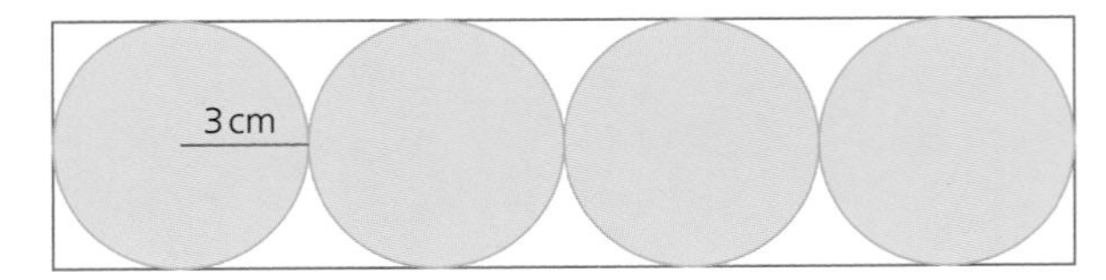

5 A garden is made up of a rectangular patch of grass and two semi-circular vegetable patches. If the dimensions of the rectangular patch are 16 m (length) and 8 m (width) respectively, calculate in exact form:

a the perimeter of the garden

b the total area of the garden.

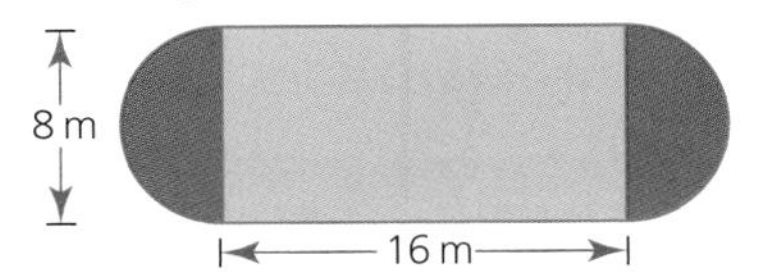

The surface area of a cuboid and a cylinder

To calculate the surface area of a **cuboid**, start by looking at its individual faces. These are either squares or rectangles. The surface area of a cuboid is the sum of the areas of its faces.

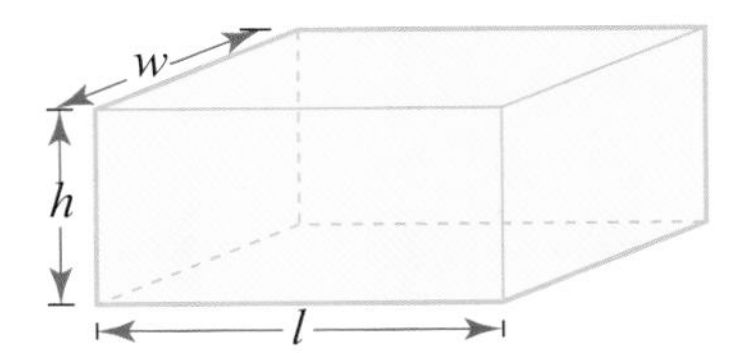

Area of top $= wl$ Area of bottom $= wl$

Area of front $= lh$ Area of back $= lh$

Area of left side $= wh$ Area of right side $= wh$

Total surface area

$= 2wl + 2lh + 2wh$

$= 2(wl + lh + wh)$

For the surface area of a **cylinder**, it is best to visualise the net of the solid. It is made up of one rectangular piece and two circular pieces:

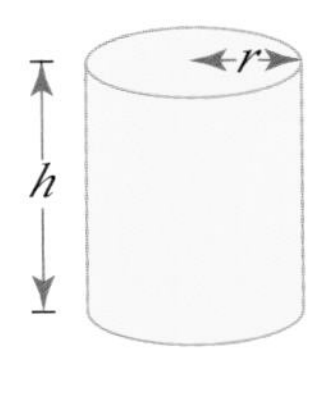

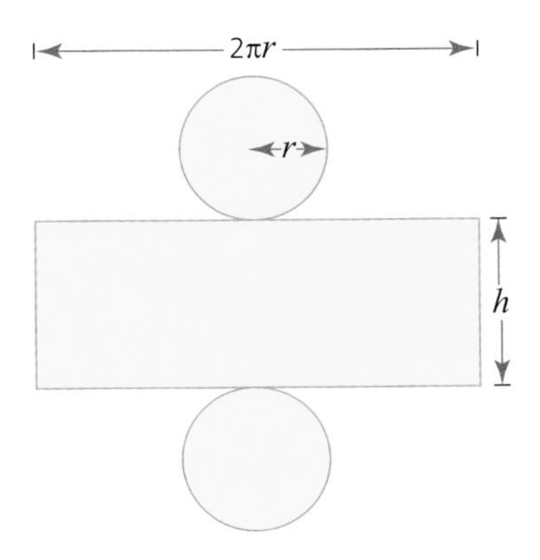

Area of circular pieces $= 2 \times \pi r^2$

Area of rectangular piece $= 2\pi r \times h$

Total surface area $= 2\pi r^2 + 2\pi rh$

$= 2\pi r(r + h)$

Worked examples

1 Calculate the surface area of the cuboid.

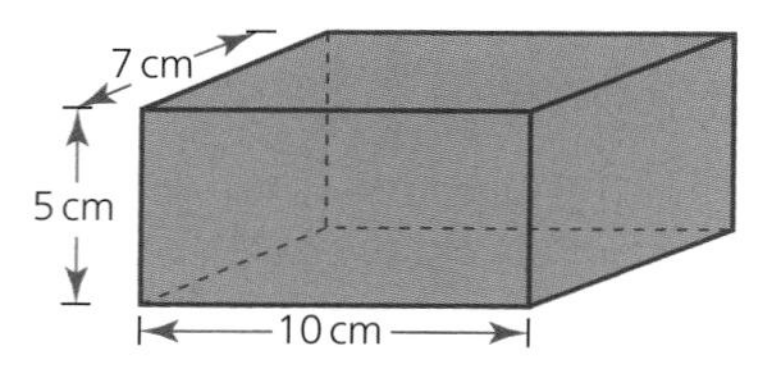

Total area of top and bottom $= 2 \times 7 \times 10 = 140\,\text{cm}^2$

Total area of front and back $= 2 \times 5 \times 10 = 100\,\text{cm}^2$

Total area of both sides $= 2 \times 5 \times 7 = 70\,\text{cm}^2$

Total surface area $= 140 + 100 + 70$

$= 310\,\text{cm}^2$

2 If the height of a cylinder is 7 cm and the radius of its circular top is 3 cm, calculate its surface area.

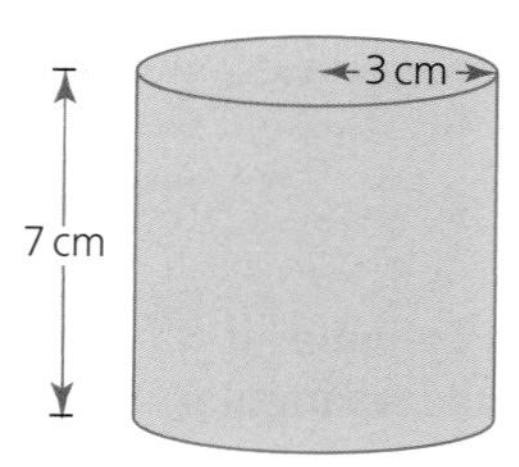

Total surface area $= 2\pi r(r + \text{h})$

$= 2\pi \times 3 \times (3 + 7)$

$= 6\pi \times 10$

$= 60\pi$

$= 188.50\,\text{cm}^2$ (2 d.p.)

The total surface area is $188.50\,\text{cm}^2$.

Exercise 23.9

1 These are the dimensions of some cuboids. Calculate the surface area of each one.

a $l = 12\,\text{cm}$, $w = 10\,\text{cm}$, $h = 5\,\text{cm}$
b $l = 4\,\text{cm}$, $w = 6\,\text{cm}$, $h = 8\,\text{cm}$
c $l = 4.2\,\text{cm}$, $w = 7.1\,\text{cm}$, $h = 3.9\,\text{cm}$
d $l = 5.2\,\text{cm}$, $w = 2.1\,\text{cm}$, $h = 0.8\,\text{cm}$

2 These are the dimensions of some cuboids. Calculate the height of each one.

a $l = 5$ cm, $w = 6$ cm, surface area $= 104\text{ cm}^2$
b $l = 2$ cm, $w = 8$ cm, surface area $= 112\text{ cm}^2$
c $l = 3.5$ cm, $w = 4$ cm, surface area $= 118\text{ cm}^2$
d $l = 4.2$ cm, $w = 10$ cm, surface area $= 226\text{ cm}^2$

3 These are the dimensions of some cylinders. Calculate the surface area of each one.

a $r = 2$ cm, $h = 6$ cm
b $r = 4$ cm, $h = 7$ cm
c $r = 3.5$ cm, $h = 9.2$ cm
d $r = 0.8$ cm, $h = 4.3$ cm

4 These are the dimensions of some cylinders. Calculate the height of each one. Give your answers to 1 d.p.

a $r = 2.0$ cm, surface area $= 40\text{ cm}^2$
b $r = 3.5$ cm, surface area $= 88\text{ cm}^2$
c $r = 5.5$ cm, surface area $= 250\text{ cm}^2$
d $r = 3.0$ cm, surface area $= 189\text{ cm}^2$

Exercise 23.10

1 Two cubes are placed next to each other. The length of each edge of the larger cube is 4 cm.

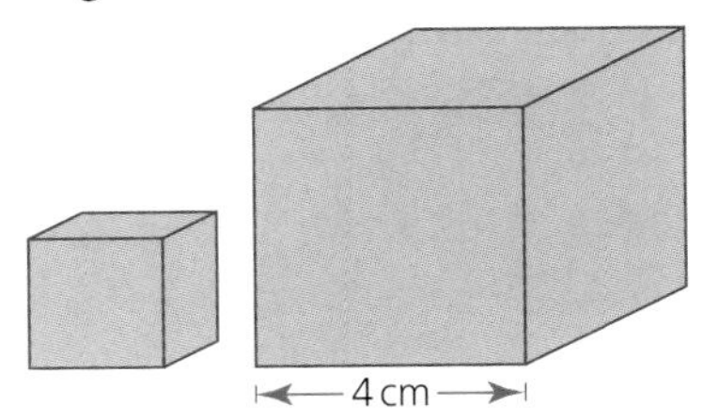

If the ratio of their surface areas is 1 : 4, calculate:

a the surface area of the small cube
b the length of an edge of the small cube.

2 A cube and a cylinder have the same surface area. If the cube has an edge length of 6 cm and the cylinder a radius of 2 cm, calculate:

a the surface area of the cube
b the height of the cylinder.

3 Two cylinders have the same surface area. The shorter of the two has a radius of 3 cm and a height of 2 cm, and the taller cylinder has a radius of 1 cm. Calculate:

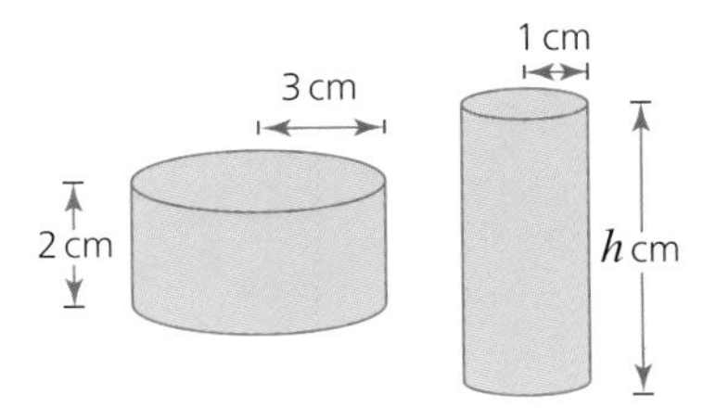

a the surface area of one of the cylinders in terms of π
b the height of the taller cylinder.

4 Two cuboids have the same surface area. The dimensions of one of them are: length = 3 cm, width = 4 cm and height = 2 cm. Calculate the height of the other cuboid if its length is 1 cm and its width is 4 cm.

The volume and surface area of a prism

A **prism** is any three-dimensional object which has a constant cross-sectional area. Some examples of common types of prism are:

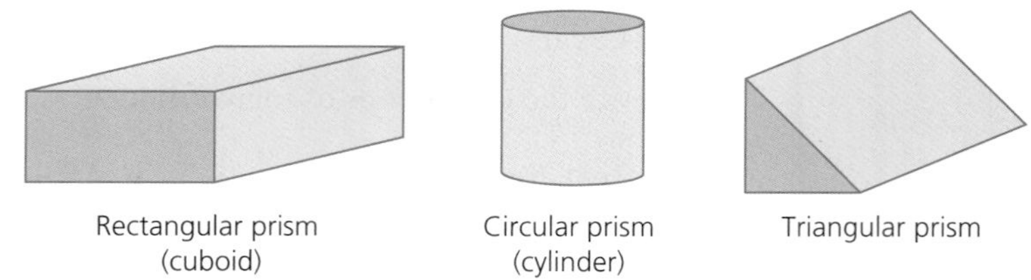

When each of the shapes is cut parallel to the shaded face, the cross-section is constant and the shape is therefore classified as a prism.

Volume of a prism = area of cross-section × length
Surface area of a prism = sum of the area of each of its faces

Worked examples

1 Calculate the volume of the cylinder in the diagram:

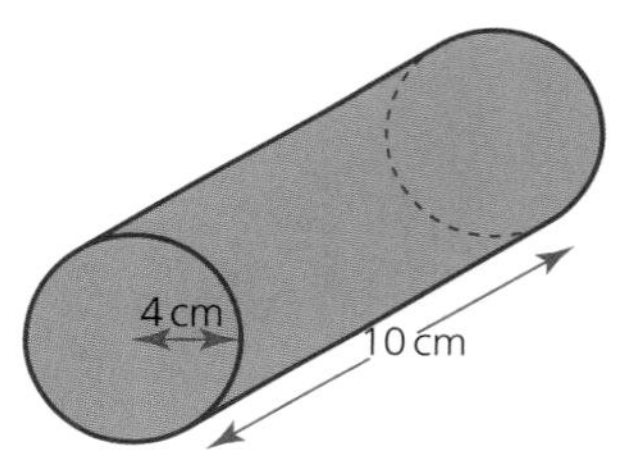

Volume = area of cross-section × length
$= \pi \times 4^2 \times 10$
Volume = $502.7\,\text{cm}^3$ (1 d.p.)
As an exact value the volume would be left as $160\pi\,\text{cm}^3$

2 For the 'L' shaped prism in the diagram, calculate:
a the volume

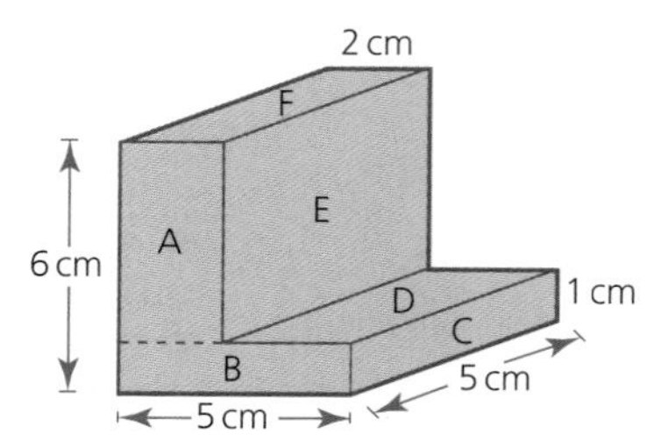

The cross-sectional area can be split into two rectangles:

Area of rectangle A = 5×2
$= 10\,\text{cm}^2$

Area of rectangle B $= 5 \times 1$
$= 5\,\text{cm}^2$

Total cross-sectional area $= 10\,\text{cm}^2 + 5\,\text{cm}^2 = 15\,\text{cm}^2$

Volume of prism $= 15 \times 5$
$= 75\,\text{cm}^3$

b the surface area.

Area of rectangle A $= 5 \times 2 = 10\,\text{cm}^2$
Area of rectangle B $= 5 \times 1 = 5\,\text{cm}^2$
Area of rectangle C $= 5 \times 1 = 5\,\text{cm}^2$
Area of rectangle D $= 3 \times 5 = 15\,\text{cm}^2$
Area of rectangle E $= 5 \times 5 = 25\,\text{cm}^2$
Area of rectangle F $= 2 \times 5 = 10\,\text{cm}^2$
Area of back is the same as area of rectangle A + area of rectangle B $= 15\,\text{cm}^2$
Area of left face is the same as area of rectangle C + area of rectangle E $= 30\,\text{cm}^2$
Area of base $= 5 \times 5 = 25\,\text{cm}^2$

Total surface area $= 10 + 5 + 5 + 15 + 25 + 10 + 15 + 30 + 25$
$= 140\,\text{cm}^2$

Exercise 23.11

1 Calculate the volume of each of the following cuboids, where w, l and h represent the width, length and height, respectively.

a $w = 2\,\text{cm}$, $l = 3\,\text{cm}$, $h = 4\,\text{cm}$
b $w = 6\,\text{cm}$, $l = 1\,\text{cm}$, $h = 3\,\text{cm}$
c $w = 6\,\text{cm}$, $l = 23\,\text{mm}$, $h = 2\,\text{cm}$
d $w = 42\,\text{mm}$, $l = 3\,\text{cm}$, $h = 0.007\,\text{m}$

2 Calculate the volume of each of the following cylinders, where r represents the radius of the circular face and h the height of the cylinder.

a $r = 4\,\text{cm}$, $h = 9\,\text{cm}$
b $r = 3.5\,\text{cm}$, $h = 7.2\,\text{cm}$
c $r = 25\,\text{mm}$, $h = 10\,\text{cm}$
d $r = 0.3\,\text{cm}$, $h = 17\,\text{mm}$

3 Calculate the volume and total surface area of each of these right-angled triangular prisms:

a

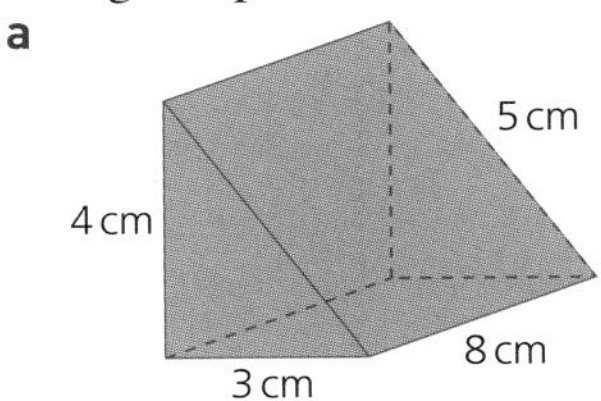

b

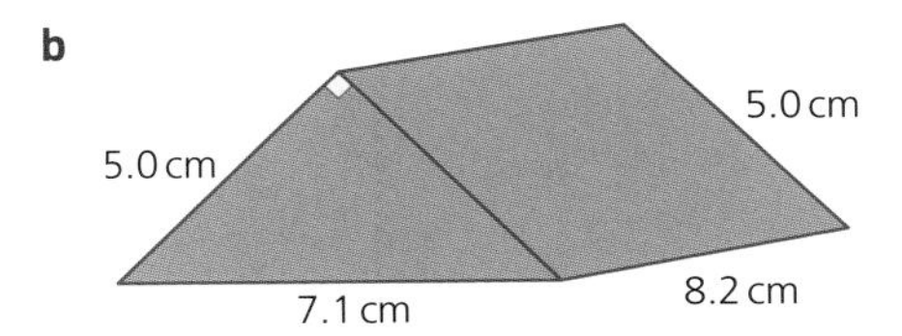

Exercise 23.11 (cont)

4 Calculate the volume of each prism. All dimensions given are centimetres.

a

b

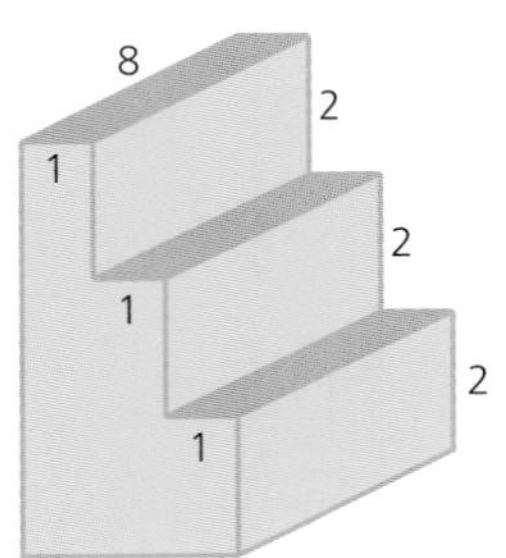

c

d

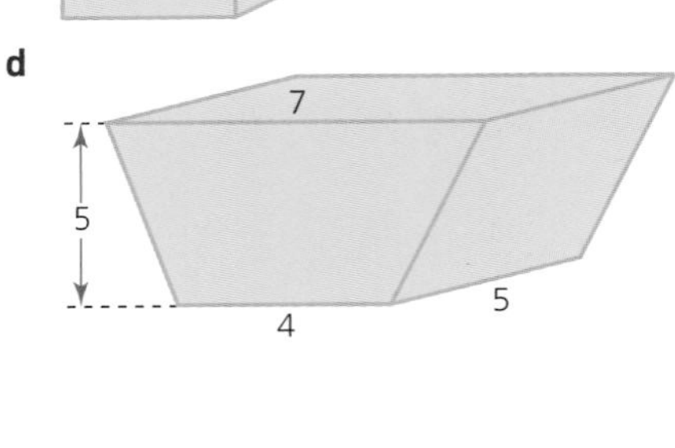

Exercise 23.12

1 The diagram shows a plan view of a cylinder inside a box the shape of a cube. The radius of the cylinder is 8 cm.

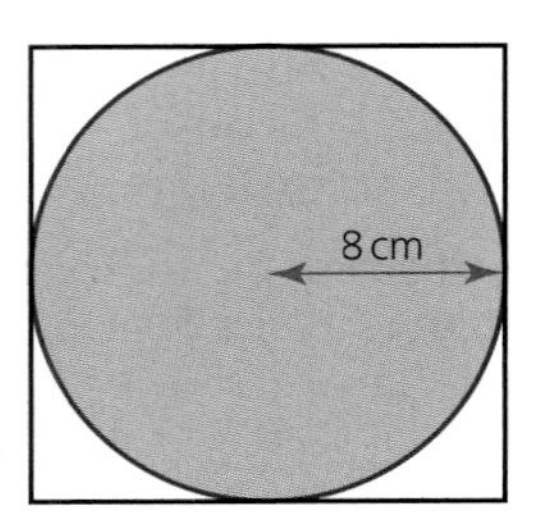

Calculate the percentage volume of the cube not occupied by the cylinder.

2 A chocolate bar is made in the shape of a triangular prism. The triangular face of the prism is equilateral and has an edge length of 4 cm and a perpendicular height of 3.5 cm. The manufacturer also sells these in special packs of six bars arranged as a hexagonal prism, as shown.

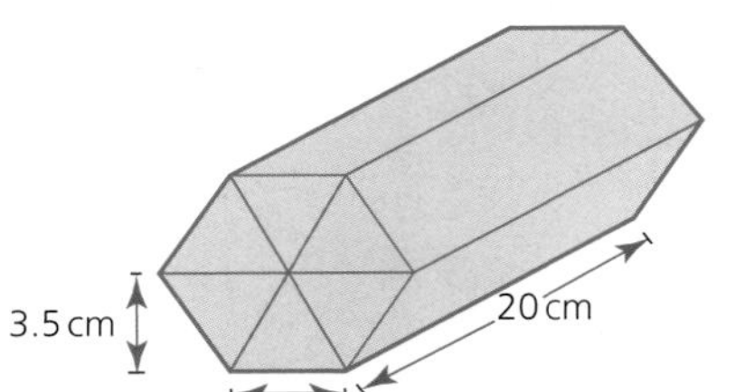

If the prisms are 20 cm long, calculate:

a the cross-sectional area of the pack

b the volume of the pack.

3 A cuboid and a cylinder have the same volume. The radius and height of the cylinder are 2.5 cm and 8 cm, respectively. If the length and width of the cuboid are each 5 cm, calculate its height to 1 d.p.

4 A section of steel pipe is shown in the diagram below. The inner radius is 35 cm and the outer radius is 36 cm. Calculate the volume of steel used in making the pipe if it has a length of 130 m. Give your answer in terms of π.

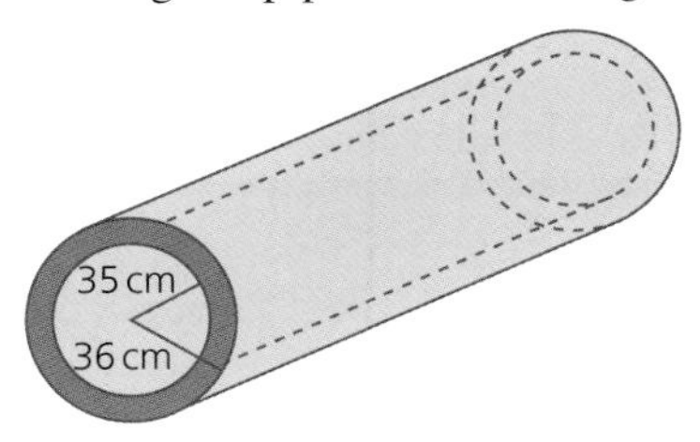

Arc length

An **arc** is part of the circumference of a circle between two radii.

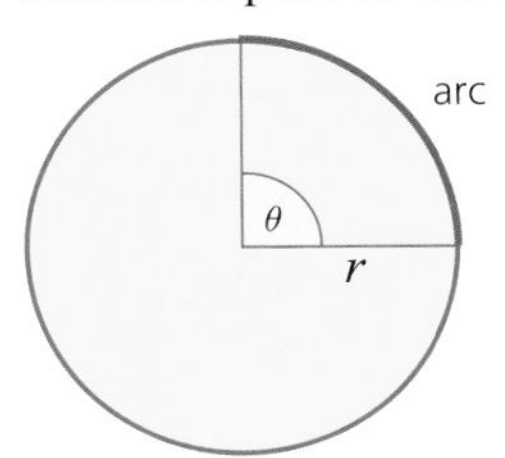

The length of an arc is proportional to the size of the angle θ between the two radii. The length of the arc as a fraction of the circumference of the whole circle is therefore equal to the fraction that θ is of 360°.

$$\text{Arc length} = \frac{\theta}{360} \times 2\pi r$$

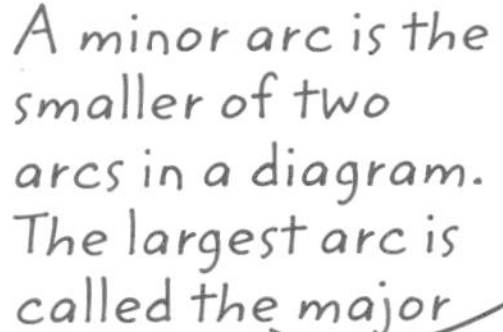

A minor arc is the smaller of two arcs in a diagram. The largest arc is called the major arc.

Worked examples

1 Find the length of the minor arc in the circle (right).

a Give your answer to 3 s.f.

$$\text{Arc length} = \frac{60}{360} \times 2 \times \pi \times 6$$
$$= 6.28\,\text{cm}$$

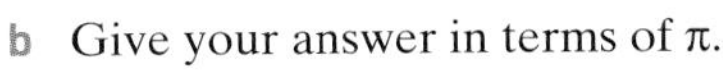

b Give your answer in terms of π.

$$\text{Arc length} = \frac{60}{360} \times 2 \times \pi \times 6$$
$$= 2\pi\,\text{cm}$$

2 In the circle, the length of the minor arc is 3π cm and the radius is 9 cm.

a Calculate the angle θ.

$$\text{Arc length} = \frac{\theta}{360} \times 2\pi r$$

$$3\pi = \frac{\theta}{360} \times 2 \times \pi \times 9$$

$$\theta = \frac{3\pi \times 360}{2 \times \pi \times 9}$$

$$= 60°$$

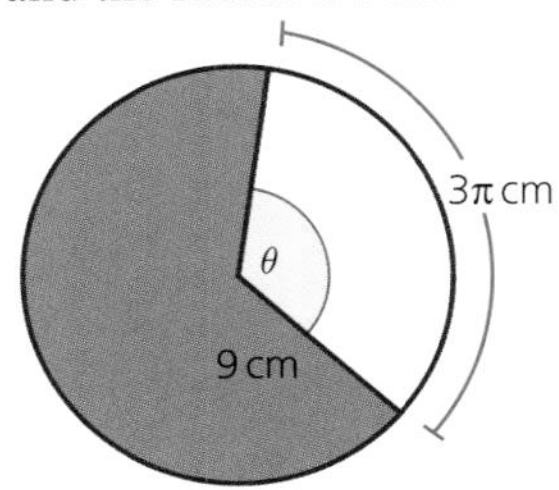

b Calculate the length of the major arc.

$C = 2\pi r$
$= 2 \times \pi \times 9 = 18\pi$

Major arc = circumference – minor arc
$= 18\pi - 3\pi = 15\pi$ cm

Exercise 23.13

A sector is the region of a circle enclosed by two radii and an arc.

1 For each of these sectors, give the length of the arc to 3 s.f. O is the centre of the circle.

a

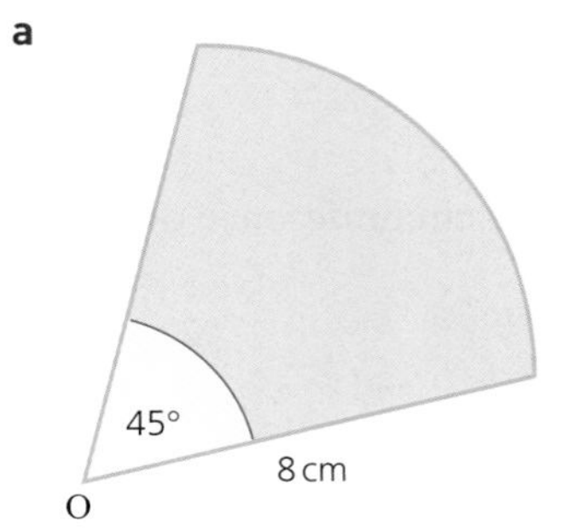

b

O
8°
15 cm

c

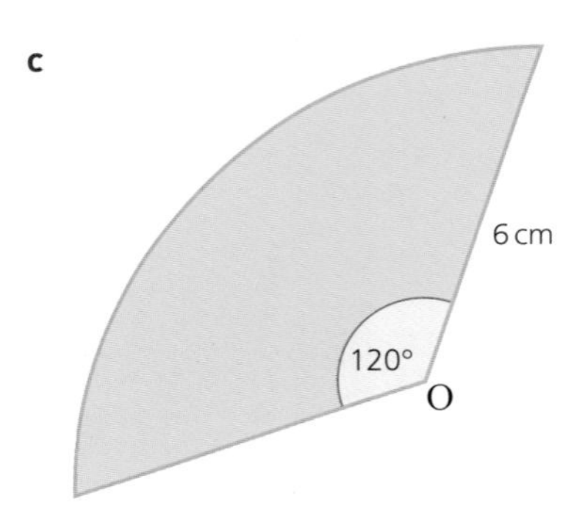

d

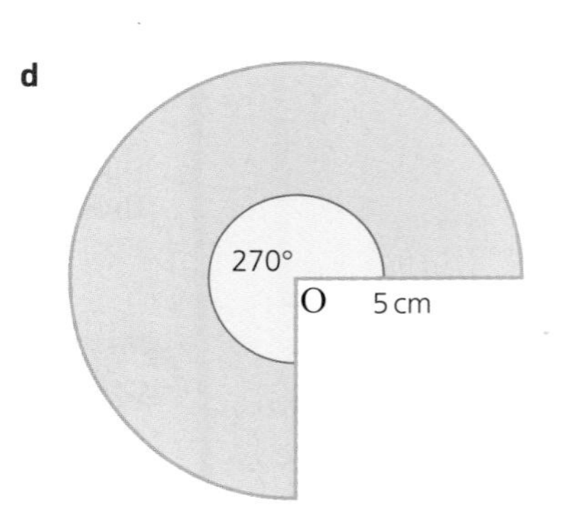

2 Calculate the angle θ for each sector. The radius r and arc length a are given in each case.

a $r = 15$ cm, $a = 3\pi$ cm
b $r = 12$ cm, $a = 6\pi$ cm
c $r = 24$ cm, $a = 8\pi$ cm
d $r = \frac{3}{2}$ cm, $a = \frac{3\pi}{3^2}$ cm

3 Calculate the radius r for each sector. The angle θ and arc length a are given in each case.

a $\theta = 90°$, $a = 16$ cm
b $\theta = 36°$, $a = 24$ cm
c $\theta = 20°$, $a = 6.5$ cm
d $\theta = 72°$, $a = 17$ cm

Exercise 23.14

1 Calculate the perimeter of each of these shapes. Give your answers in exact form.

a

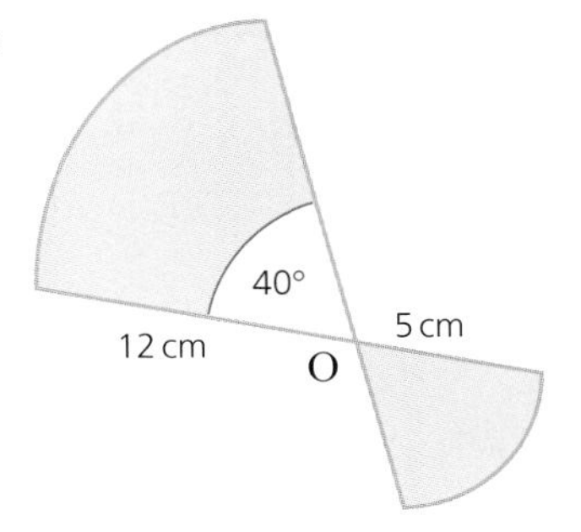

b

12 cm

36 cm

O

2 For the diagram, calculate:
  a the radius of the smaller sector
  b the perimeter of the shape
  c the angle θ.

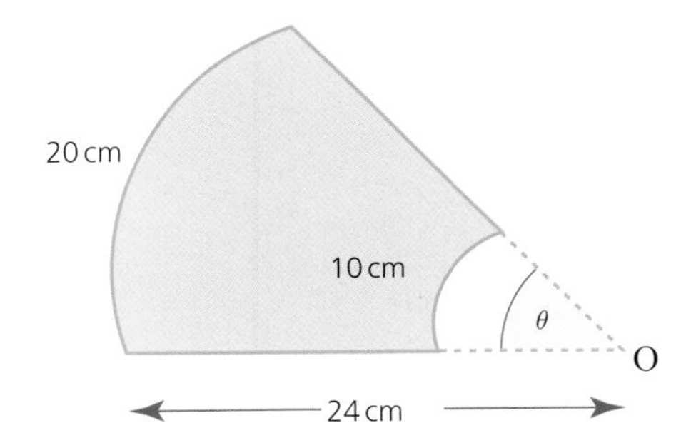

The area of a sector

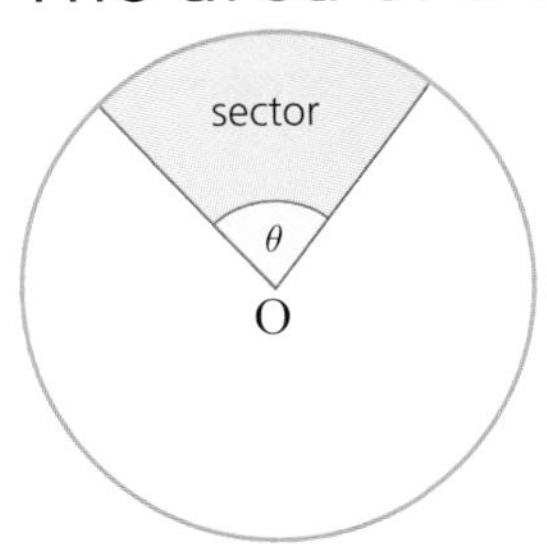

A **sector** is the region of a circle enclosed by two radii and an arc. Its area is proportional to the size of the angle θ between the two radii. The area of the sector as a fraction of the area of the whole circle is therefore equal to the fraction that θ is of 360°.

Area of sector $= \frac{\theta}{360}\pi r^2$

Worked examples

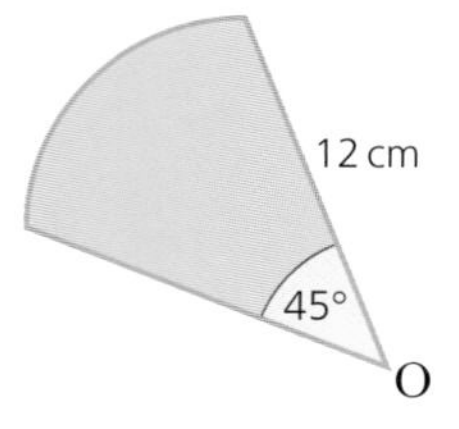

1 Calculate the area of the sector, giving your answer:

a to 3 s.f.

$$\text{Area} = \frac{\theta}{360} \times \pi r^2$$

$$= \frac{45}{360} \times \pi \times 12^2$$

$$= 56.5\,\text{cm}^2$$

b in terms of π.

$$\text{Area} = \frac{45}{360} \times \pi \times 12^2$$

$$= 18\pi\,\text{cm}^2$$

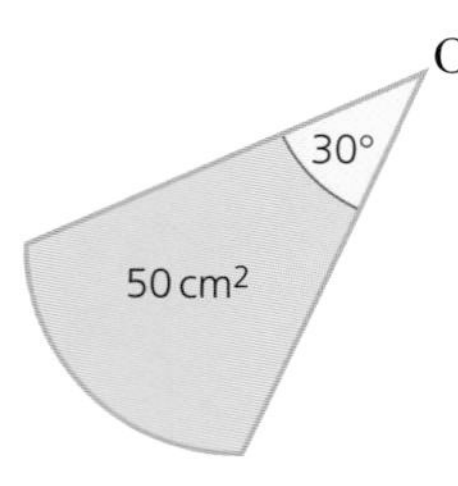

2 Calculate the radius of the sector, giving your answer to 3 s.f.

$$\text{Area} = \frac{\theta}{360} \times \pi r^2$$

$$50 = \frac{30}{360} \times \pi \times r^2$$

$$\frac{50 \times 360}{30\pi} = r^2$$

$$r = 13.8$$

The radius is 13.8 cm.

Exercise 23.15

1 Calculate the area of each of the sectors described in the table, using the values of the angle θ and radius r:

	a	b	c	d
θ	60	120	2	72
r (cm)	8	14	18	14

2 Calculate the radius for each of the sectors described in the table, using the values of the angle θ and the area A:

	a	b	c	d
θ	40	12	72	18
A (cm²)	120	42	4	400

3 Calculate the value of the angle θ, to the nearest degree, for each of the sectors described in the table, using the values of the radius r and area A:

	a	**b**	**c**	**d**
r	12 cm	20 cm	2 cm	0.5 m
A	$8\pi\,\text{cm}^2$	$40\pi\,\text{cm}^2$	$\frac{\pi}{2}\,\text{cm}^2$	$\frac{\pi}{60}\,\text{m}^2$

Exercise 23.16

1 A rotating sprinkler is placed in one corner of a garden. It has a reach of 8 m and rotates through an angle of 30°. Calculate the area of garden not being watered. Give your answer in terms of π.

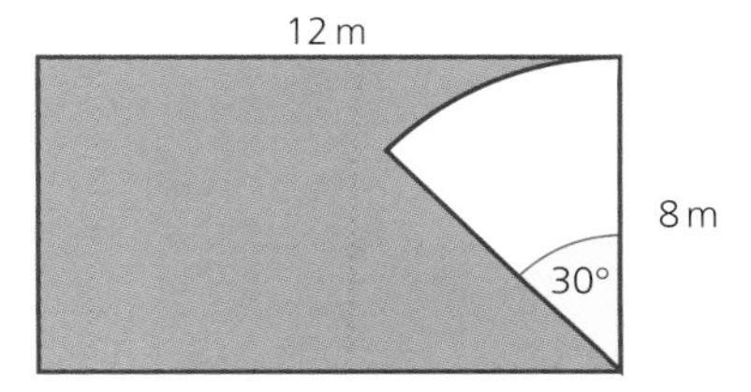

2 Two sectors AOB and COD share the same centre O. The area of AOB is three times the area of COD. Calculate:

a the area of sector AOB
b the area of sector COD
c the radius r cm of sector COD.

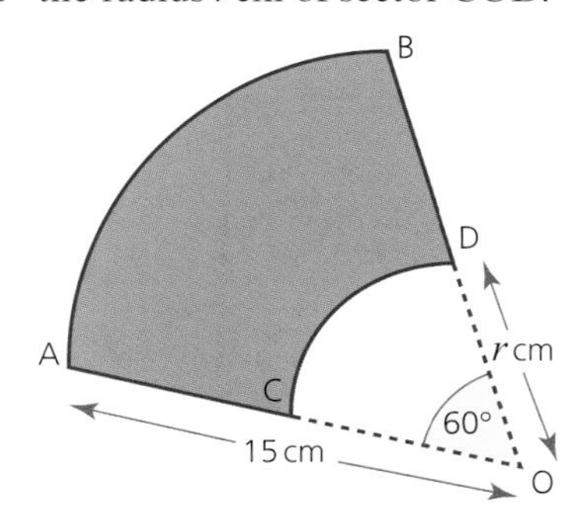

3 A circular cake is cut. One of the slices is shown. Calculate:

a the length a cm of the arc
b the total surface area of all the sides of the slice
c the volume of the slice.

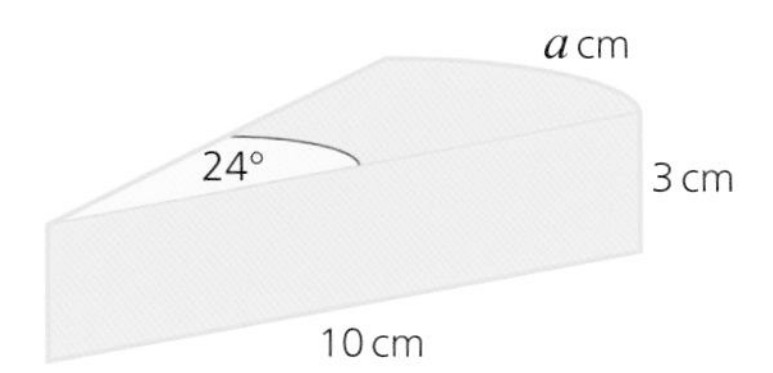

The volume of a sphere

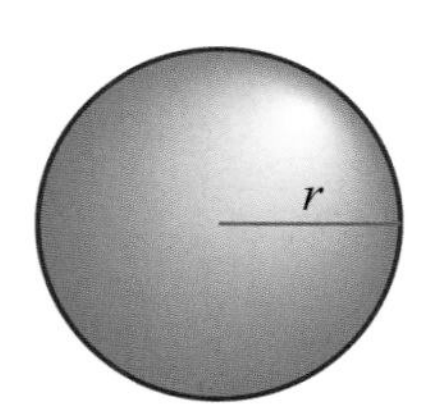

Volume of sphere $= \frac{4}{3}\pi r^3$

Worked examples

1 Calculate the volume of the sphere, giving your answer:

a to 3 s.f.

b in terms of π.

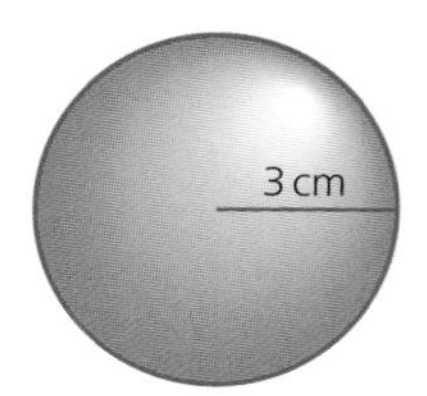

a

$$\text{Volume of sphere} = \frac{4}{3}\pi r^3$$

$$= \frac{4}{3} \times \pi \times 3^3$$

$$= 113.1$$

The volume is $113\,\text{cm}^3$.

b

$$\text{Volume of sphere} = \frac{4}{3}\pi \times 3^3$$

$$= 36\pi\,\text{cm}^3$$

2 Given that the volume of a sphere is $150\,\text{cm}^3$, calculate its radius to 3 s.f.

$$V = \frac{4}{3}\pi r^3$$

$$r^3 = \frac{3 \times V}{4 \times \pi}$$

$$r^3 = \frac{3 \times 150}{4 \times \pi}$$

$$r = \sqrt[3]{35.8} = 3.30$$

The radius is 3.30 cm.

Exercise 23.17

1 The radius of four spheres is given. Calculate the volume of each one.

a $r = 6\,\text{cm}$

b $r = 9.5\,\text{cm}$

c $r = 8.2\,\text{cm}$

d $r = 0.7\,\text{cm}$

2 The volume of four spheres is given. Calculate the radius of each one, giving your answers in centimetres and to 1 d.p.

a $V = 130\,\text{cm}^3$

b $V = 720\,\text{cm}^3$

c $V = 0.2\,\text{m}^3$

d $V = 1000\,\text{mm}^3$

Exercise 23.18

1 Given that sphere B has twice the volume of sphere A, calculate the radius of sphere B. Give your answer to 1 d.p.

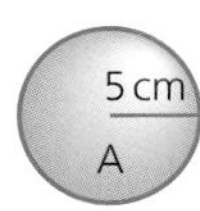

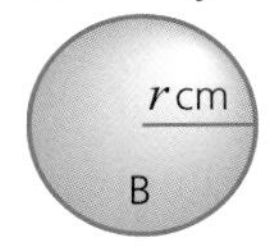

2 Calculate the volume of material used to make the hemispherical bowl in the diagram, given the inner radius of the bowl is 5 cm and its outer radius 5.5 cm. Give your answer in terms of π.

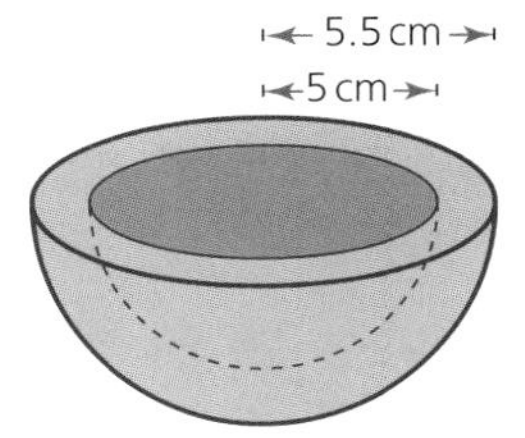

3 The volume of material used to make the sphere and hemispherical bowl is the same. Given that the radius of the sphere is 7 cm and the inner radius of the bowl is 10 cm, calculate, to 1 d.p., the outer radius r cm of the bowl.

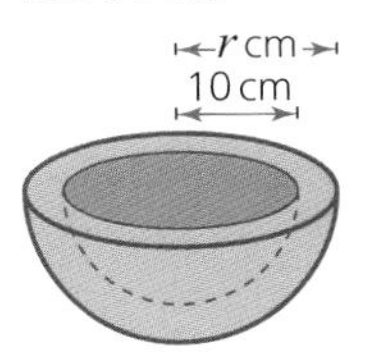

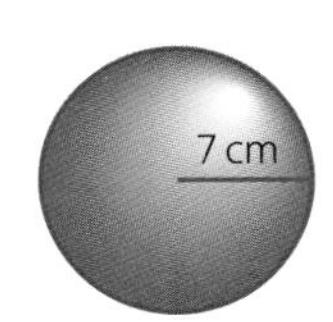

4 A ball is placed inside a box into which it will fit tightly. If the radius of the ball is 10 cm, calculate the percentage volume of the box not occupied by the ball.

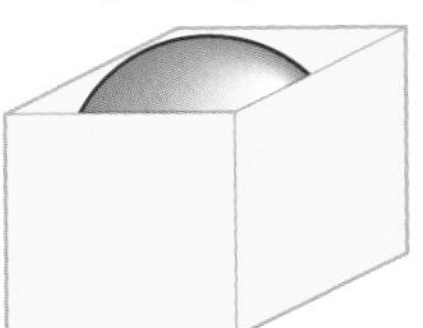

The surface area of a sphere

Surface area of sphere $= 4\pi r^2$

Exercise 23.19

1 The radius of four spheres is given. Calculate the surface area of each one.

a $r = 6$ cm

b $r = 4.5$ cm

c $r = 12.25$ cm

d $r = \frac{1}{3}$ cm

Exercise 23.19 (cont)

2 The surface area of four spheres is given. Calculate the radius of each one.

a $A = 50\,\text{cm}^2$ **b** $A = 16.5\,\text{cm}^2$

c $A = 120\,\text{mm}^2$ **d** $A = \pi\,\text{cm}^2$

3 Sphere A has a radius of 8 cm and sphere B has a radius of 16 cm. Calculate the ratio of their surface areas in the form $1 : n$.

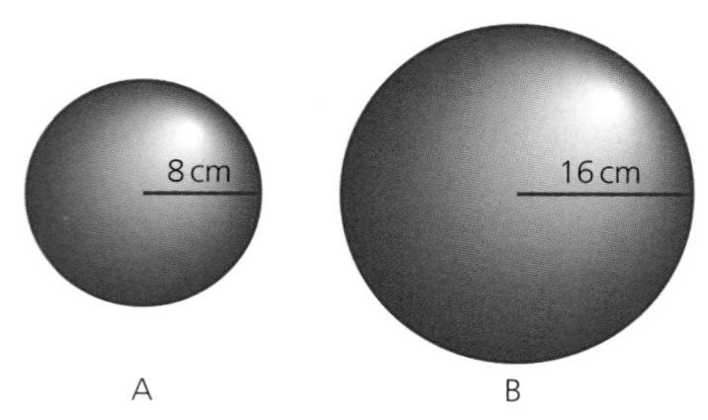

4 A hemisphere of diameter 10 cm is attached to a cylinder of equal diameter as shown. If the total length of the shape is 20 cm, calculate the surface area of the whole shape.

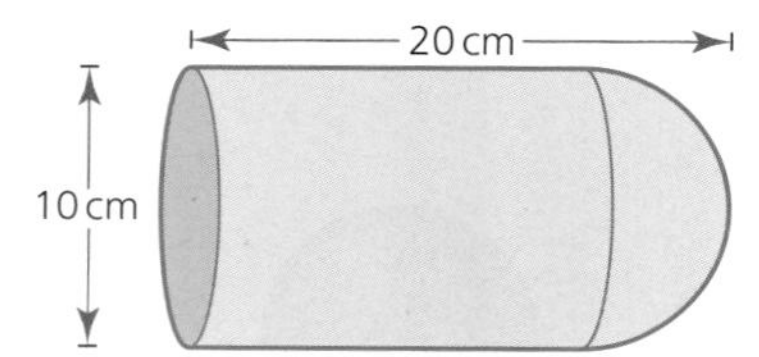

The volume of a pyramid

A **pyramid** is a three-dimensional shape in which each of its faces must be plane. A pyramid has a polygon for its base and the other faces are triangles with a common vertex, known as the **apex**. Its individual name is taken from the shape of the base.

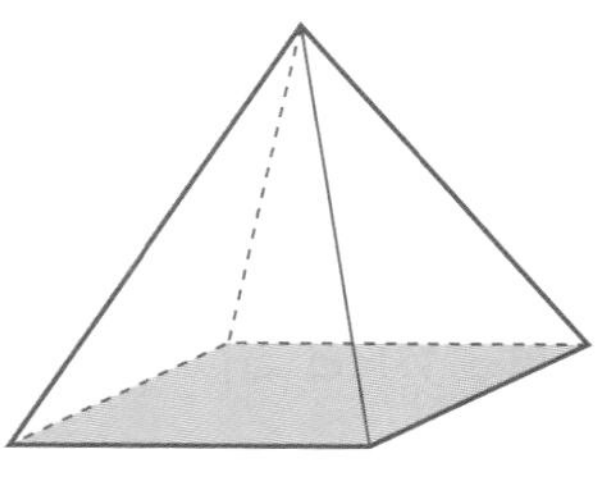

Square-based pyramid

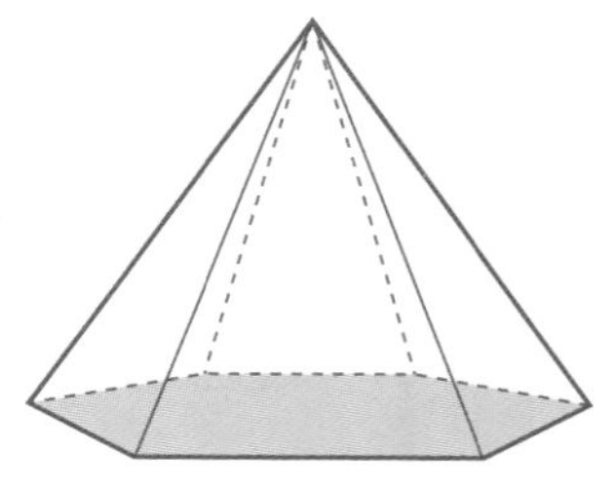

Hexagonal-based pyramid

$$\text{Volume of any pyramid} = \tfrac{1}{3} \times \text{area of base} \times \text{perpendicular height}$$

Worked examples

1 A rectangular-based pyramid has a perpendicular height of 5 cm and base dimensions as shown. Calculate the volume of the pyramid.

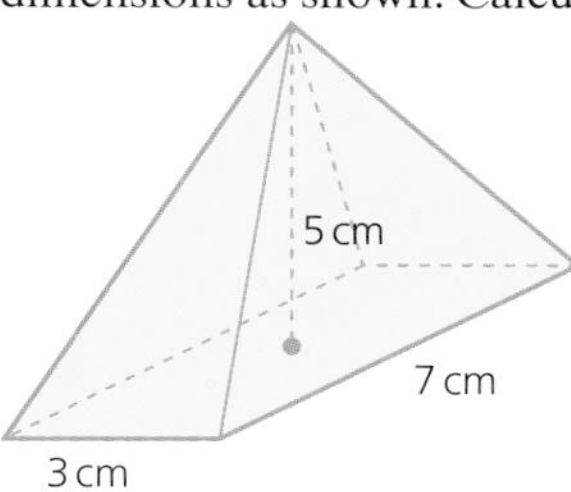

$$\text{Volume} = \frac{1}{3} \times \text{base area} \times \text{height}$$

$$= \frac{1}{3} \times 3 \times 7 \times 5 = 35$$

The volume is $35\,\text{cm}^3$.

2 The pyramid shown has a volume of $60\,\text{cm}^3$. Calculate its perpendicular height h cm.

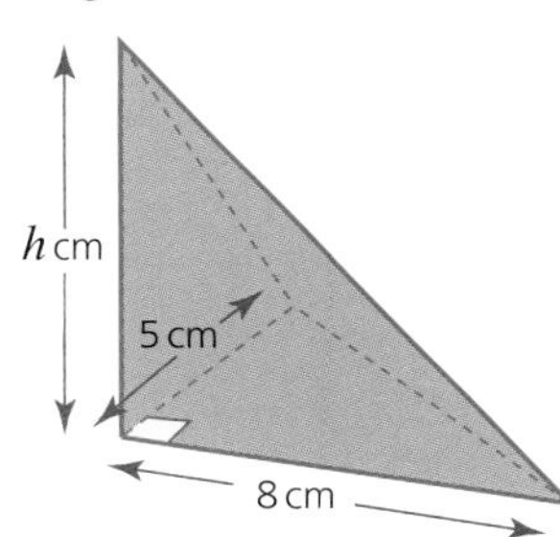

$$\text{Volume} = \frac{1}{3} \times \text{base area} \times \text{height}$$

$$\text{Height} = \frac{3 \times \text{volume}}{\text{base area}}$$

$$h = \frac{3 \times 60}{\frac{1}{2} \times 8 \times 5}$$

$$h = 9$$

The height is 9 cm.

Exercise 23.20

Find the volume of each of the following pyramids:

a

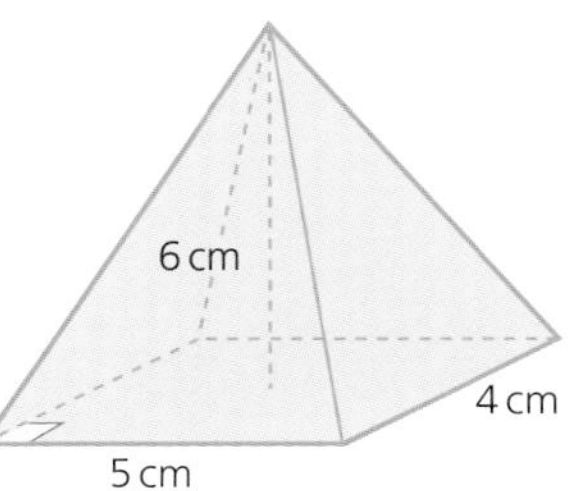

b

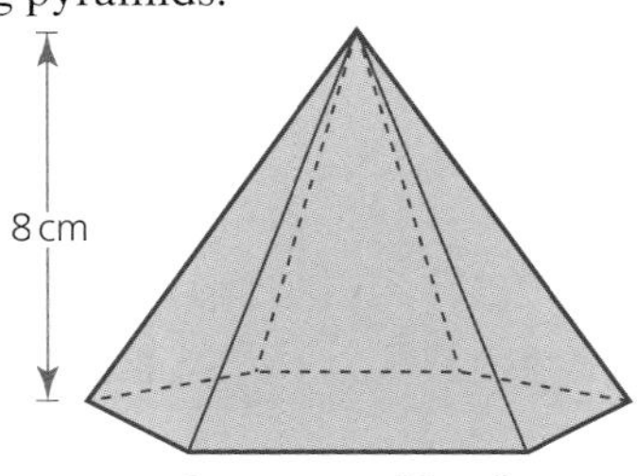

c

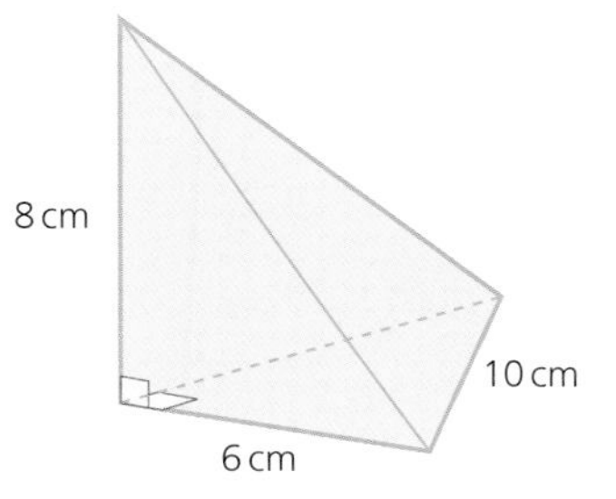

d

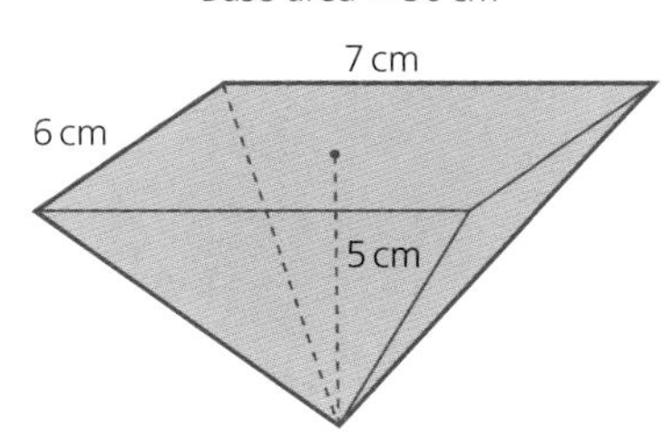

The surface area of a pyramid

The surface area of a pyramid is found by adding together the areas of all faces.

Exercise 23.21

1 Calculate the surface area of a regular tetrahedron with edge length 2 cm.

2 cm

2 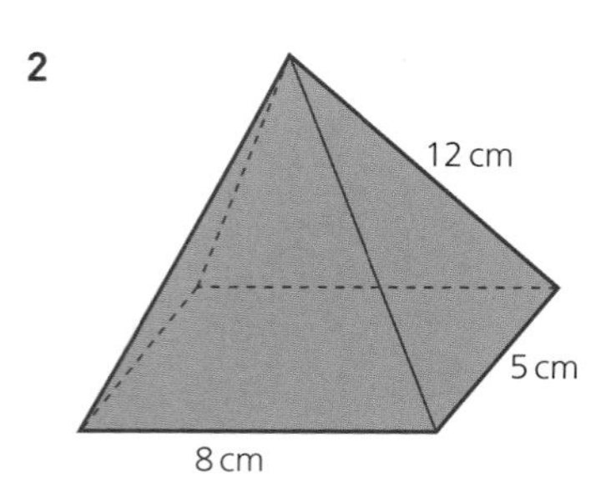

The rectangular-based pyramid shown has a sloping edge length of 12 cm. Calculate its surface area.

3 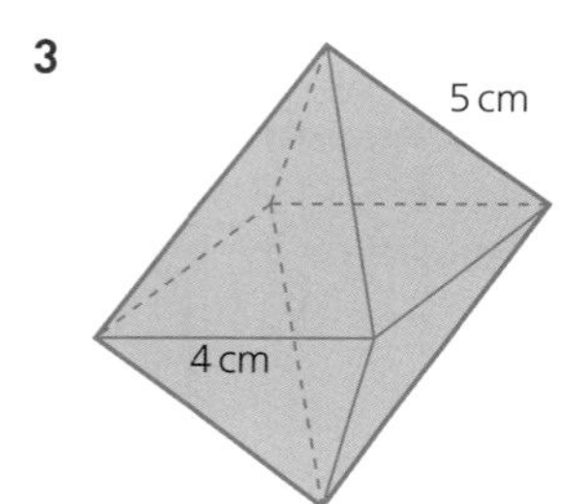

Two square-based pyramids are glued together as shown. Given that all the triangular faces are identical, calculate the surface area of the whole shape.

The volume of a cone

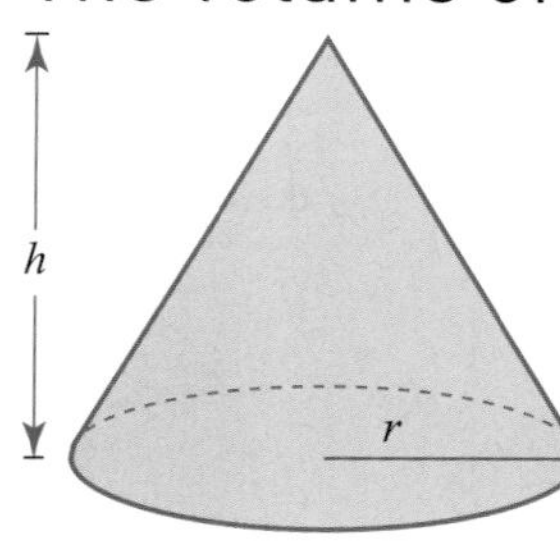

A **cone** is a pyramid with a circular base. The formula for its volume is therefore the same as for any other pyramid.

$$\text{Volume} = \tfrac{1}{3} \times \text{base area} \times \text{height}$$

$$= \tfrac{1}{3}\pi r^2 h$$

The surface area of a cone

The surface area of a cone comprises the area of the circular base and the area of the curved face. The area of the curved face is equal to the area of the sector from which it is formed.

Worked example

Calculate the total surface area of the cone.

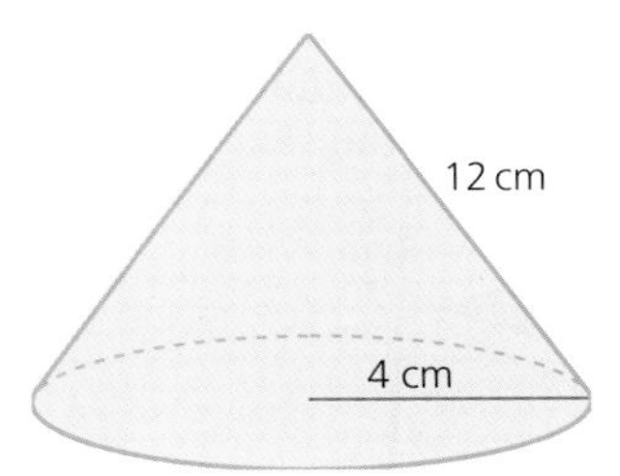

Surface area of base = $\pi r^2 = 16\pi\,\text{cm}^2$

The curved surface area can best be visualised if drawn as a sector as shown in the diagram. The radius of the sector is equivalent to the slant height of the cone. The curved perimeter of the sector is equivalent to the base circumference of the cone.

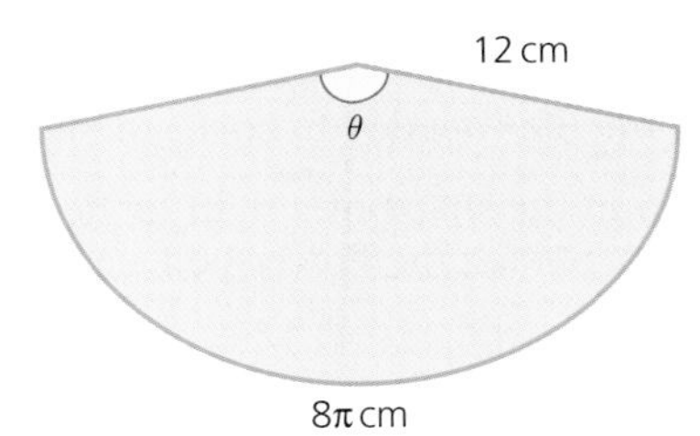

$\frac{\theta}{360} = \frac{8\pi}{24\pi}$

Therefore $\theta = 120°$

Area of sector = $\frac{120}{360} \times \pi \times 12^2 = 48\pi\,\text{cm}^2$

Total surface area = $48\pi + 16\pi$

$= 64\pi$

$= 201$ (3 s.f.)

The total surface area is $201\,\text{cm}^2$.

Exercise 23.24

1 Calculate the surface area of the cones. Give your answers in exact form.

a

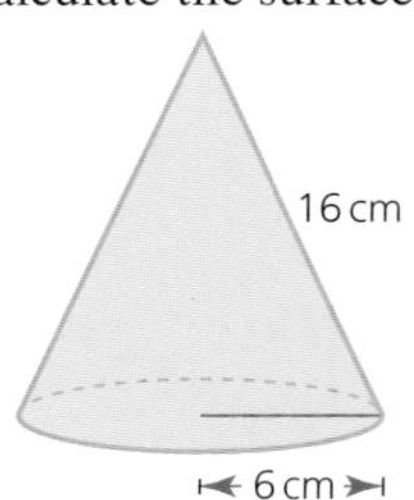

b

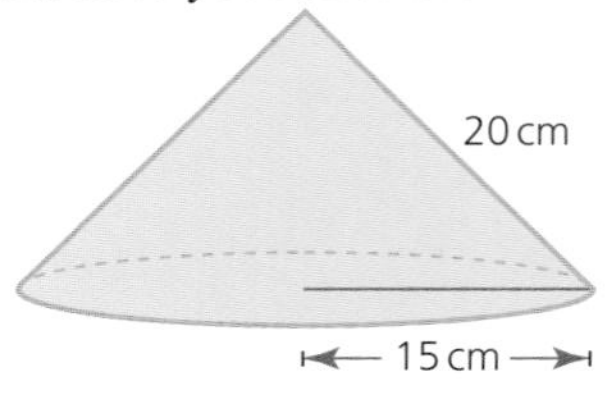

2 Two cones with the same base radius are stuck together as shown. Calculate the surface area of the shape giving your answer in exact form.

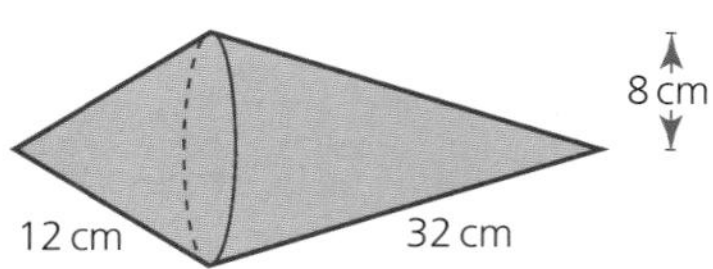

Student assessment 1

1 A rowing lake, rectangular in shape, is 2.5 km long by 500 m wide. Calculate the surface area of the water in km^2.

2 A rectangular floor 12 m long by 8 m wide is to be covered in ceramic tiles 40 cm long by 20 cm wide.
   a Calculate the number of tiles required to cover the floor.
   b The tiles are bought in boxes of 24 at a cost of $70 per box. What is the cost of the tiles needed to cover the floor?

3 A flower bed is in the shape of a right-angled triangle of sides 3 m, 4 m and 5 m. Sketch the flower bed, and calculate its area and perimeter.

4 A drawing of a building shows a rectangle 50 cm high and 10 cm wide with a triangular tower 20 cm high and 10 cm wide at the base on top of it. Find the area of the drawing of the building.

5 The squares of a chessboard are each of side 7.5 cm. What is the area of the chessboard?

Student assessment 2

1 Calculate the circumference and area of each of the following circles. Give your answers to 1 d.p.

a

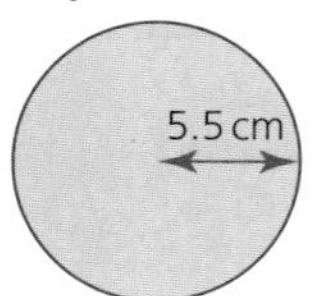

b

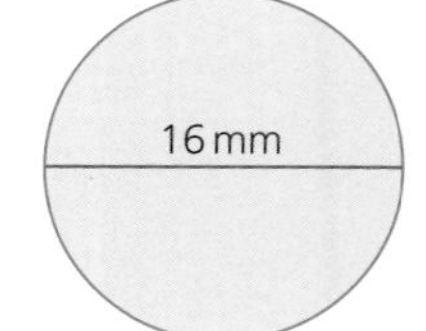

2 A semi-circular shape is cut out of the side of a rectangle as shown. Calculate the shaded area to 1 d.p.

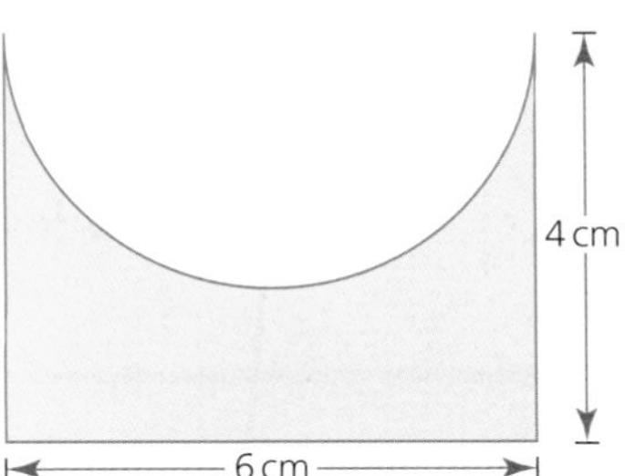

3 For the shape shown in the diagram, calculate the area of:
a the semi-circle
b the trapezium
c the whole shape.

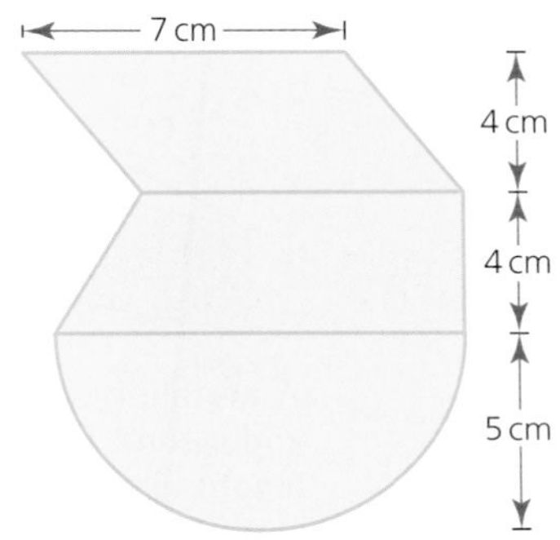

4 A cylindrical tube has an inner diameter of 6 cm, an outer diameter of 7 cm and a length of 15 cm.

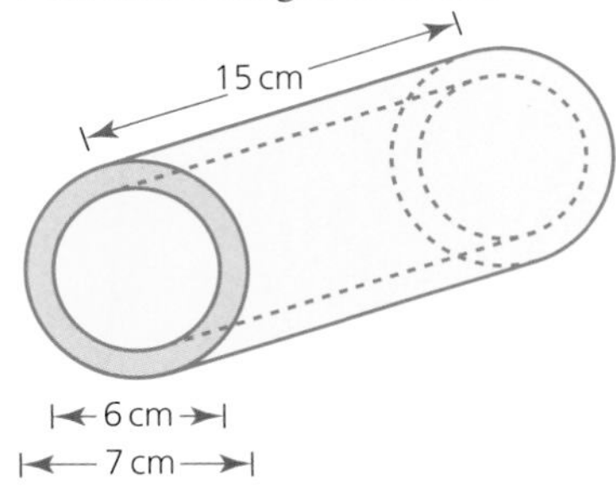

Calculate the following to 1 d.p.:
a the surface area of the shaded end
b the inside surface area of the tube
c the total surface area of the tube.

5 Calculate the volume of each of the following cylinders:

a

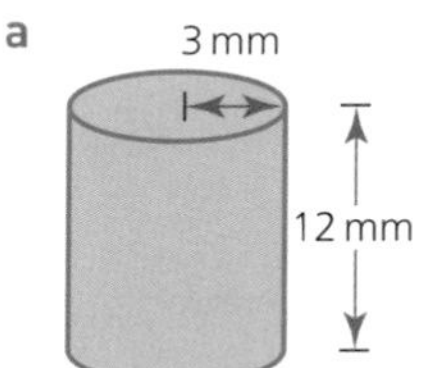

b

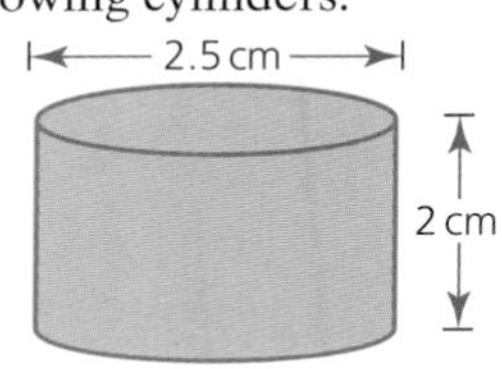

Student assessment 3

1 Calculate the area of the sector shown below:

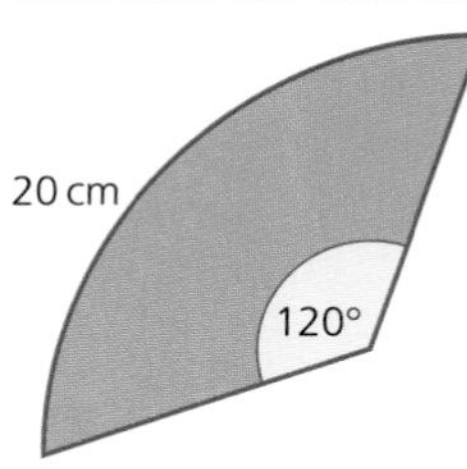

2 A hemisphere has a radius of 8 cm. Calculate to 1 d.p.:

a its total surface area

b its volume.

3 A cone has its top cut as shown. Calculate the volume of the truncated cone.

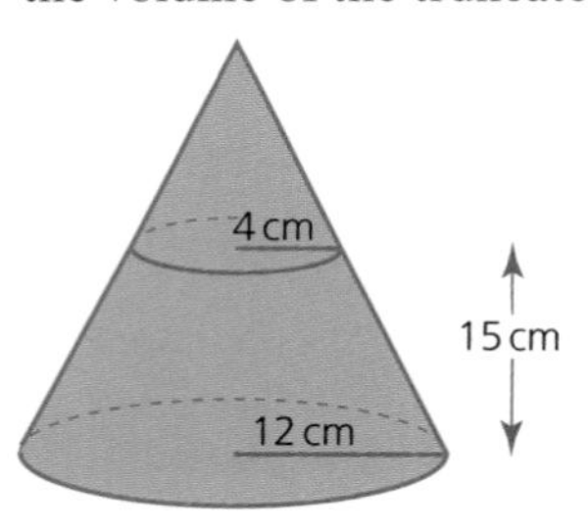

4 The prism has a cross-sectional area in the shape of a sector. Calculate:

a the radius r cm

b the cross-sectional area of the prism

c the total surface area of the prism

d the volume of the prism.

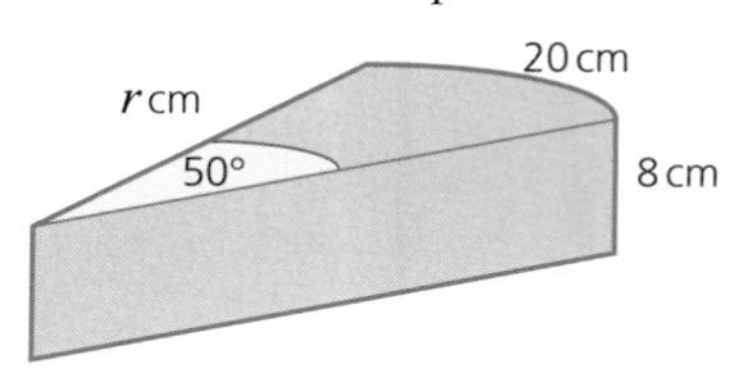

5 A metal object is made from a hemisphere and a cone, both of base radius 12 cm. The height of the object when upright is 36 cm. Calculate:

a the volume of the hemisphere

b the volume of the cone

c the curved surface area of the hemisphere

d the total surface area of the object.

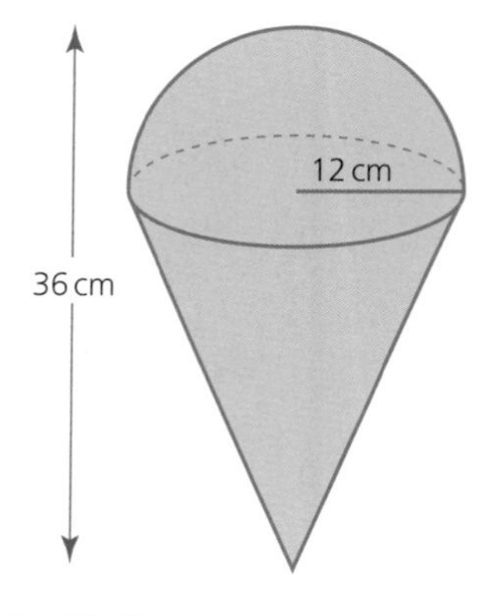

TOPIC 5

Mathematical investigations and ICT

Metal trays

A rectangular sheet of metal measures 30 cm by 40 cm.

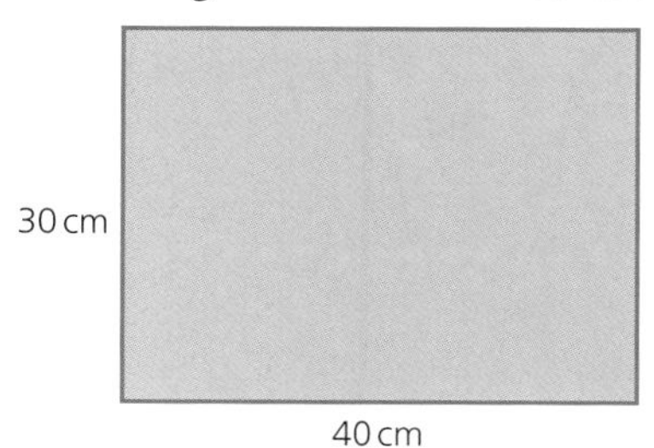

The sheet has squares of equal size cut from each corner. It is then folded to form a metal tray as shown.

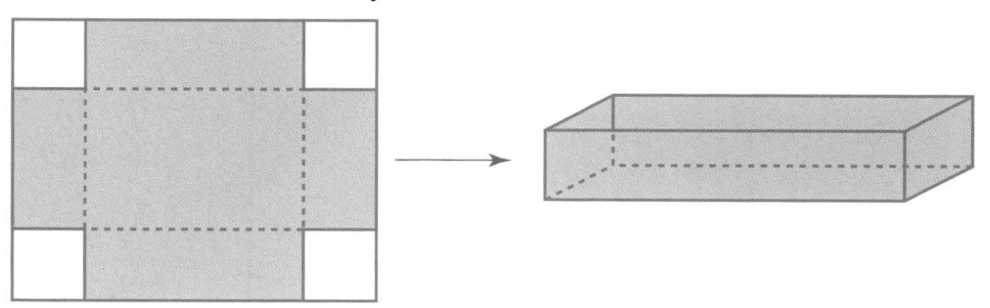

1 a Calculate the length, width and height of the tray if a square of side length 1 cm is cut from each corner of the sheet of metal.
  b Calculate the volume of this tray.

2 a Calculate the length, width and height of the tray if a square of side length 2 cm is cut from each corner of the sheet of metal.
  b Calculate the volume of this tray.

3 Using a spreadsheet if necessary, investigate the relationship between the volume of the tray and the size of the square cut from each corner. Enter your results in an ordered table.

4 Calculate, to 1 d.p., the side length of the square that produces the tray with the greatest volume.

5 State the greatest volume to the nearest whole number.

TOPIC 6

Trigonometry

Contents

Course

C6.1
Interpret and use three-figure bearings.

C6.2
Apply Pythagoras' theorem and the sine, cosine and tangent ratios for acute angles to the calculation of a side or of an angle of a right-angled triangle.

C6.3
Extended curriculum only.

C6.4
Extended curriculum only.

C6.5
Extended curriculum only.

The development of trigonometry

In about 2000BCE, astronomers in Sumer in ancient Mesopotamia introduced angle measure. They divided the circle into 360 degrees. They and the ancient Babylonians studied the ratios of the sides of similar triangles. They discovered some properties of these ratios. However, they did not develop these into a method for finding sides and angles of triangles, what we now call trigonometry.

The ancient Greeks, among them Euclid and Archimedes, developed trigonometry further. They studied the properties of chords in circles and produced proofs of the trigonometric formulae we use today.

The modern sine function was first defined in an ancient Hindu text, the *Surya Siddhanta*, and further work was done by the Indian mathematician and astronomer Aryabhata in the 5th century.

Aryabhata (476–550)

By the 10th century, Islamic mathematicians were using all six trigonometric functions (sine, cosine, tangent and their reciprocals). They made tables of trigonometric values and were applying them to problems in the geometry of the sphere.

As late as the 16th century, trigonometry was not well known in Europe. Nicolaus Copernicus decided it was necessary to explain the basic concepts of trigonometry in his book to enable people to understand his theory that the Earth went around the Sun.

Soon after, however, the need for accurate maps of large areas for navigation meant that trigonometry grew into a major branch of mathematics.

24 Bearings

NB: All diagrams are not drawn to scale.

Bearings

N
NW NE
W E
SW SE
S

In the days when sailing ships travelled the oceans of the world, compass bearings like the ones in the diagram (left) were used.

As the need for more accurate direction arose, extra points were added to N, S, E, W, NE, SE, SW and NW. Midway between North and North East was North North East, and midway between North East and East was East North East, and so on. This gave 16 points of the compass. This was later extended even further, eventually to 64 points.

As the speed of travel increased, a new system was required. The new system was the **three-figure bearing** system. North was given the **bearing** zero. 360° in a clockwise direction was one full rotation.

Exercise 24.1

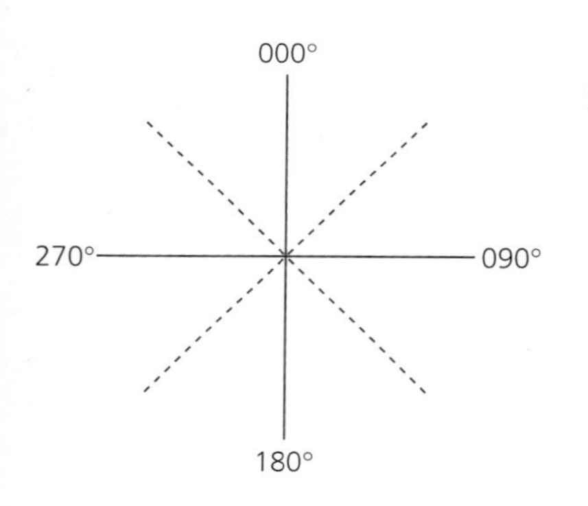

1 Copy the three-figure bearing diagram (left). On your diagram, mark the bearings for the compass points North East, South East, South West and North West.

2 Draw diagrams to show the following compass bearings and journeys. Use a scale of 1 cm : 1 km. North can be taken to be a line vertically up the page.
   a Start at point A. Travel 7 km on a bearing of 135° to point B. From B, travel 12 km on a bearing of 250° to point C. Measure the distance and bearing of A from C.
   b Start at point P. Travel 6.5 km on a bearing of 225° to point Q. From Q, travel 7.8 km on a bearing of 105° to point R. From R, travel 8.5 km on a bearing of 090° to point S. What are the distance and bearing of P from S?
   c Start from point M. Travel 11.2 km on a bearing of 270° to point N. From point N, travel 5.8 km on a bearing of 170° to point O. What are the bearing and distance of M from O?

Back bearings

Worked examples

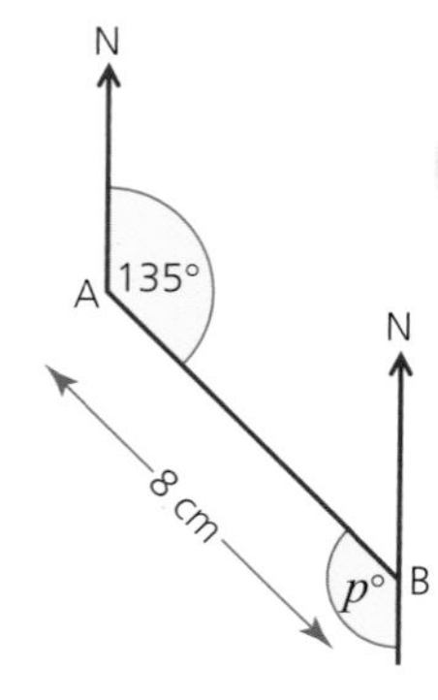

1 The bearing of B from A is 135° and the distance from A to B is 8 cm, as shown. The bearing of A from B is called the **back bearing**.

Since the two North lines are parallel:

$p = 135°$ (alternate angles), so the back bearing is $(180 + 135)°$.

That is, 315°.

(There are a number of methods of solving this type of problem.)

2 The bearing of B from A is 245°. What is the bearing of A from B?

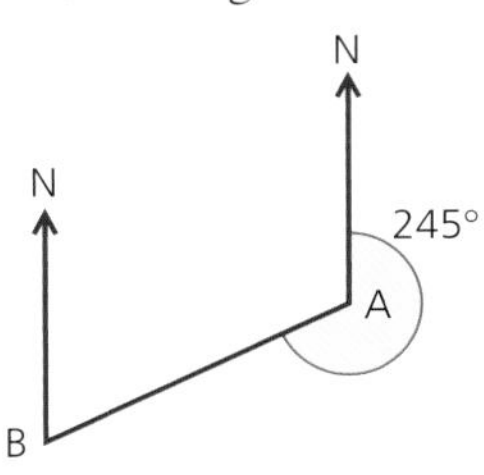

Since the two North lines are parallel:

$b = (245 - 180)° = 65°$ (alternate angles), so the bearing is 065°.

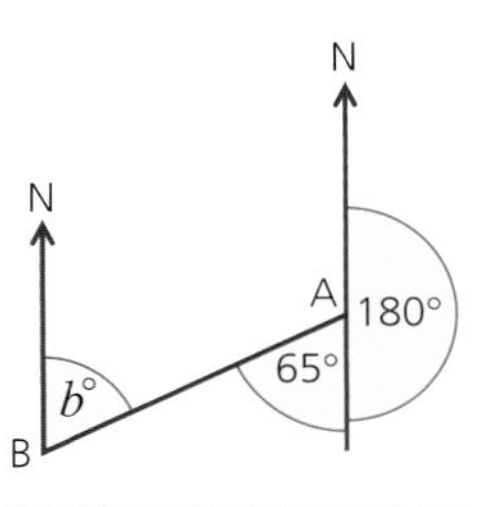

Exercise 24.2

1 Given the following bearings of point B from point A, draw diagrams and use them to calculate the bearing of A from B.

a bearing 130° **b** bearing 145°
c bearing 220° **d** bearing 200°
e bearing 152° **f** bearing 234°
g bearing 163° **h** bearing 214°

2 Given the following bearings of point D from point C, draw diagrams and use them to calculate the bearing of C from D.

a bearing 300° **b** bearing 320°
c bearing 290° **d** bearing 282°

Student assessment 1

1 From the top of a tall building in a town it is possible to see five towns. The bearing and distance of each one are given in the table:

Town	Distance (km)	Bearing
Bourn	8	070°
Catania	12	135°
Deltaville	9	185°
Etta	7.5	250°
Freetown	11	310°

Choose an appropriate scale and draw a diagram to show the position of each town. What are the distance and bearing of the following?

a Bourn from Deltaville
b Etta from Catania

2 A coastal radar station picks up a distress call from a ship. It is 50 km away on a bearing of 345°. The radar station contacts a lifeboat at sea which is 20 km away on a bearing of 220°.

Make a scale drawing and use it to find the distance and bearing of the ship from the lifeboat.

3 A climber gets to the top of Mont Blanc. He can see in the distance a number of ski resorts. He uses his map to find the bearing and distance of the resorts, and records them in a table:

Resort	Distance (km)	Bearing
Val d'Isère	30	082°
Les Arcs	40	135°
La Plagne	45	205°
Méribel	35	320°

Choose an appropriate scale and draw a diagram to show the position of each resort. What are the distance and bearing of the following?

a Val d'Isère from La Plagne

b Méribel from Les Arcs

4 An aircraft is seen on radar at airport X. The aircraft is 210 km away from the airport on a bearing of 065°. The aircraft is diverted to airport Y, which is 130 km away from airport X on a bearing of 215°. Use an appropriate scale and make a scale drawing to find the distance and bearing of airport Y from the aircraft.

25 Right-angled triangles

NB: All diagrams are not drawn to scale.

Trigonometric ratios

There are three basic trigonometric ratios: **sine**, **cosine** and **tangent**.

Each of these relates an angle of a right-angled triangle to a ratio of the lengths of two of its sides.

The sides of the triangle have names, two of which are dependent on their position in relation to a specific angle.

The longest side (always opposite the right angle) is called the **hypotenuse**. The side opposite the angle is called the **opposite** side and the side next to the angle is called the **adjacent** side.

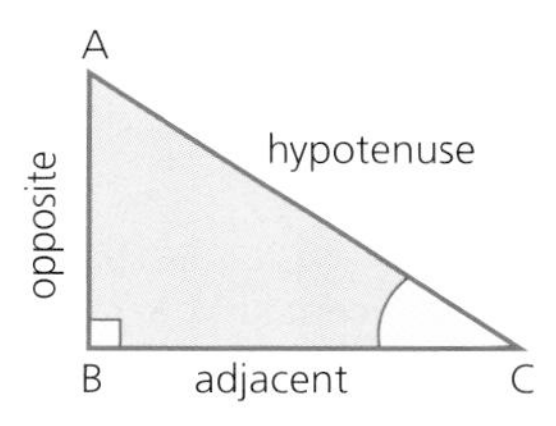

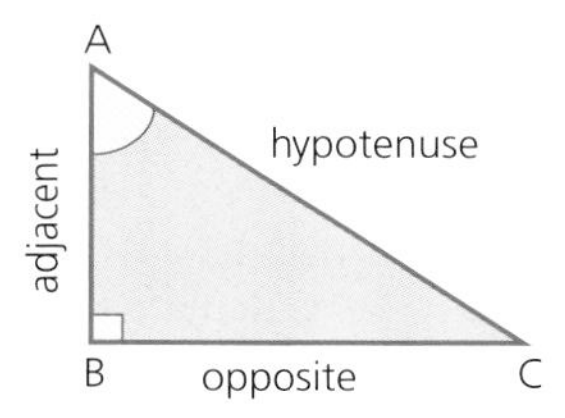

Note that, when the chosen angle is at A, the sides labelled opposite and adjacent change (above right).

Tangent

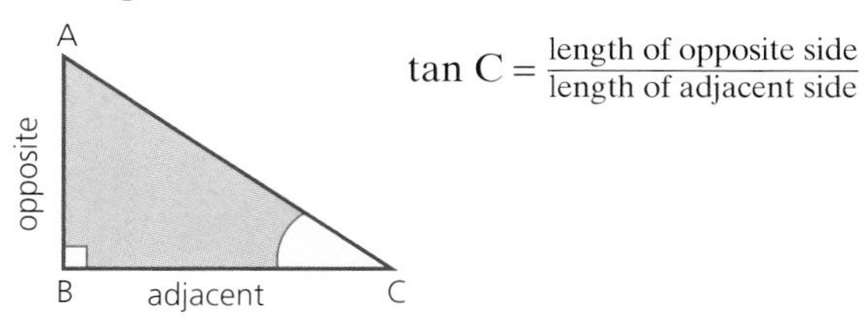

$$\tan C = \frac{\text{length of opposite side}}{\text{length of adjacent side}}$$

Worked examples

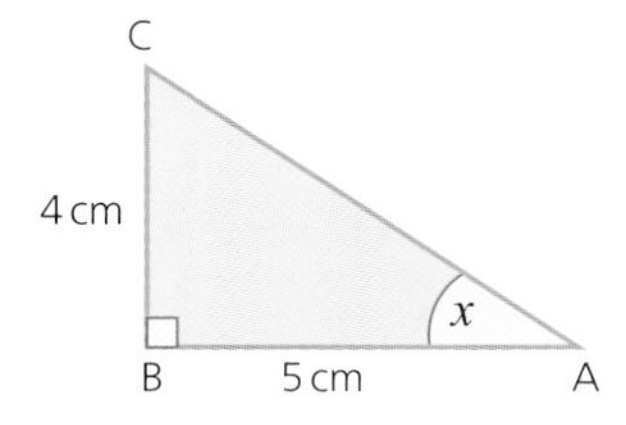

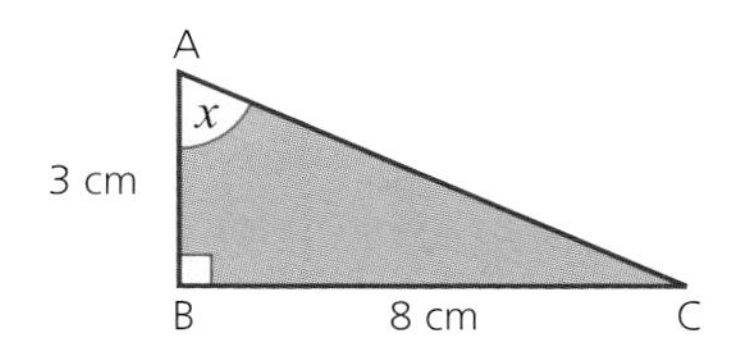

1 Calculate the size of angle BAC in each triangle.

Triangle 1: $\tan x = \frac{\text{opposite}}{\text{adjacent}} = \frac{4}{5}$

$$x = \tan^{-1}\left(\frac{4}{5}\right)$$

$$x = 38.7 \text{ (3 s.f.)}$$

angle $BAC = 38.7°$ (3 s.f.)

Triangle 2: $\tan x = \frac{8}{3}$

$$x = \tan^{-1}\left(\frac{8}{3}\right)$$

$$x = 69.4 \text{ (3 s.f.)}$$

angle $BAC = 69.4°$ (3 s.f.)

2 Calculate the length of the opposite side QR (right).

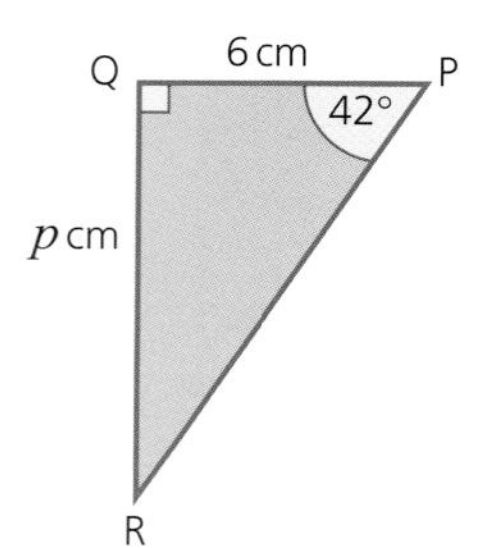

$$\tan 42° = \frac{p}{6}$$

$$6 \times \tan 42° = p$$

$$p = 5.40 \text{ (3 s.f.)}$$

QR = 5.40 cm (3 s.f.)

3 Calculate the length of the adjacent side XY.

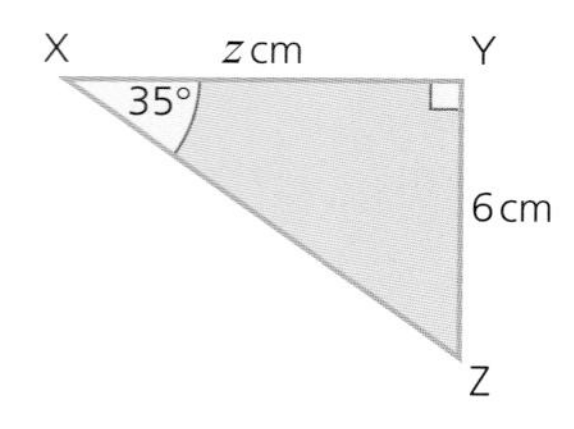

$$\tan 35° = \frac{6}{z}$$

$$z \times \tan 35° = 6$$

$$z = \frac{6}{\tan 35°}$$

$$z = 8.57 \text{ (3 s.f.)}$$

XY = 8.57 cm (3 s.f.)

Exercise 25.1

Calculate the length of the side marked x cm in each of the diagrams in Q.1 and 2. Answer to 1 d.p.

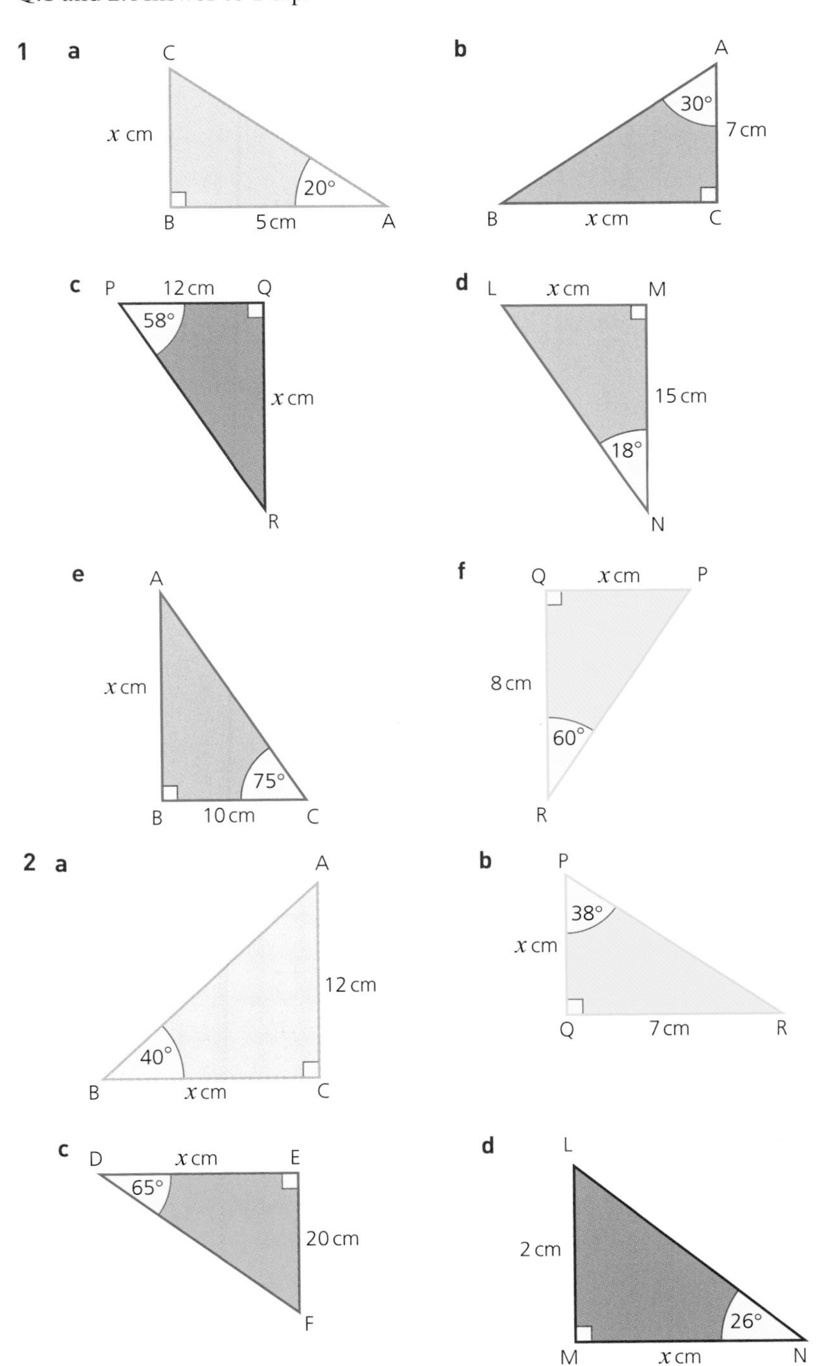

Exercise 25.1 (cont)

e

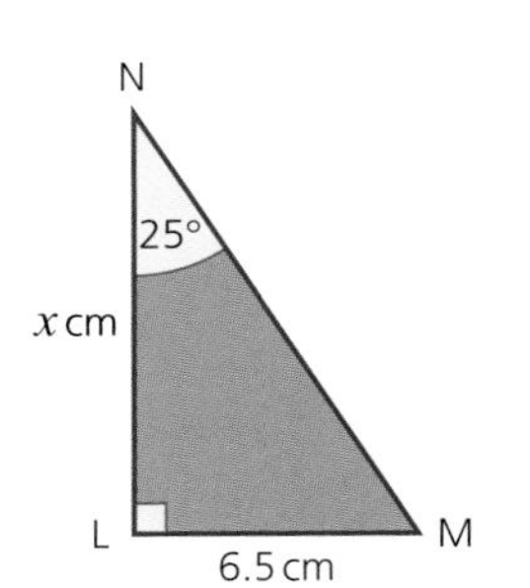

f

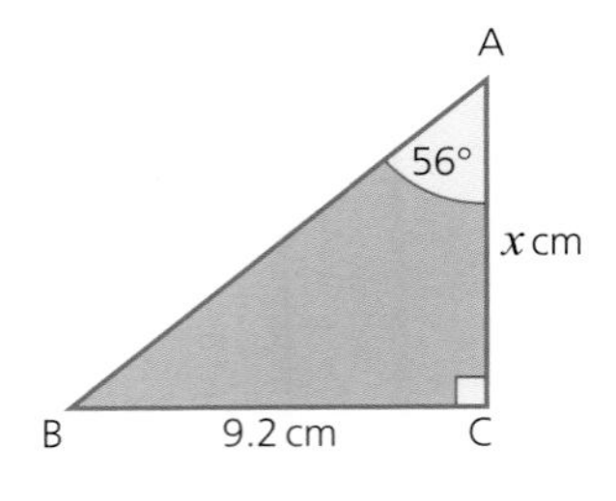

3 Calculate the size of the marked angle $x°$ in each diagram. Give your answers to 1 d.p.

a

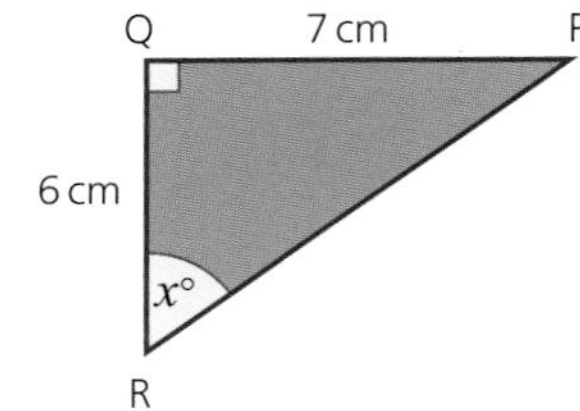

b

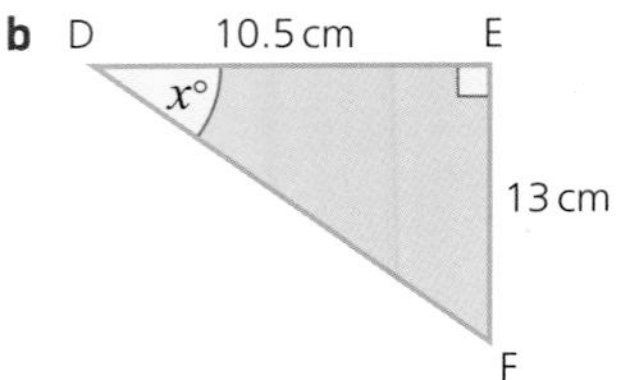

c

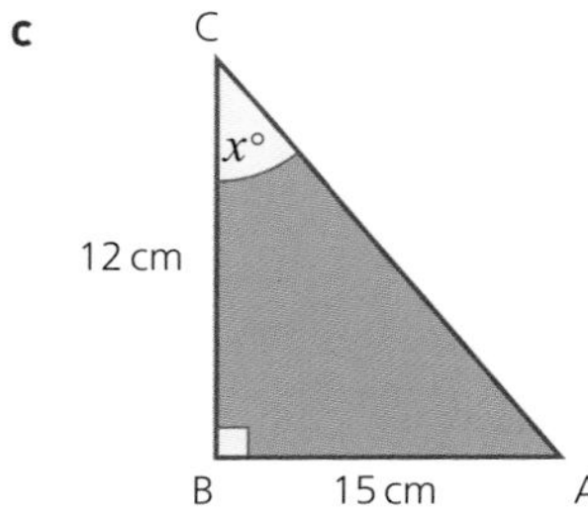

d

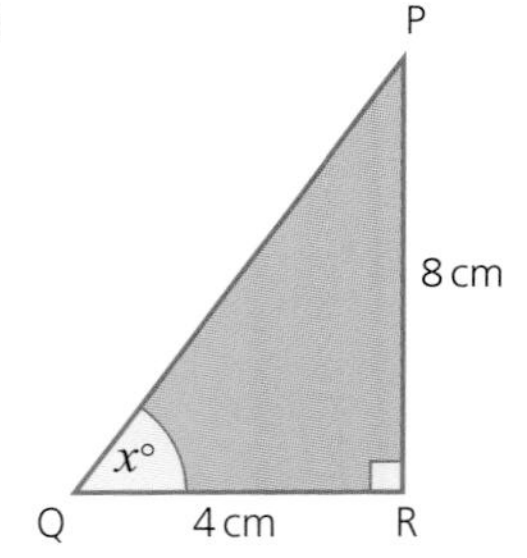

e

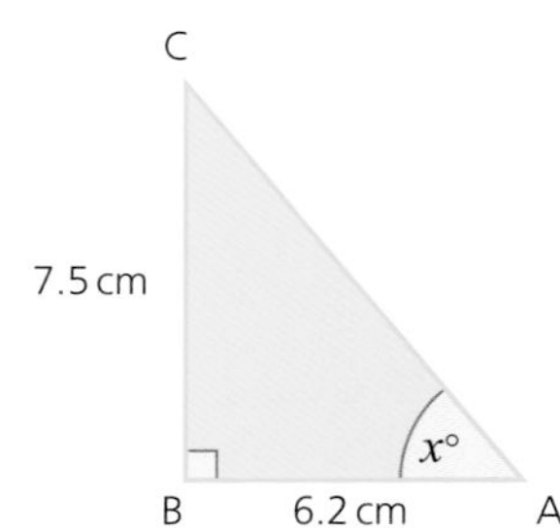

f

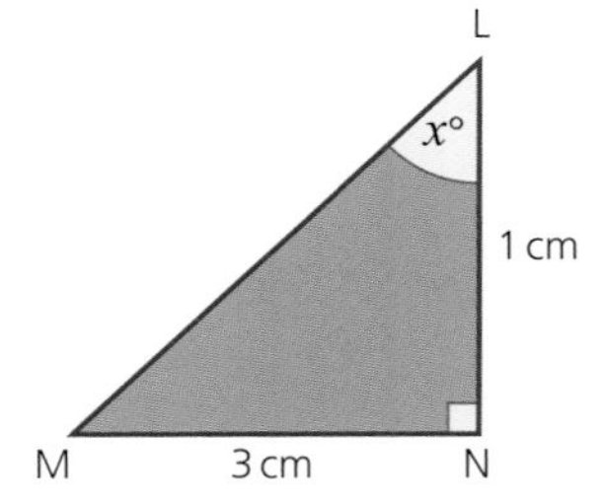

Sine

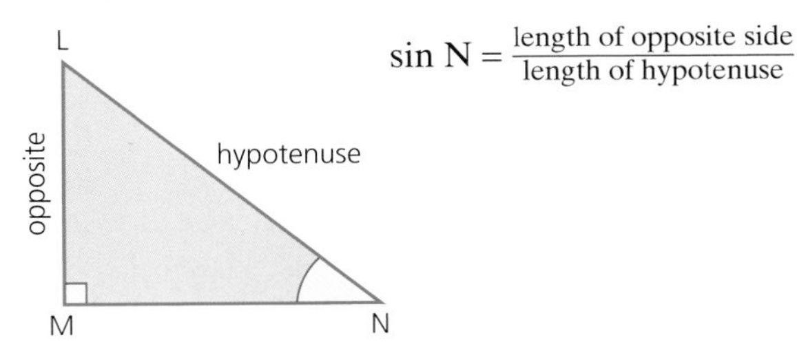

$$\sin N = \frac{\text{length of opposite side}}{\text{length of hypotenuse}}$$

Worked examples

1 Calculate the size of angle *BAC*.

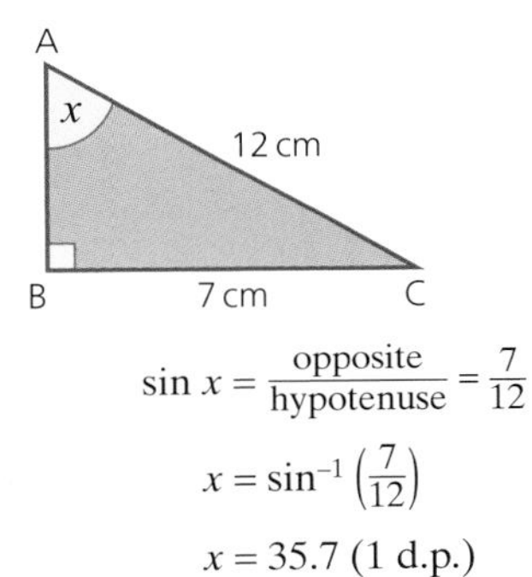

$$\sin x = \frac{\text{opposite}}{\text{hypotenuse}} = \frac{7}{12}$$

$$x = \sin^{-1}\left(\frac{7}{12}\right)$$

$$x = 35.7 \text{ (1 d.p.)}$$

angle $BAC = 35.7°$ (1 d.p.)

2 Calculate the length of the hypotenuse PR.

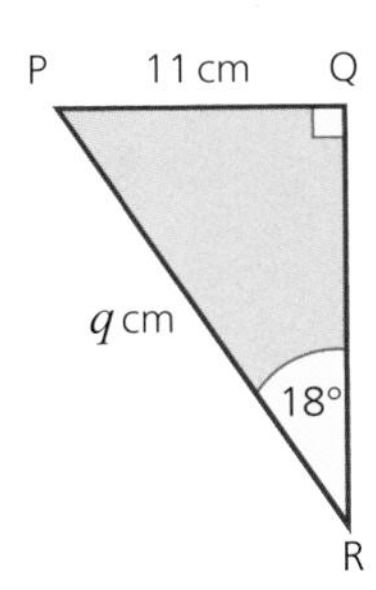

$$\sin 18° = \frac{11}{q}$$

$$q \times \sin 18° = 11$$

$$q = \frac{11}{\sin 18°}$$

$$q = 35.6 \text{ (3 s.f.)}$$

$$\text{PR} = 35.6 \text{ cm (3 s.f.)}$$

Exercise 25.2

1 Calculate the length of the marked side in each diagram. Give your answers to 1 d.p.

a

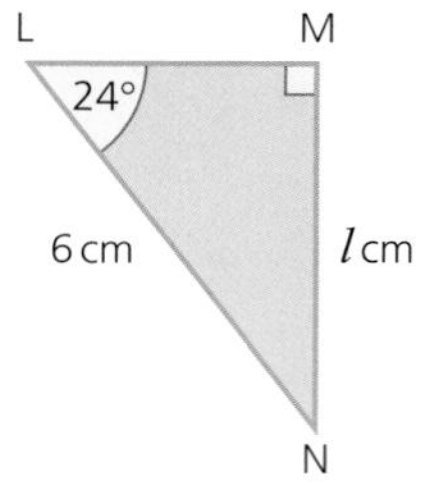

b

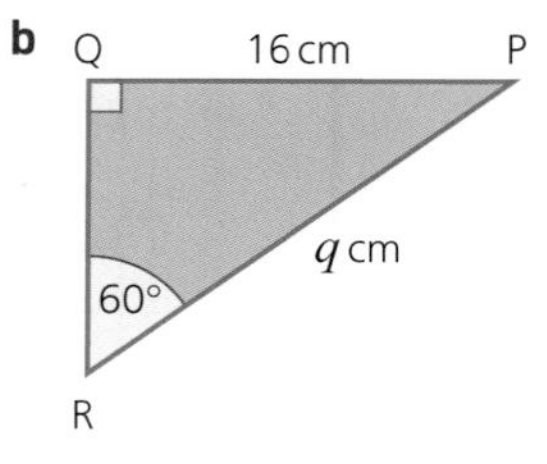

c

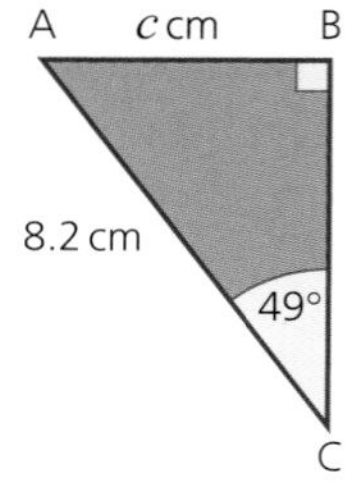

d

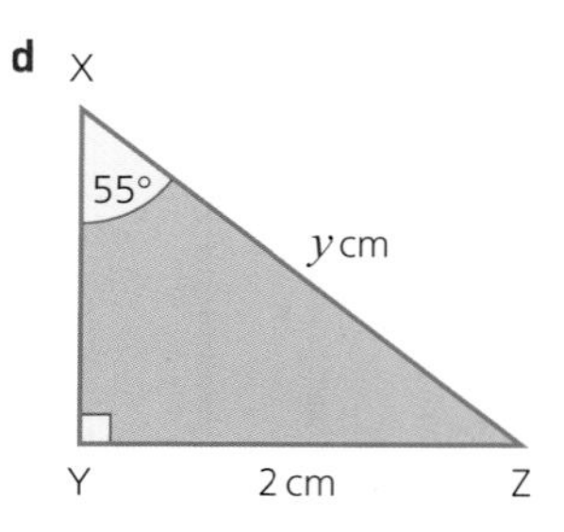

e

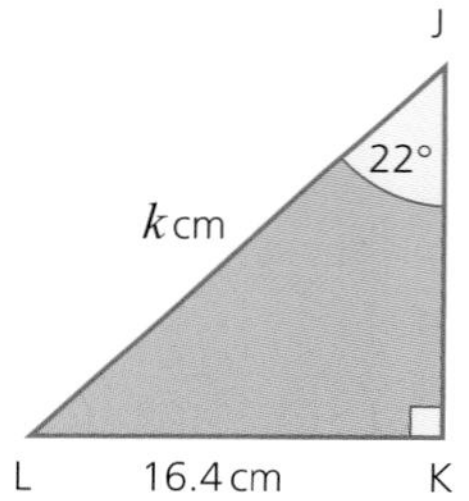

f

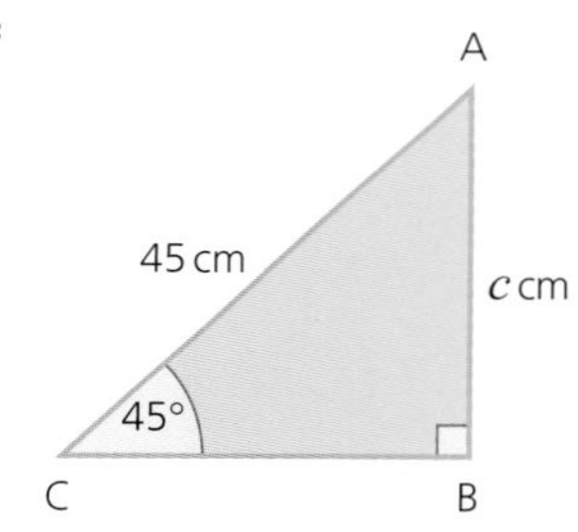

2 Calculate the size of the angle marked $x°$ in each diagram. Give your answers to 1 d.p.

a

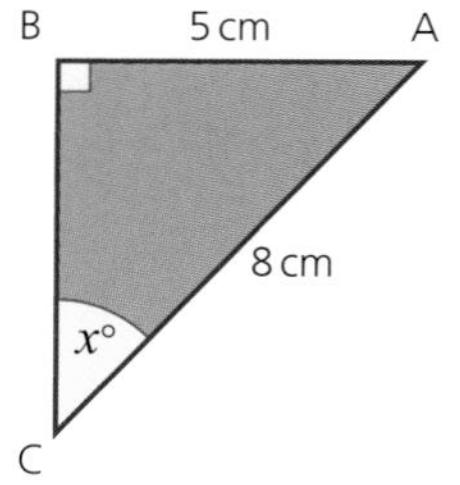

b

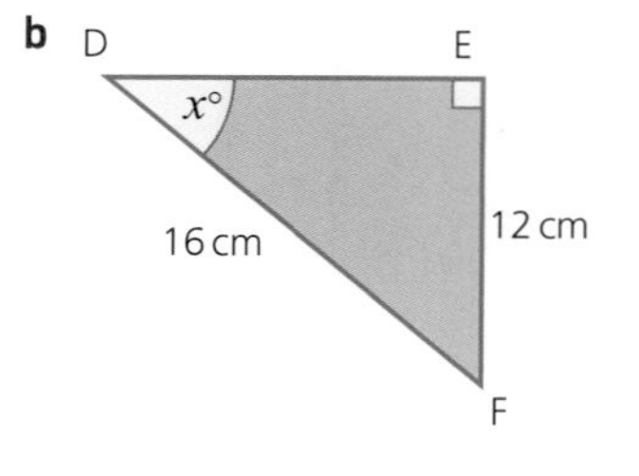

c

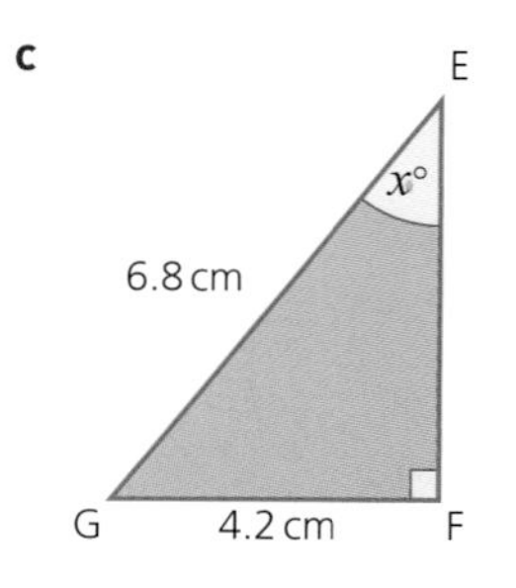

d

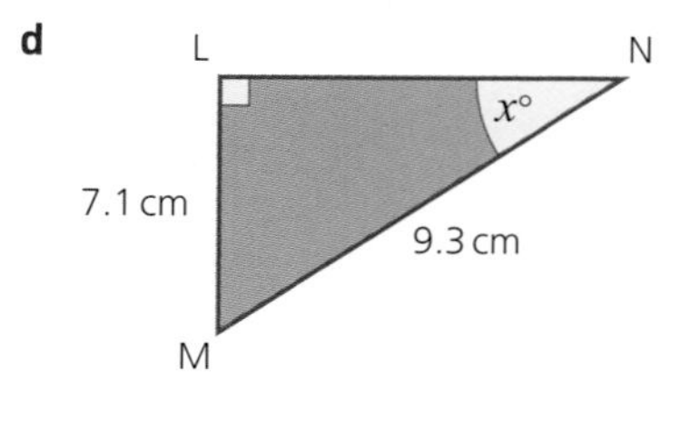

3 Calculate the length of the side marked q cm in each diagram. Give your answers correct to 1 d.p.

a

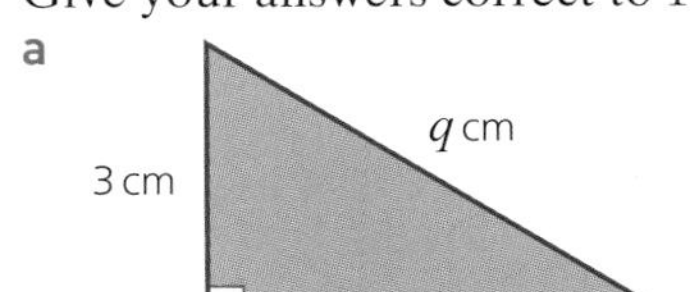

b

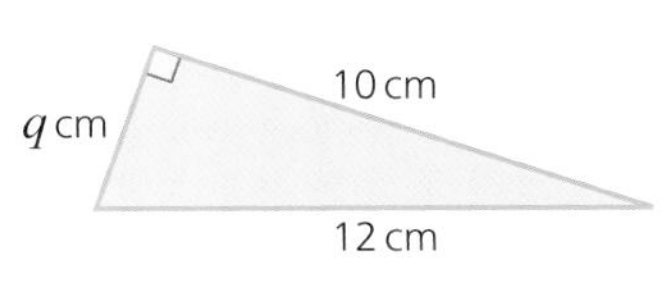

c

3 cm
65°
6 cm
q cm

d

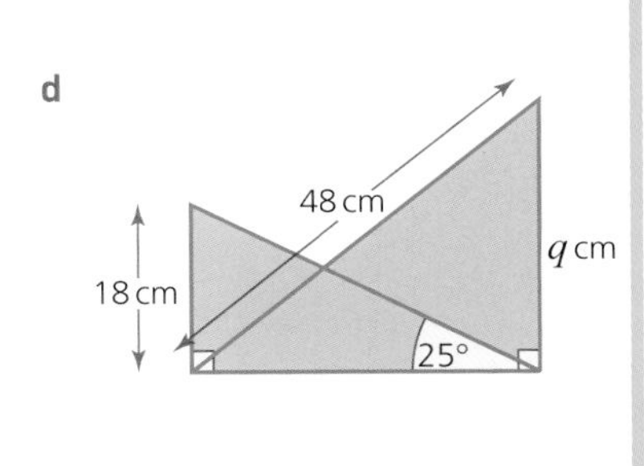

4 A table measures 3.9 m by 2.4 m. Calculate the distance between the opposite corners. Give your answer correct to 1 d.p.

Student assessment 2

1 A map shows three towns A, B and C. Town A is due North of C. Town B is due East of A. The distance AC is 75 km and the bearing of C from B is 245°. Calculate, giving your answers to the nearest 100 m:

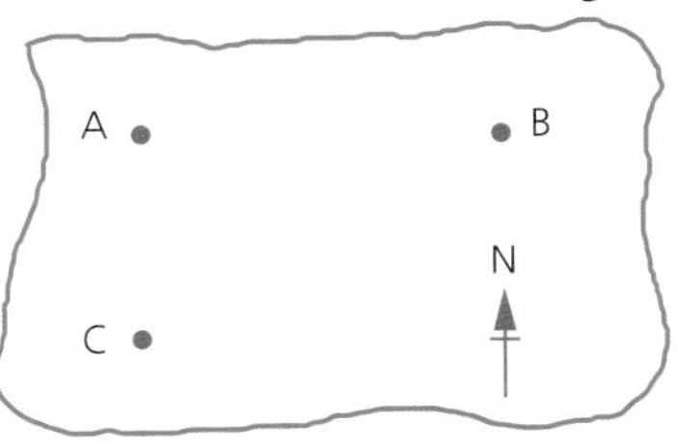

a the distance AB

b the distance BC.

2 Two trees stand 16 m apart. Their tops make an angle of $\theta°$ at point A on the ground.

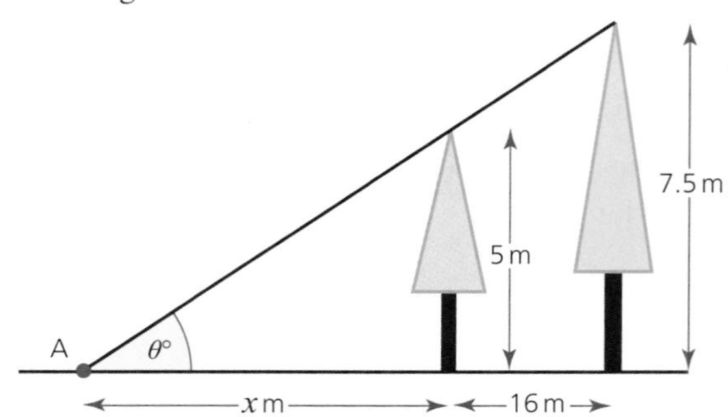

a Express $\theta°$ in terms of the height of the shorter tree and its distance x metres from point A.
b Express $\theta°$ in terms of the height of the taller tree and its distance from A.
c Form an equation in terms of x.
d Calculate the value of x.
e Calculate the value of θ.

3 Two boats X and Y, sailing in a race, are shown in the diagram:

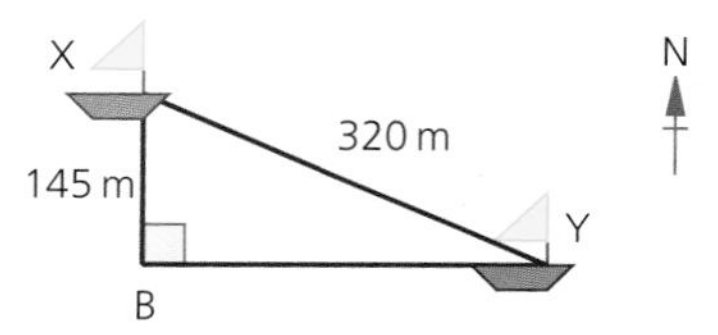

Boat X is 145 m due North of a buoy B. Boat Y is due East of buoy B. Boats X and Y are 320 m apart. Calculate:
a the distance BY
b the bearing of Y from X
c the bearing of X from Y.

TOPIC 6

Mathematical investigations and ICT

Pythagoras and circles

The explanation for Pythagoras' theorem usually shows a right-angled triangle with squares drawn on each of its three sides, as in the diagram.

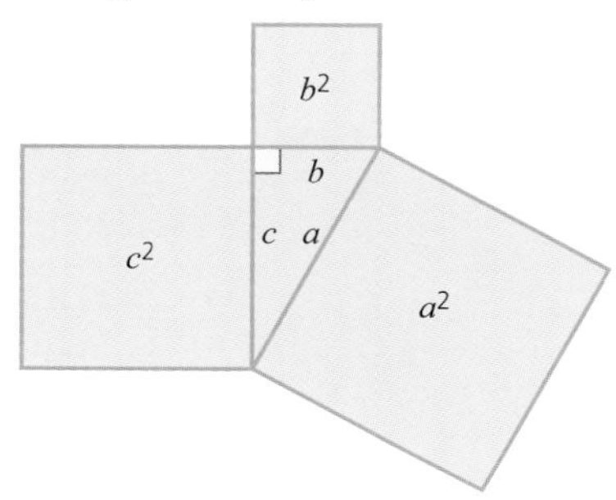

In this example, the area of the square on the hypotenuse, a^2, is equal to the sum of the areas of the squares on the other two sides, $b^2 + c^2$.

This gives the formula $a^2 = b^2 + c^2$.

1 Draw a right-angled triangle.
2 Using a pair of compasses, construct a semi-circle off each side of the triangle. Your diagram should look similar to the one below.

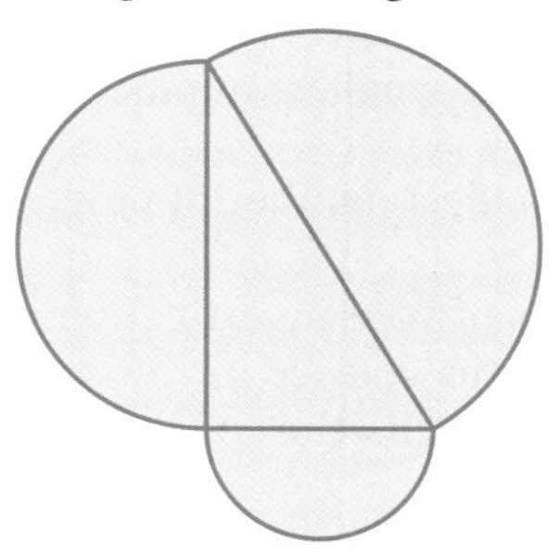

3 By measuring the diameter of each semi-circle, calculate their areas.
4 Is the area of the semi-circle on the hypotenuse the sum of the areas of the semi-circles drawn on the other two sides? Does Pythagoras' theorem still hold for semi-circles?
5 Does Pythagoras' theorem still hold if equilateral triangles are drawn on each side?
6 Investigate for other regular polygons.

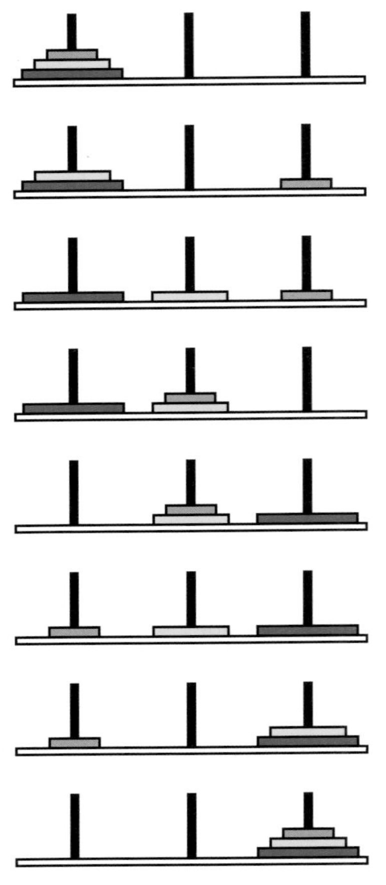

Towers of Hanoi

This investigation is based on an old Vietnamese legend. The legend is as follows:

> At the beginning of time a temple was created by the Gods. Inside the temple stood three giant rods. On one of these rods, 64 gold discs, all of different diameters, were stacked in descending order of size, i.e. the largest at the bottom rising to the smallest at the top. Priests at the temple were responsible for moving the discs onto the remaining two rods until all 64 discs were stacked in the same order on one of the other rods. When this task was completed, time would cease and the world would come to an end.

The discs however could only be moved according to certain rules. These were:

- Only one disc could be moved at a time.
- A disc could only be placed on top of a larger one.

The diagrams (left) show the smallest number of moves required to transfer three discs from the rod on the left to the rod on the right.

With three discs, the smallest number of moves is seven.

1 What is the smallest number of moves needed for two discs?
2 What is the smallest number of moves needed for four discs?
3 Investigate the smallest number of moves needed to move different numbers of discs.
4 Display the results of your investigation in an ordered table.
5 Describe any patterns you see in your results.
6 Predict, from your results, the smallest number of moves needed to move ten discs.
7 Determine a formula for the smallest number of moves for n discs.
8 Assume the priests have been transferring the discs at the rate of one per second and assume the Earth is approximately 4.54 billion years old (4.54×10^9 years). According to the legend, is the world coming to an end soon? Justify your answer with relevant calculations.

ICT activity

In this activity you will need to use a graphical calculator to investigate the relationship between different trigonometric ratios.

1 a Using the graphical calculator or graphing software, plot the graph of $y = \sin x$ for $0^\circ \leqslant x \leqslant 180^\circ$. The graph should look similar to the one shown below:

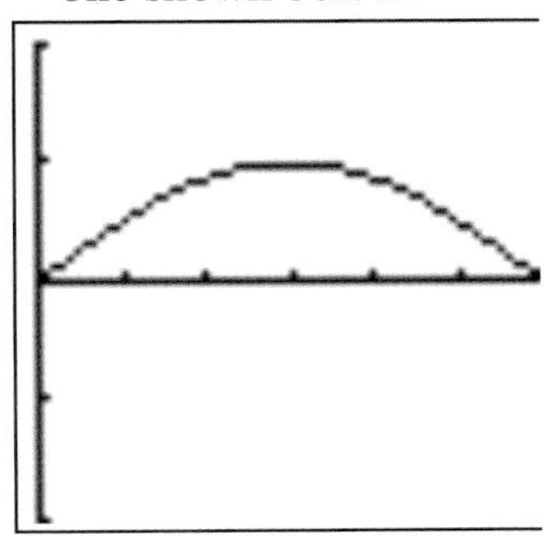

b Using the equation solving facility, evaluate sin 70°.

c Referring to the graph, explain why $\sin x = 0.7$ has two solutions between 0° and 180°.

d Use the graph to solve the equation $\sin x = 0.5$.

2 a On the same axes as before, plot $y = \cos x$.

b How many solutions are there to the equation $\sin x = \cos x$ between 0° and 180°?

c What is the solution to the equation $\sin x = \cos x$ between 0° and 180°?

3 By plotting appropriate graphs, solve the following for $0^\circ \leqslant x \leqslant 180^\circ$.

a $\sin x = \tan x$

b $\cos x = \tan x$

TOPIC 7

Vectors and transformations

Contents

Course

C7.1

Describe a translation by using a vector represented by e.g. $\begin{pmatrix} x \\ y \end{pmatrix}$, $\overrightarrow{AB}$ or $\mathbf{a}$.

Add and subtract vectors.

Multiply a vector by a scalar.

C7.2

Reflect simple plane figures in horizontal or vertical lines.

Rotate simple plane figures about the origin, vertices or midpoints of edges of the figures, through multiples of 90°.

Construct given translations and enlargements of simple plane figures.

Recognise and describe reflections, rotations, translations and enlargements.

C7.3

Extended curriculum only.

The development of vectors

The study of vectors arose from coordinates in two dimensions. Around 1636, René Descartes and Pierre de Fermat founded analytic geometry by linking the solutions to an equation with two variables with points on a curve.

In 1804, Czech mathematician Bernhard Bolzano worked on the mathematics of points, lines and planes and his work formed the beginnings of work on vectors. Later in the 19th century, this was further developed by German mathematician August Möbius and Italian mathematician Giusto Bellavitis.

Bernhard Bolzano (1781–1848)

26 Vectors

Translations

A **translation** (a sliding movement) can be described using a **column vector**. A column vector describes the movement of the object in both the x direction and the y direction.

Worked examples

1 Describe the translation from A to B in the diagram in terms of a column vector.

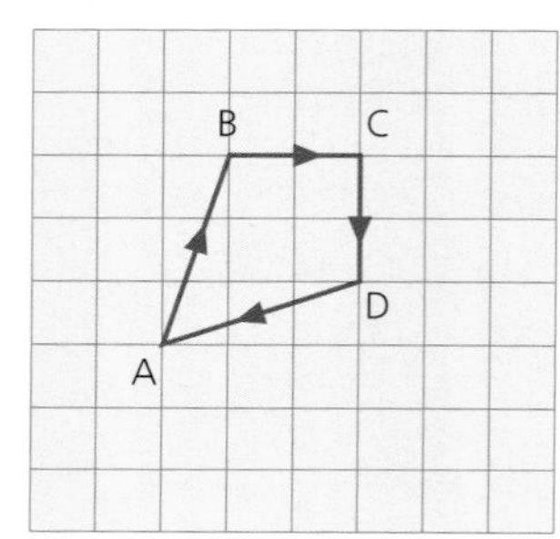

$\overrightarrow{AB} = \begin{pmatrix} 1 \\ 3 \end{pmatrix}$

i.e. 1 unit in the x direction, 3 units in the y direction

2 Describe $\overrightarrow{BC}$ in terms of a column vector.

$\overrightarrow{BC} = \begin{pmatrix} 2 \\ 0 \end{pmatrix}$

3 Describe $\overrightarrow{CD}$ in terms of a column vector.

$\overrightarrow{CD} = \begin{pmatrix} 0 \\ -2 \end{pmatrix}$

4 Describe $\overrightarrow{DA}$ in terms of a column vector.

$\overrightarrow{DA} = \begin{pmatrix} -3 \\ -1 \end{pmatrix}$

Translations can also be named by a single letter. The direction of the arrow indicates the direction of the translation.

Worked example

Describe **a** and **b** in the diagram (right) using column vectors.

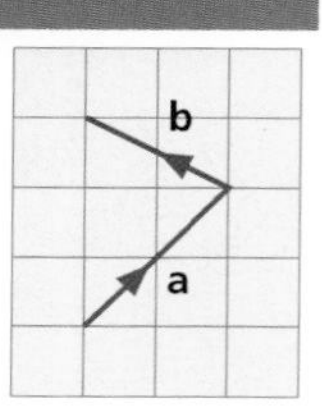

$\mathbf{a} = \begin{pmatrix} 2 \\ 2 \end{pmatrix}$

$\mathbf{b} = \begin{pmatrix} -2 \\ 1 \end{pmatrix}$

Note: When you represent vectors by single letters, e.g. **a**, in handwritten work, you should write them as $\underset{\sim}{a}$.

If $\mathbf{a} = \begin{pmatrix} 2 \\ 5 \end{pmatrix}$ and $\mathbf{b} = \begin{pmatrix} -3 \\ -2 \end{pmatrix}$, they can be represented diagrammatically as shown:

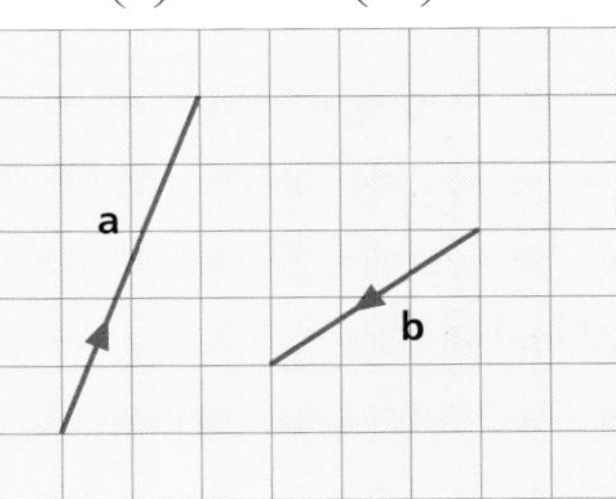

The diagrammatic representation of –**a** and –**b** is shown below.

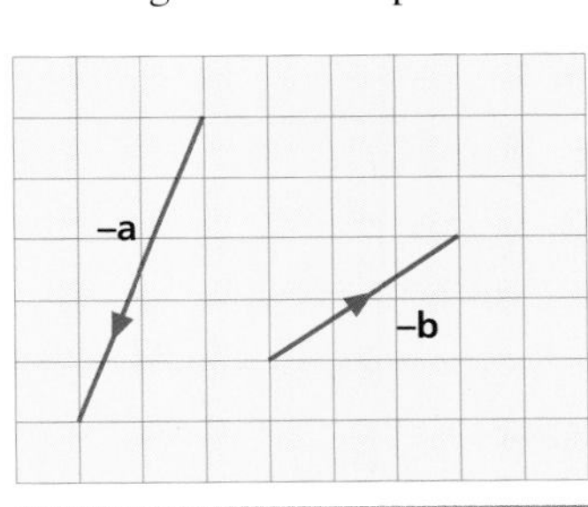

It can be seen from the diagram that:

$-\mathbf{a} = \begin{pmatrix} -2 \\ -5 \end{pmatrix}$ and $-\mathbf{b} = \begin{pmatrix} 3 \\ 2 \end{pmatrix}$

Exercise 26.1

In Q.1 and Q.2 describe each translation using a column vector.

1 a $\overrightarrow{AB}$
b $\overrightarrow{BC}$
c $\overrightarrow{CD}$
d $\overrightarrow{DE}$
e $\overrightarrow{EA}$
f $\overrightarrow{AE}$
g $\overrightarrow{DA}$
h $\overrightarrow{CA}$
i $\overrightarrow{DB}$

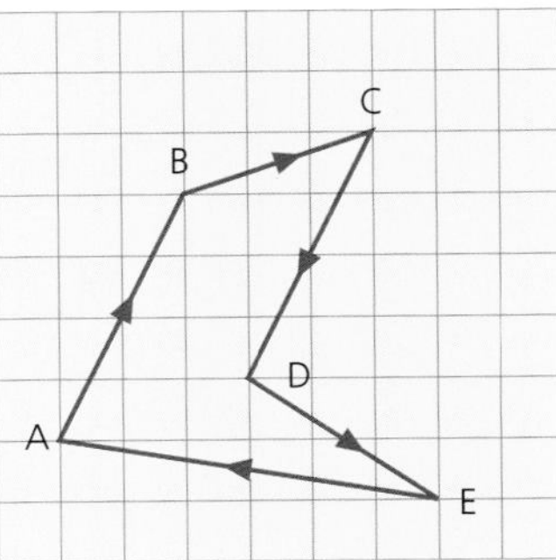

2 a **a**
b **b**
c **c**
d **d**
e **e**
f –**b**
g –**c**
h –**d**
i –**a**

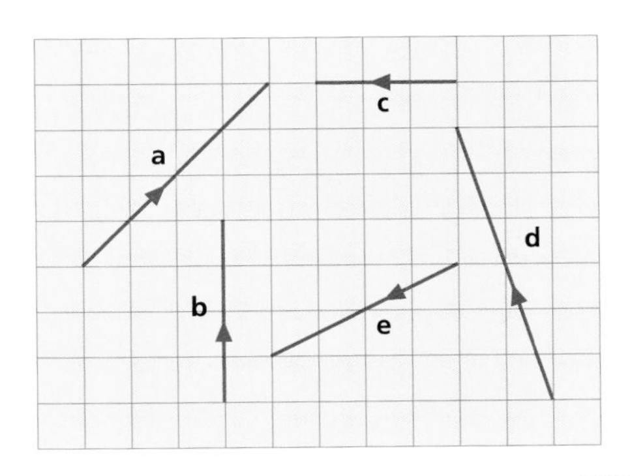

Exercise 26.1 (cont)

3 Draw and label the following vectors on a square grid:

a $\mathbf{a} = \begin{pmatrix} 2 \\ 4 \end{pmatrix}$ **b** $\mathbf{b} = \begin{pmatrix} -3 \\ 6 \end{pmatrix}$ **c** $\mathbf{c} = \begin{pmatrix} 3 \\ -5 \end{pmatrix}$

d $\mathbf{d} = \begin{pmatrix} -4 \\ -3 \end{pmatrix}$ **e** $\mathbf{e} = \begin{pmatrix} 0 \\ -6 \end{pmatrix}$ **f** $\mathbf{f} = \begin{pmatrix} -5 \\ 0 \end{pmatrix}$

g $\mathbf{g} = -\mathbf{c}$ **h** $\mathbf{h} = -\mathbf{b}$ **i** $\mathbf{i} = -\mathbf{f}$

Addition of vectors

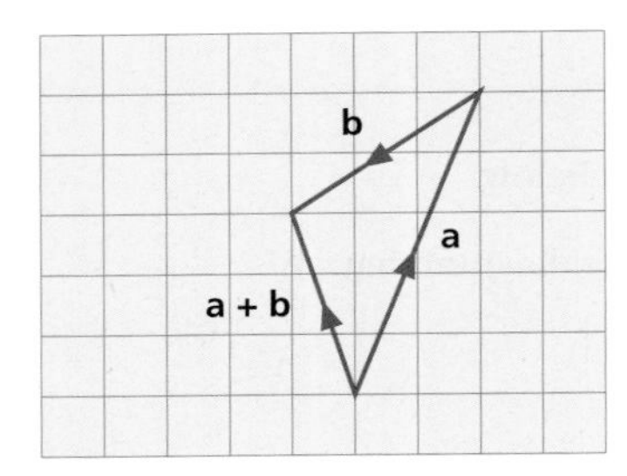

Vectors can be added together and represented diagrammatically, as shown in the diagram.

The translation represented by **a** followed by **b** can be written as a single transformation **a** + **b**:

i.e. $\begin{pmatrix} 2 \\ 5 \end{pmatrix} + \begin{pmatrix} -3 \\ -2 \end{pmatrix} = \begin{pmatrix} -1 \\ 3 \end{pmatrix}$

Exercise 26.2

In the following questions,

$\mathbf{a} = \begin{pmatrix} 3 \\ 4 \end{pmatrix}$ $\mathbf{b} = \begin{pmatrix} -2 \\ 1 \end{pmatrix}$ $\mathbf{c} = \begin{pmatrix} -4 \\ -3 \end{pmatrix}$ $\mathbf{d} = \begin{pmatrix} 3 \\ -2 \end{pmatrix}$

1 Draw vector diagrams to represent:

a $\mathbf{a} + \mathbf{b}$ **b** $\mathbf{b} + \mathbf{a}$ **c** $\mathbf{a} + \mathbf{d}$

d $\mathbf{d} + \mathbf{a}$ **e** $\mathbf{b} + \mathbf{c}$ **f** $\mathbf{c} + \mathbf{b}$

2 What conclusions can you draw from your answers to Q.1?

Multiplying a vector by a scalar

Look at the two vectors in the diagram:

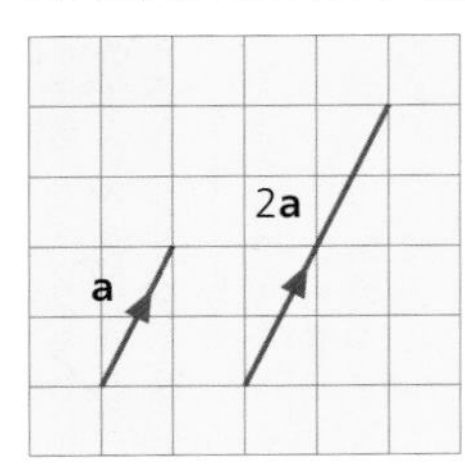

$\mathbf{a} = \begin{pmatrix} 1 \\ 2 \end{pmatrix}$ $2\mathbf{a} = 2\begin{pmatrix} 1 \\ 2 \end{pmatrix} = \begin{pmatrix} 2 \\ 4 \end{pmatrix}$

Worked example

If $\mathbf{a} = \begin{pmatrix} 2 \\ -4 \end{pmatrix}$, express the vectors **b**, **c**, **d** and **e** in terms of **a**.

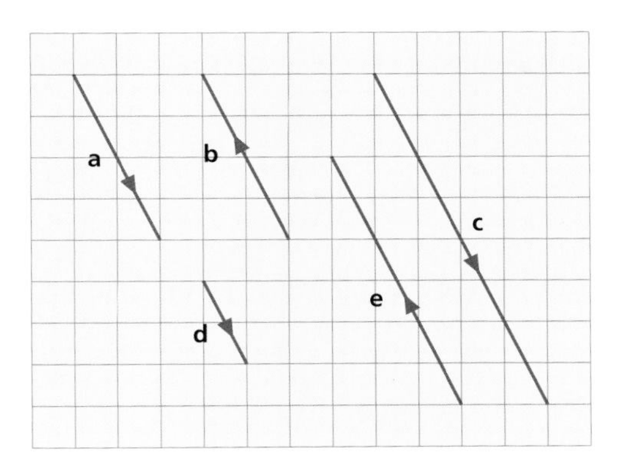

$\mathbf{b} = -\mathbf{a}$ $\quad$ $\mathbf{c} = 2\mathbf{a}$ $\quad$ $\mathbf{d} = \frac{1}{2}\mathbf{a}$ $\quad$ $\mathbf{e} = -\frac{3}{2}\mathbf{a}$

Exercise 26.3

1 $\mathbf{a} = \begin{pmatrix} 1 \\ 4 \end{pmatrix}$ $\quad$ $\mathbf{b} = \begin{pmatrix} -4 \\ -2 \end{pmatrix}$ $\quad$ $\mathbf{c} = \begin{pmatrix} -4 \\ 6 \end{pmatrix}$

Express the following vectors in terms of either **a**, **b** or **c**.

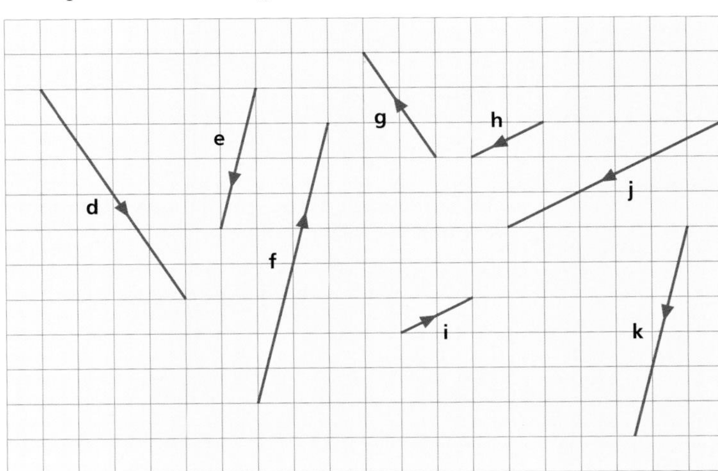

Exercise 26.3 (cont)

2 $\mathbf{a} = \begin{pmatrix} 2 \\ 3 \end{pmatrix}$ $\mathbf{b} = \begin{pmatrix} -4 \\ -1 \end{pmatrix}$ $\mathbf{c} = \begin{pmatrix} -2 \\ 4 \end{pmatrix}$

Represent each of the following as a single column vector:

a $2\mathbf{a}$ b $3\mathbf{b}$ c $-\mathbf{c}$
d $\mathbf{a} + \mathbf{b}$ e $\mathbf{b} - \mathbf{c}$ f $3\mathbf{c} - \mathbf{a}$
g $2\mathbf{b} - \mathbf{a}$ h $\frac{1}{2}(\mathbf{a} + \mathbf{b})$ i $2\mathbf{a} + 3\mathbf{c}$

3 $\mathbf{a} = \begin{pmatrix} -2 \\ 3 \end{pmatrix}$ $\mathbf{b} = \begin{pmatrix} 0 \\ -3 \end{pmatrix}$ $\mathbf{c} = \begin{pmatrix} 4 \\ -1 \end{pmatrix}$

Express each of the following vectors in terms of **a**, **b** and **c**:

a $\begin{pmatrix} -4 \\ 6 \end{pmatrix}$ b $\begin{pmatrix} 0 \\ 3 \end{pmatrix}$ c $\begin{pmatrix} 4 \\ -4 \end{pmatrix}$ d $\begin{pmatrix} 8 \\ -2 \end{pmatrix}$

Student assessment 1

1 Using the diagram, describe the following translations using column vectors.

a $\overrightarrow{AB}$ b  $\overrightarrow{DA}$ c  $\overrightarrow{CA}$

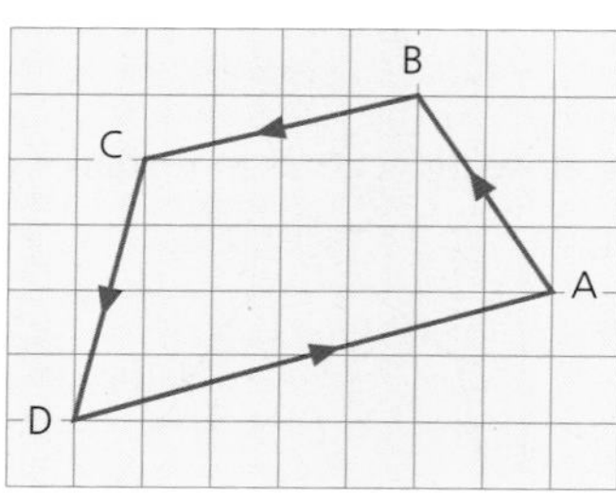

2 Describe each of the translations **a** to **e** shown in the diagram using column vectors.

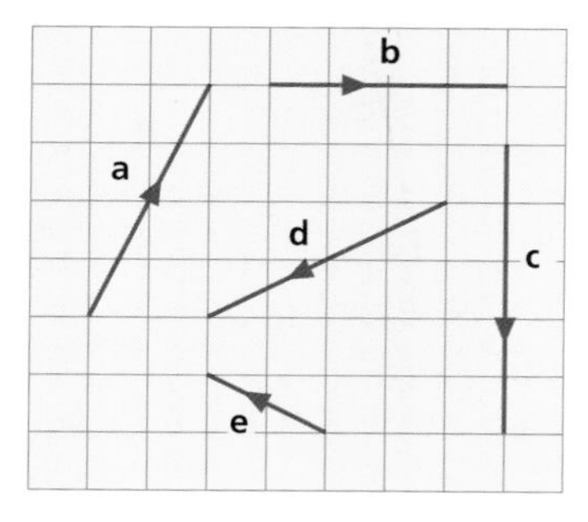

3 Using the vectors from Q.2, draw diagrams to represent:

a $\mathbf{a} + \mathbf{b}$ b $\mathbf{e} - \mathbf{d}$ c $\mathbf{c} - \mathbf{e}$ d $2\mathbf{e} + \mathbf{b}$

Exercise 27.4

For questions 1–6, the object (unshaded) and image (shaded) have been drawn. Copy each diagram then:

a mark the centre of rotation

b calculate the angle and direction of rotation.

1

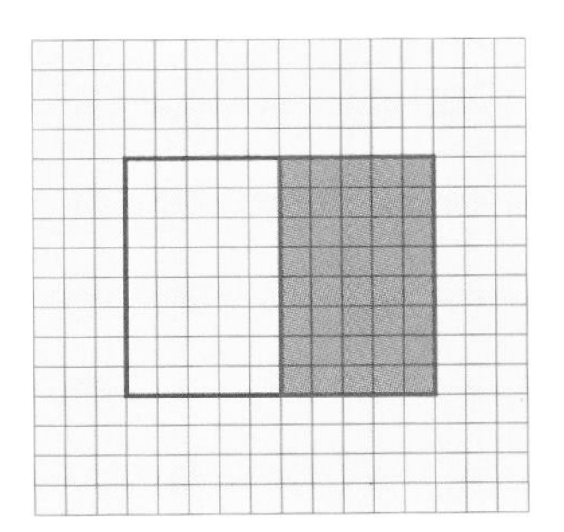

2

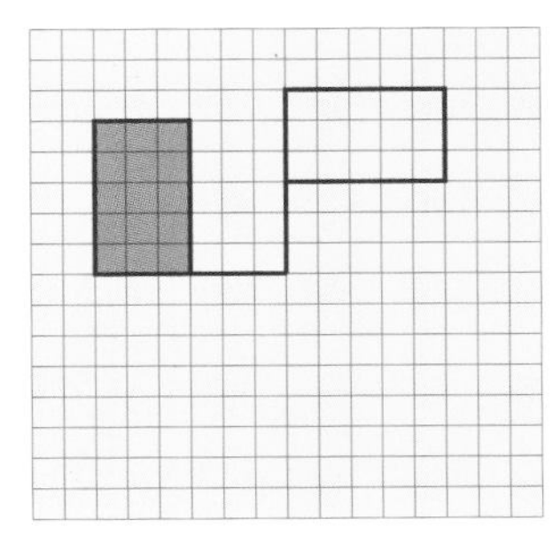

3

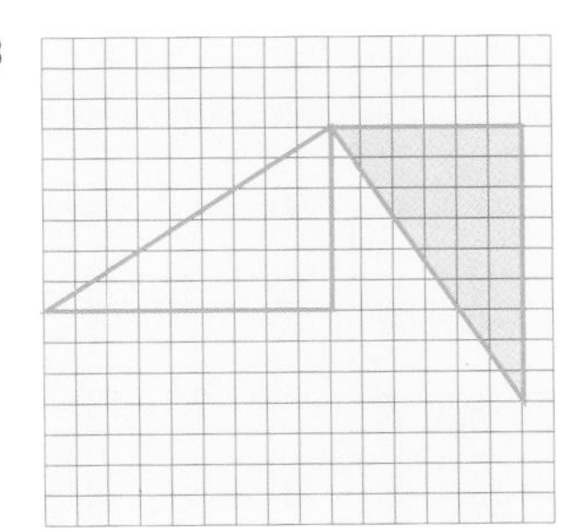

4

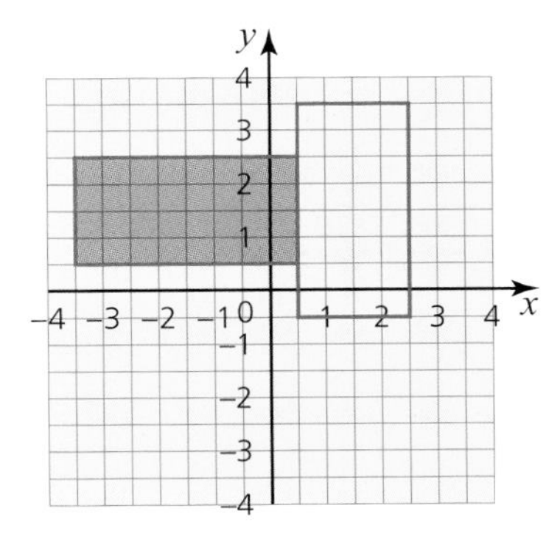

5

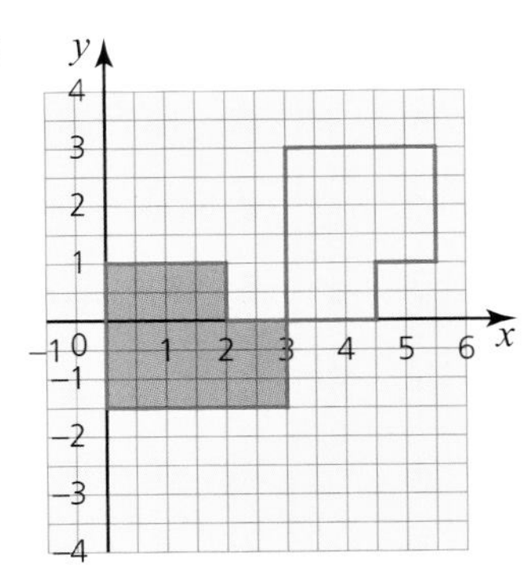

6

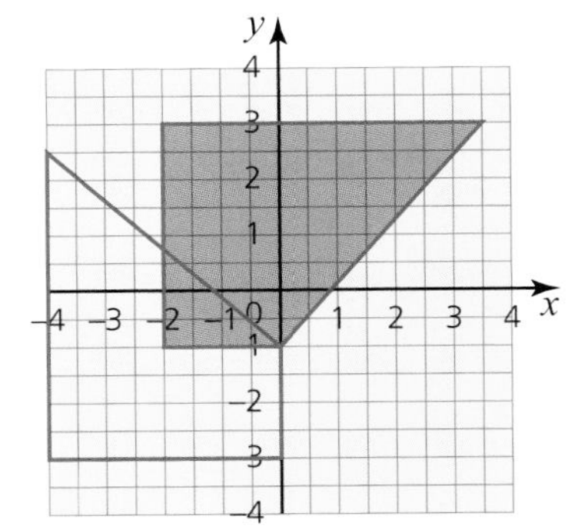

Translation

When an object is translated, it undergoes a 'straight sliding' movement. To describe a translation, it is necessary to give the **translation vector**. As no rotation is involved, each point on the object moves in the same way to its corresponding point on the image, e.g.

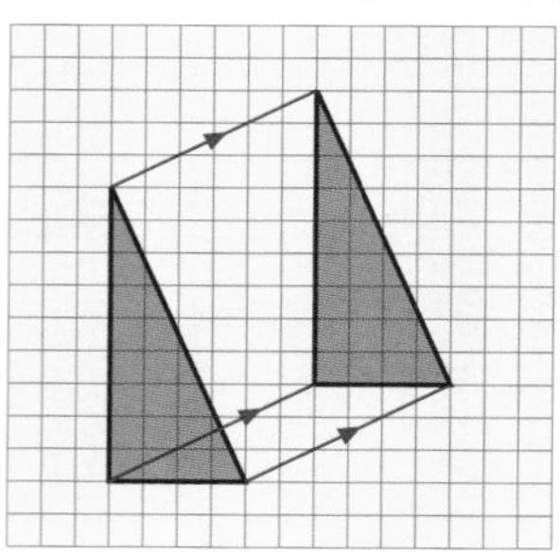

Vector $= \begin{pmatrix} 6 \\ 3 \end{pmatrix}$

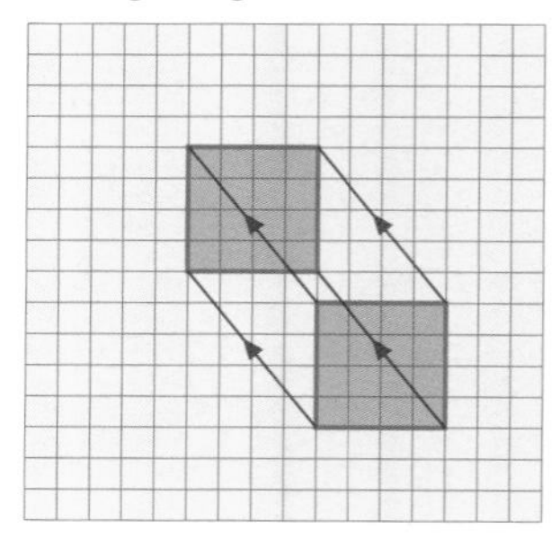

Vector $= \begin{pmatrix} -4 \\ 5 \end{pmatrix}$

Exercise 27.5

For questions 1–4, object A has been translated to each of images B and C. Give the translation vectors in each case.

1

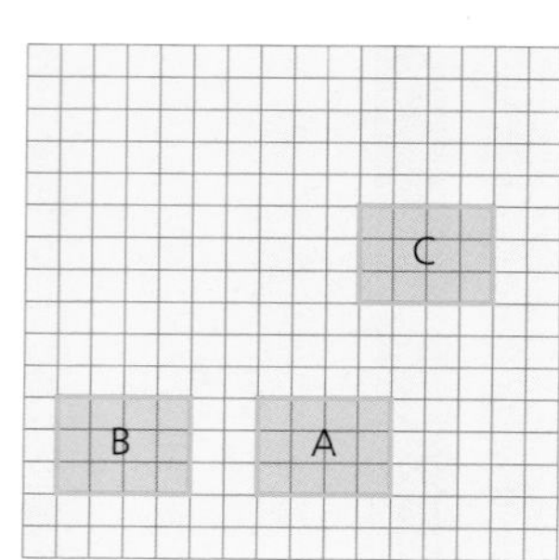

2

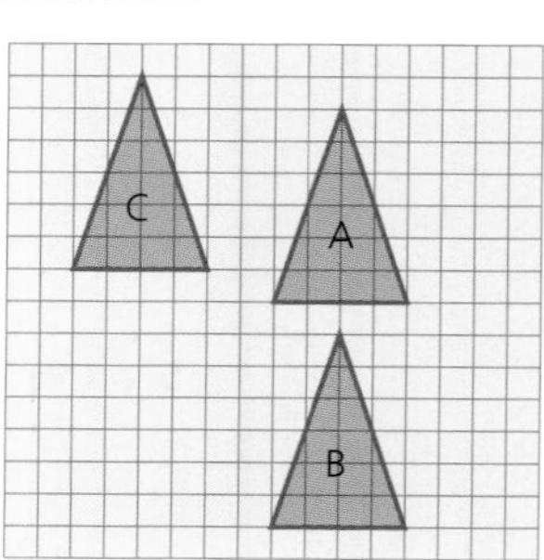

3

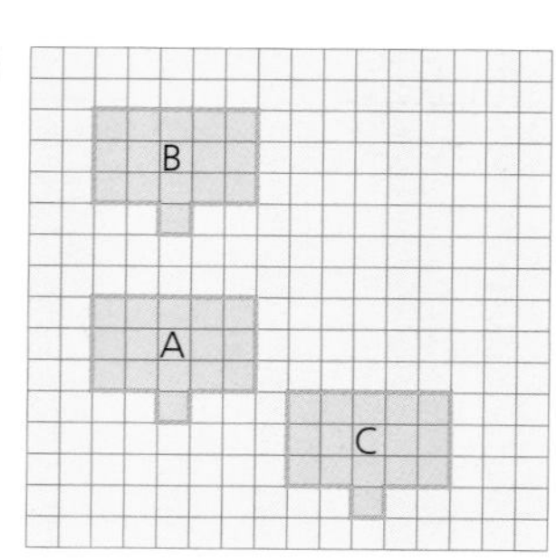

4

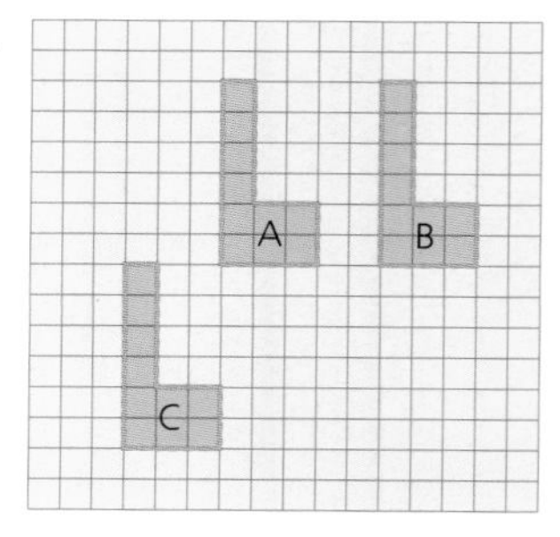

Exercise 27.6

Copy the diagrams for questions 1–6 and draw the object. Translate the object by the vector given in each case and draw the image in its new position. (Note: a bigger grid than the one shown may be needed.)

1

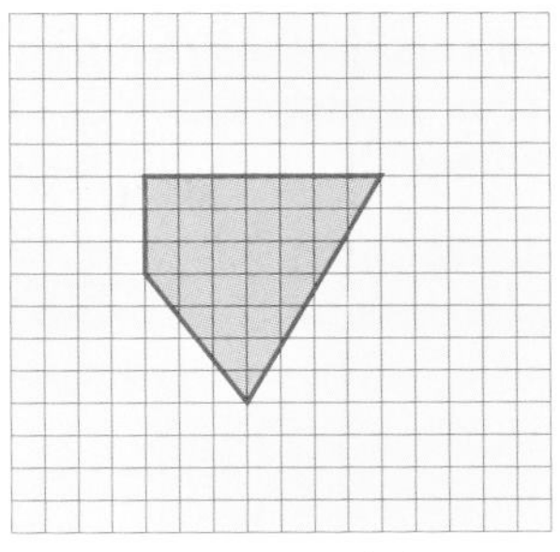

Vector = $\begin{pmatrix} 3 \\ 5 \end{pmatrix}$

2

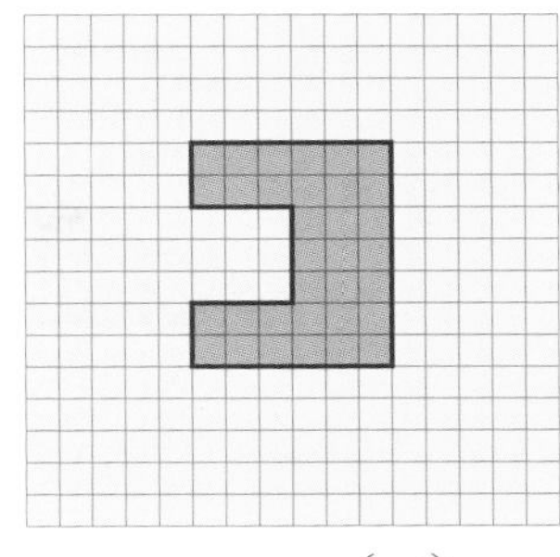

Vector = $\begin{pmatrix} 5 \\ -4 \end{pmatrix}$

3

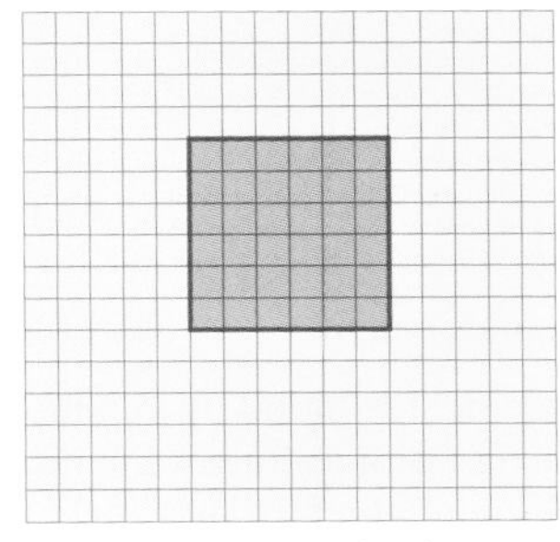

Vector = $\begin{pmatrix} -4 \\ 6 \end{pmatrix}$

4

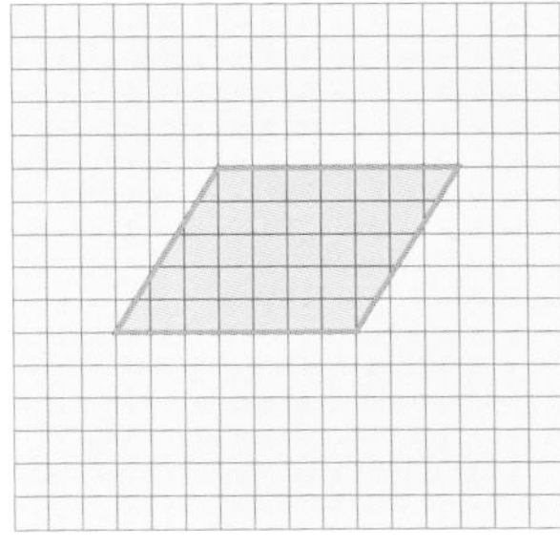

Vector = $\begin{pmatrix} -2 \\ -5 \end{pmatrix}$

5

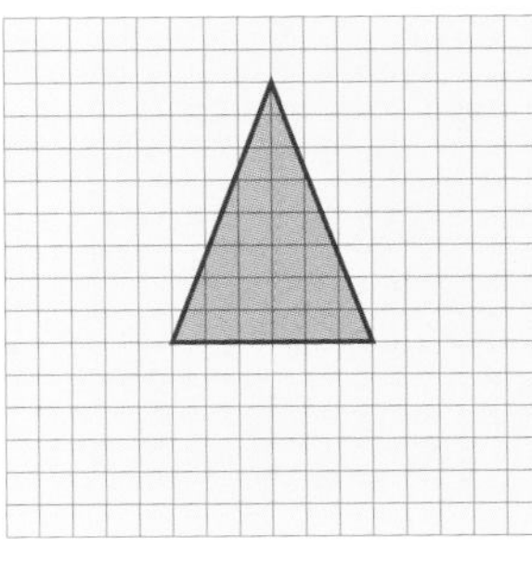

Vector = $\begin{pmatrix} -6 \\ 0 \end{pmatrix}$

6

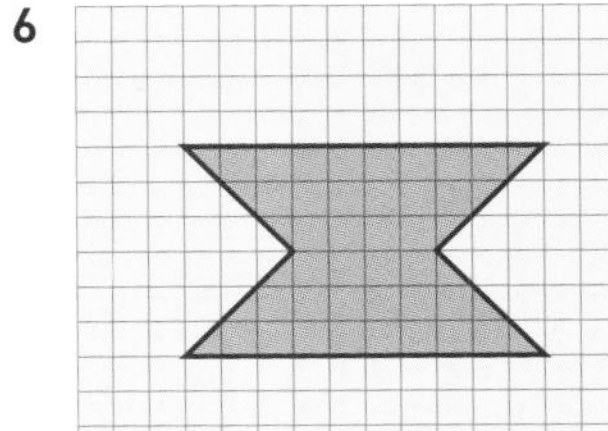

Vector = $\begin{pmatrix} 0 \\ -1 \end{pmatrix}$

Enlargement

When an object is enlarged, the result is an image which is mathematically similar to the object but is a different size. The image can be either larger or smaller than the original object. To describe an enlargement, two pieces of information need to be given: the position of the **centre of enlargement** and the **scale factor of enlargement**.

Worked examples

1 In the diagram, triangle ABC is enlarged to form triangle A'B'C'.

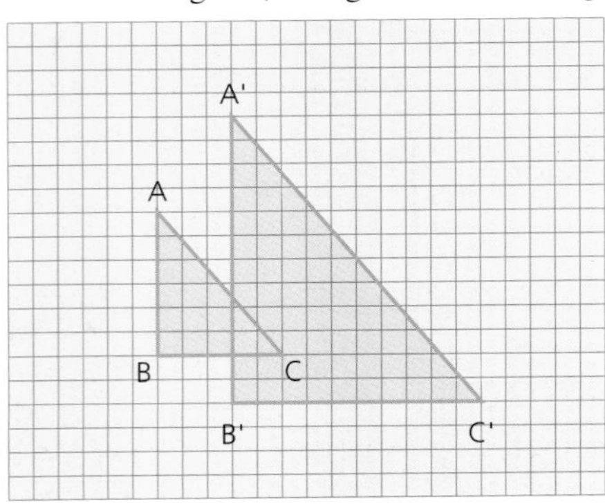

a Find the centre of enlargement.

The centre of enlargement is found by joining corresponding points on the object and on the image with a straight line. These lines are then extended until they meet. The point at which they meet is the centre of enlargement O.

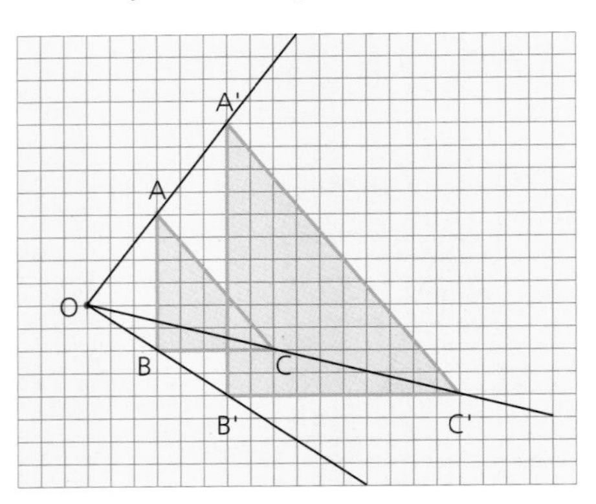

b Calculate the scale factor of enlargement.

The scale factor of enlargement can be calculated in two ways. From the previous diagram it can be seen that the distance OA' is twice the distance OA. Similarly, OC' and OB' are twice OC and OB respectively, so the scale factor of enlargement is 2.

Alternatively, the scale factor can be found by considering the ratio of the length of a side on the image to the length of the corresponding side on the object, i.e.

$$\frac{A'B'}{AB} = \frac{12}{6} = 2$$

So the scale factor of enlargement is 2.

2 In the diagram, rectangle ABCD undergoes a transformation to form rectangle A'B'C'D'.

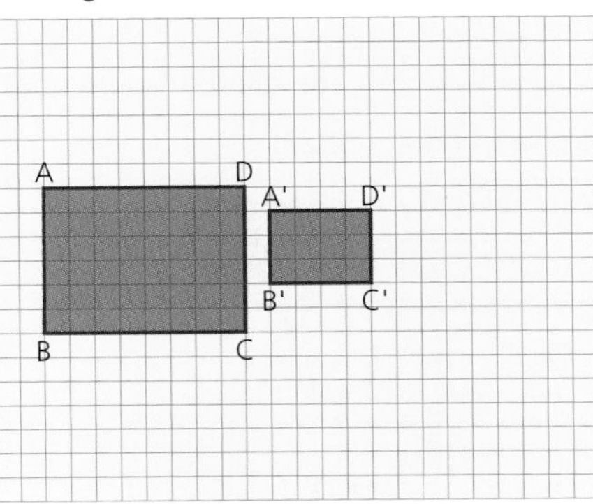

a Find the centre of enlargement.

By joining corresponding points on both the object and the image, the centre of enlargement is found at O:

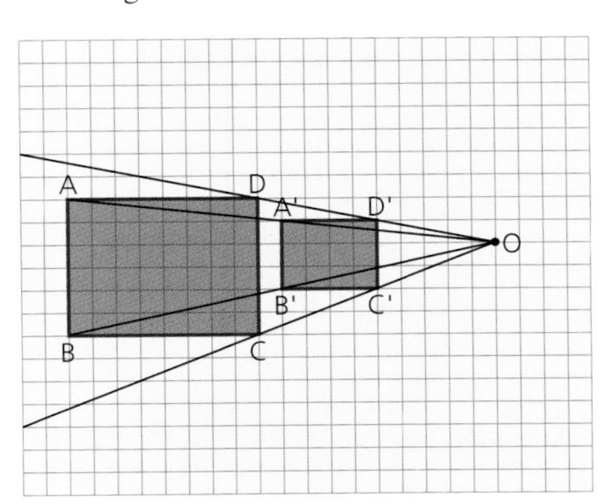

b Calculate the scale factor of enlargement.

The scale factor of enlargement $\frac{A'B'}{AB} = \frac{3}{6} = \frac{1}{2}$

Note

If the scale factor of enlargement is greater than 1, then the image is larger than the object. If the scale factor lies between 0 and 1, then the resulting image is smaller than the object. In these cases, although the image is smaller than the object, the transformation is still known as an enlargement.

Exercise 27.7 For questions 1–5, copy the diagrams and find:

a the centre of enlargement

b the scale factor of enlargement.

1

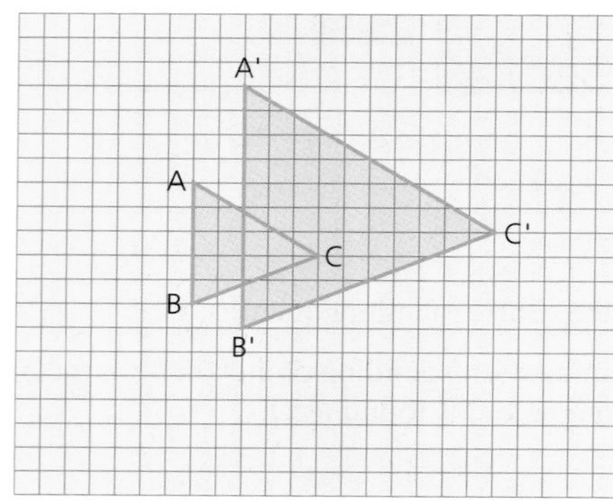

2

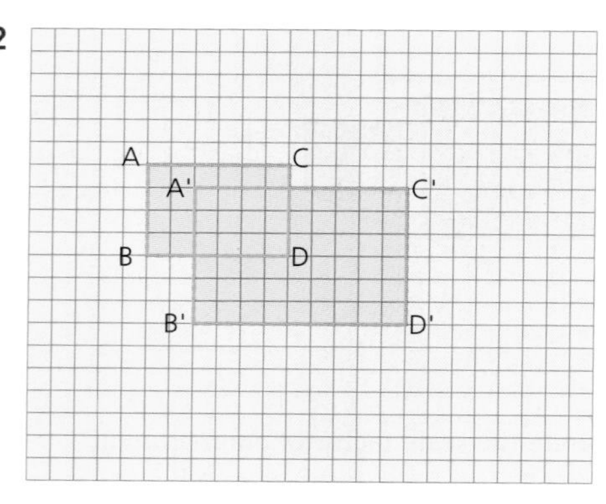

3

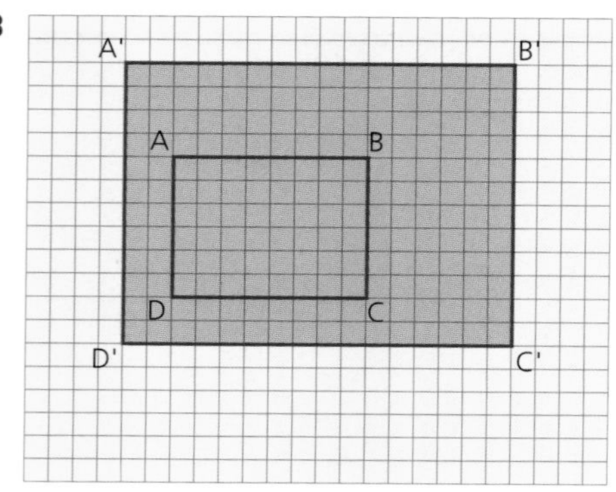

4

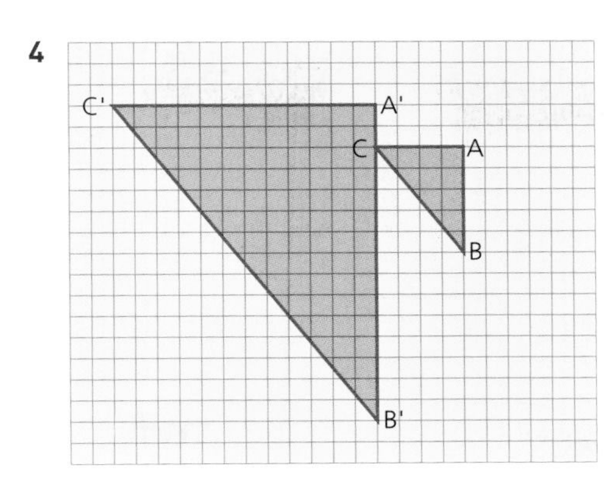

5

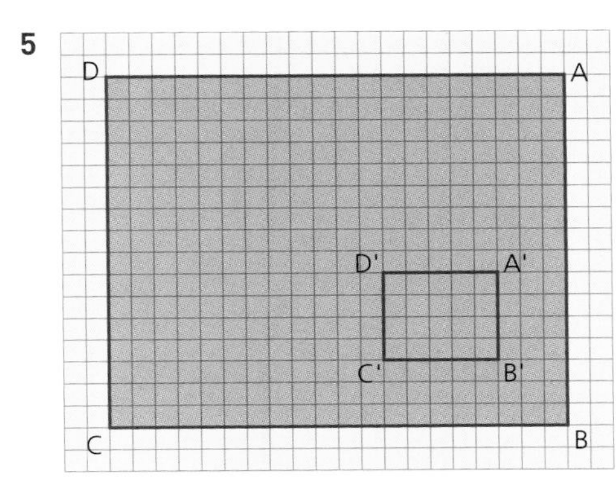

Exercise 27.8

For questions 1–4, copy the diagrams. Enlarge the objects by the scale factor given and from the centre of enlargement shown. (Note: grids larger than those shown may be needed.)

1

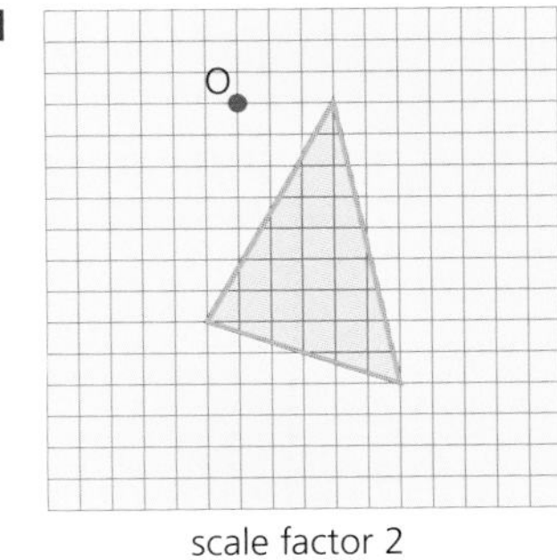

scale factor 2

2

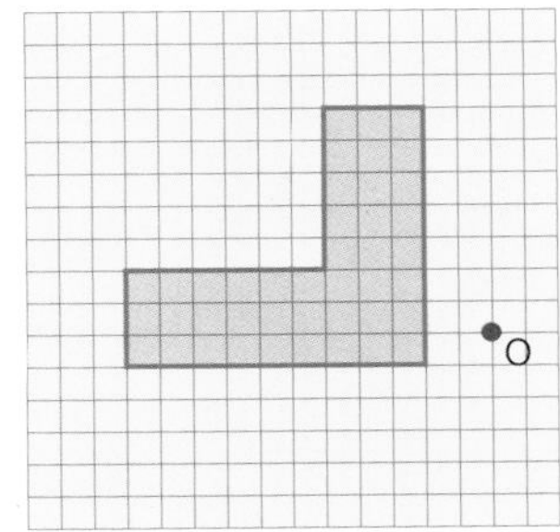

scale factor 2

Exercise 27.8 (cont)

3

O

scale factor 3

4

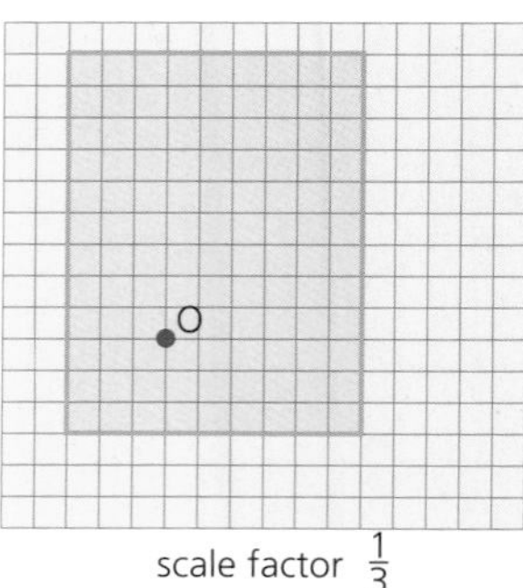

scale factor $\frac{1}{3}$

Student assessment 1

1 Copy the diagram and draw the reflection of the object in the mirror line.

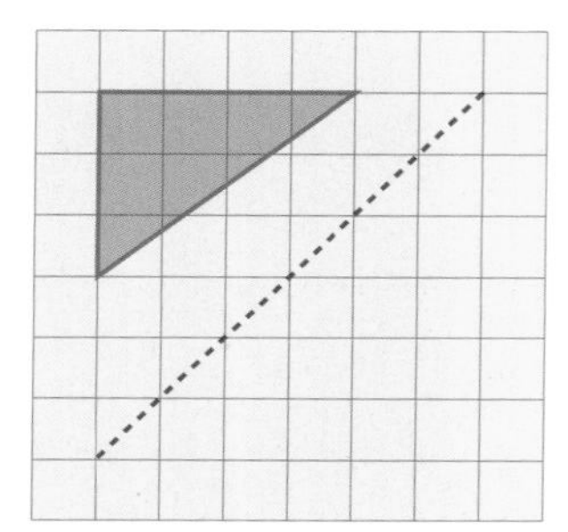

2 Copy the diagram and rotate the object 90° anticlockwise about the origin.

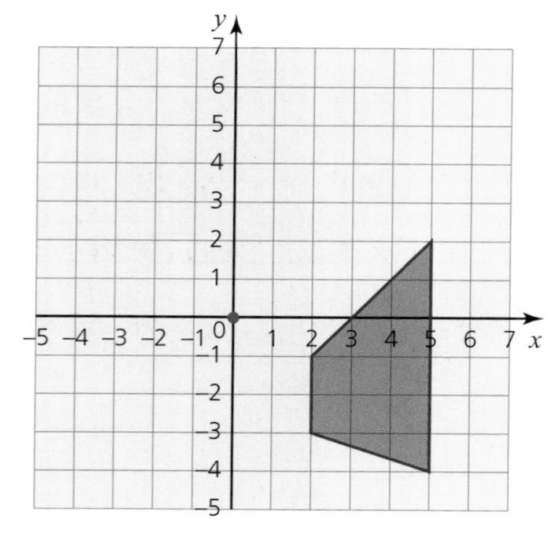

3 Write down the column vector of the translation which maps:

a rectangle A to rectangle B

b rectangle B to rectangle C.

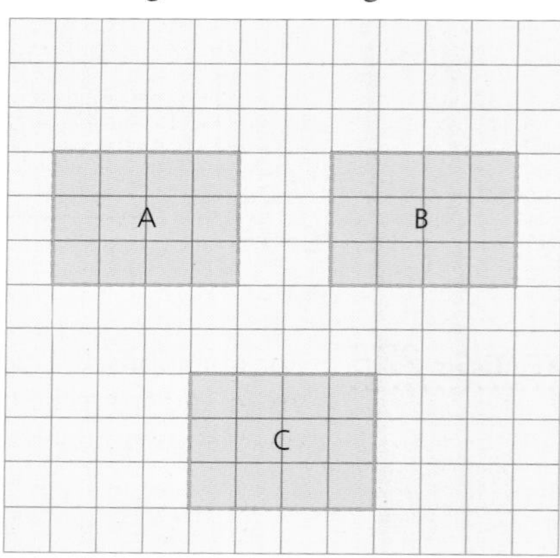

4 Enlarge the triangle by scale factor 2 and from the centre of enlargement O.

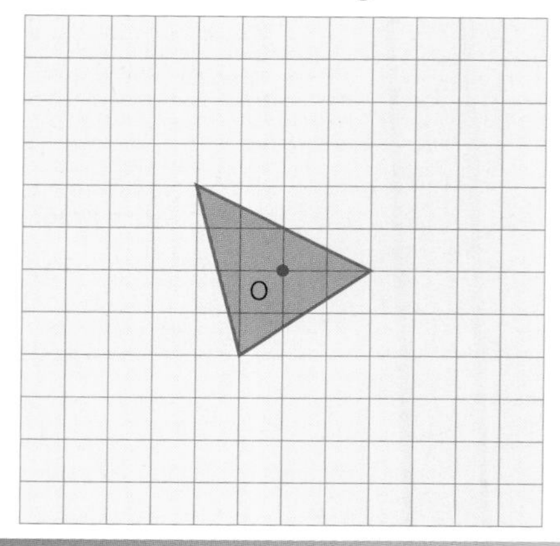

TOPIC 7

Mathematical investigations and ICT

A painted cube

A $3 \times 3 \times 3$ cm cube is painted on the outside, as shown in the left-hand diagram below:

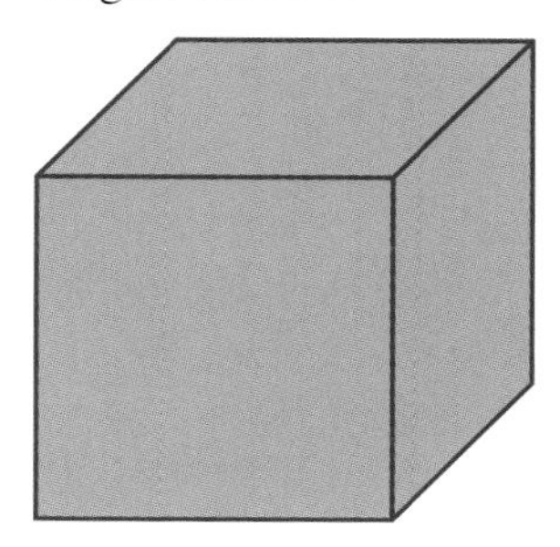

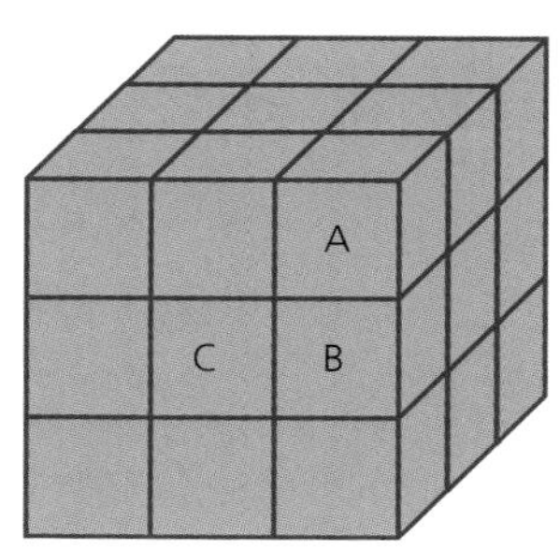

The large cube is then cut up into 27 smaller cubes, each 1 cm × 1 cm × 1 cm, as shown on the right.

$1 \times 1 \times 1$ cm cubes with 3 painted faces are labelled type A.

$1 \times 1 \times 1$ cm cubes with 2 painted faces are labelled type B.

$1 \times 1 \times 1$ cm cubes with 1 face painted are labelled type C.

$1 \times 1 \times 1$ cm cubes with no faces painted are labelled type D.

1 a How many of the 27 cubes are type A?
  b How many of the 27 cubes are type B?
  c How many of the 27 cubes are type C?
  d How many of the 27 cubes are type D?
2 Consider a $4 \times 4 \times 4$ cm cube cut into $1 \times 1 \times 1$ cm cubes. How many of the cubes are type A, B, C and D?
3 How many type A, B, C and D cubes are there when a $10 \times 10 \times 10$ cm cube is cut into $1 \times 1 \times 1$ cm cubes?
4 Generalise for the number of type A, B, C and D cubes in an $n \times n \times n$ cube.
5 Generalise for the number of type A, B, C and D cubes in a cuboid l cm long, w cm wide and h cm high.

Triangle count

The diagram below shows an isosceles triangle with a vertical line drawn from its apex to its base.

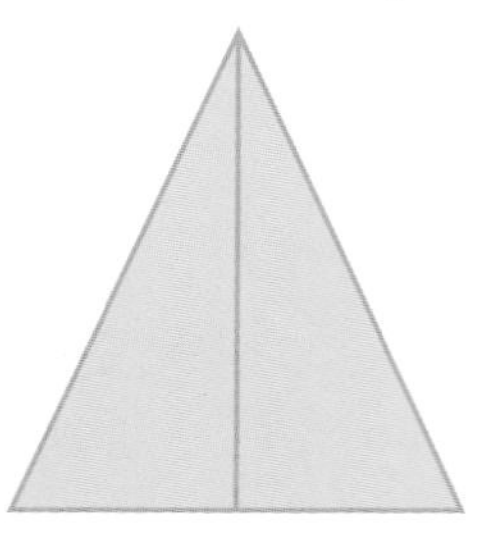

There is a total of 3 triangles in this diagram.

If a horizontal line is drawn across the triangle, it will look as shown:

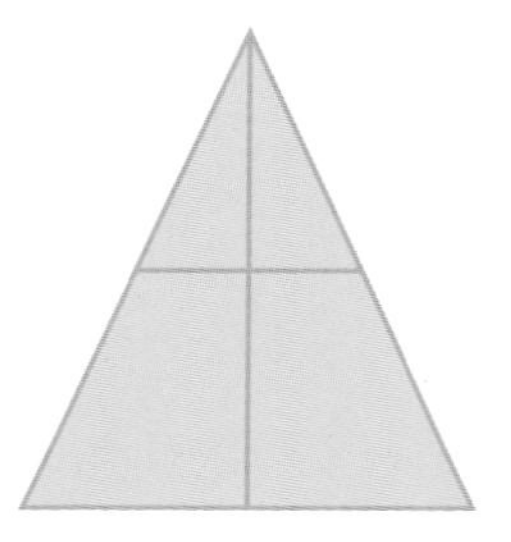

There is a total of 6 triangles in this diagram.

When one more horizontal line is added, the number of triangles increases further:

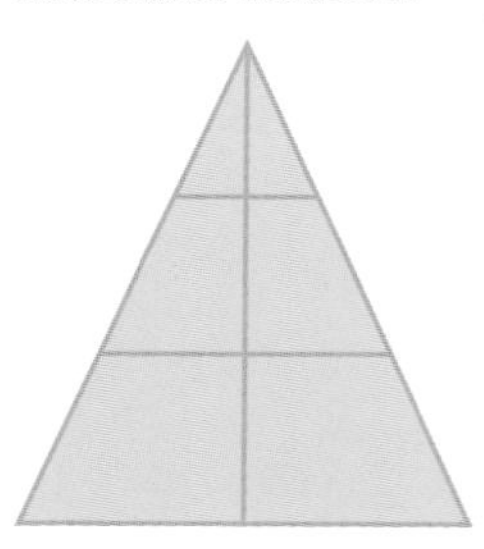

1. Calculate the total number of triangles in the diagram above with the two inner horizontal lines.
2. Investigate the relationship between the total number of triangles (t) and the number of inner horizontal lines (h). Enter your results in an ordered table.
3. Write an algebraic rule linking the total number of triangles and the number of inner horizontal lines.

ICT activity

In this activity, you will be using a geometry package to investigate enlargements.

1 Using a geometry package, draw an object and enlarge it by a scale factor of 2. An example is shown below:

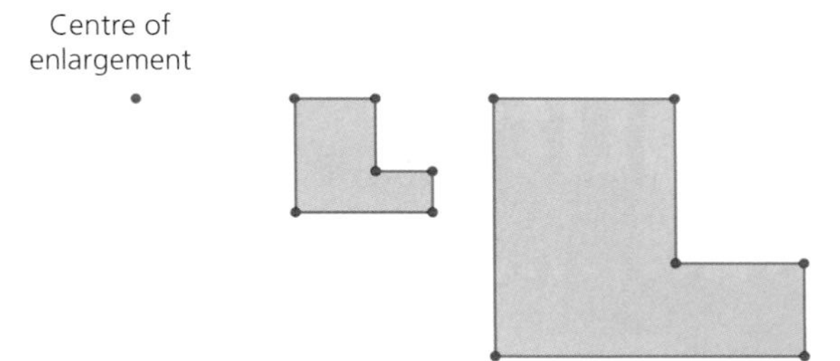

2 Describe the position of the centre of enlargement used to produce the following diagrams:

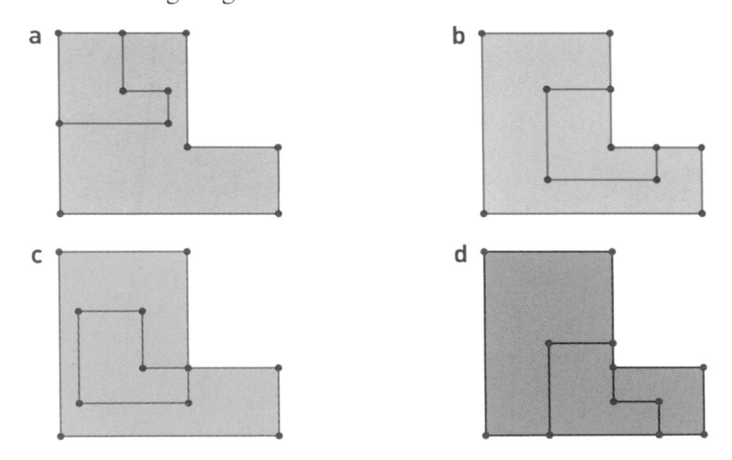

3 Move the position of the centre of enlargement to test your answers to Q.2 above.

TOPIC 8

Probability

Contents

He decides to do a simple experiment by spinning the coin lots of times. His results are shown in the table:

Number of trials	Number of heads	Relative frequency
100	40	0.4
200	90	0.45
300	142	
400	210	
500	260	
600	290	
700	345	
800	404	
900	451	
1000	499	

The relative frequency $= \frac{\text{number of successful trials}}{\text{total number of trials}}$

In the 'long run', that is after a large number of trials, did the coin appear to be fair?

Note

The greater the number of trials the better the estimated probability or relative frequency is likely to be. The key idea is that increasing the number of trials gives a better estimate of the probability and the closer the result obtained by experiment will be to that obtained by calculation.

Exercise 28.9

1. Copy and complete the table above. Draw a graph with Relative frequency as the y-axis and Number of trials as the x-axis. What do you notice?
2. Conduct a similar experiment using a dice to see if the number of sixes you get is the same as the theory of probability would make you expect.
3. Make a hexagonal spinner. Conduct an experiment to see if it is fair.
4. Ask a friend to put some coloured beads in a bag. Explain how you could use relative frequency in an experiment to find out what fraction of each colour are in the bag.

Worked examples

1 There is a group of 250 people in a hall. A girl calculates that the probability of randomly picking someone that she knows from the group is 0.032. Calculate the number of people in the group that the girl knows.

$$\text{Probability} = \frac{\text{number of favourable results } (F)}{\text{number of possible results}}$$

$$0.032 = \frac{F}{250}$$

$$250 \times 0.032 = F$$

$$8 = F$$

The girl knows 8 people in the group.

2 A boy enters 8 short stories into a writing competition. His father knows how many short stories have been entered into the competition, and tells his son that he has a probability of 0.016 of winning the first prize (assuming all the entries have an equal chance). How many short stories were entered into the competition?

$$\text{Probability} = \frac{\text{number of favourable results}}{\text{number of possible results } (T)}$$

$$0.016 = \frac{8}{T}$$

$$T = \frac{8}{0.016}$$

$$T = 500$$

So, 500 short stories were entered into the competition.

Exercise 28.10

1 A boy calculates that he has a probability of 0.004 of winning the first prize in a photography competition if the selection is made at random. If 500 photographs are entered into the competition, how many photographs did the boy enter?

2 The probability of getting any particular number on a spinner game is given as 0.04. How many numbers are there on the spinner?

3 A bag contains 7 red counters, 5 blue, 3 green and 1 yellow. If one counter is picked at random, what is the probability that it is:

a yellow
b red
c blue or green
d red, blue or green
e not blue?

4 A boy collects marbles. He has the following colours in a bag: 28 red, 14 blue, 25 yellow, 17 green and 6 purple. If he picks one marble from the bag at random, what is the probability that it is:

a red
b blue
c yellow or blue
d purple
e not purple?

5 The probability of a boy randomly picking a marble of one of the following colours from another bag of marbles is:

blue 0.25 red 0.2 yellow 0.15 green 0.35 white 0.05

If there are 49 green marbles, how many of each other colour does he have in his bag?

6 There are six red sweets in a bag. If the probability of randomly picking a red sweet is 0.02, calculate the number of sweets in the bag.

7 The probability of getting a bad egg in a batch of 400 is 0.035. How many bad eggs are there likely to be in a batch?

8 A sports arena has 25 000 seats, some of which are VIP seats. For a charity event all the seats are allocated randomly. The probability of getting a VIP seat is 0.008. How many VIP seats are there?

9 The probability of Juan's favourite football team winning 4–0 is 0.05. How many times are they likely to win by this score in a season of 40 matches?

Student assessment 1

1 An octagonal spinner has the numbers 1 to 8 on it as shown:

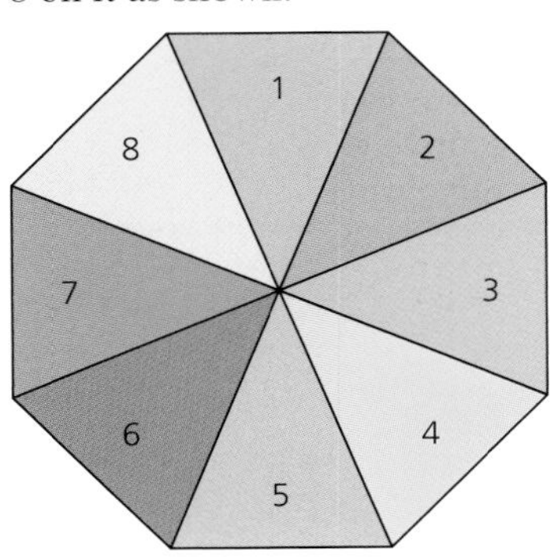

What is the probability of spinning:

a a 7
b not a 7
c a factor of 12
d a 9?

2 A game requires the use of all the playing cards in a normal pack from 6 to King inclusive.

a How many cards are used in the game?
b What is the probability of randomly picking:
   i a 6
   ii a picture card
   iii a club
   iv a prime number
   v an 8 or a spade?

3 180 students in a school are offered the chance to attend a football match for free. If the students are chosen at random, what is the chance of being picked to go if the following numbers of tickets are available?

a 1
b 9
c 15
d 40
e 180

4 A bag contains 11 white, 9 blue, 7 green and 5 red counters. What is the probability that a single counter drawn will be:

a blue
b red or green
c not white?

5 The probability of randomly picking a red, blue or green marble from a bag containing 320 marbles is:

red 0.4 blue 0.25 green 0.35

If there are no other colours in the bag, how many marbles of each colour are there?

6 Students in a class conducted a survey to see how many friends they have on a social media site. The results were grouped and are shown in the pie chart below.

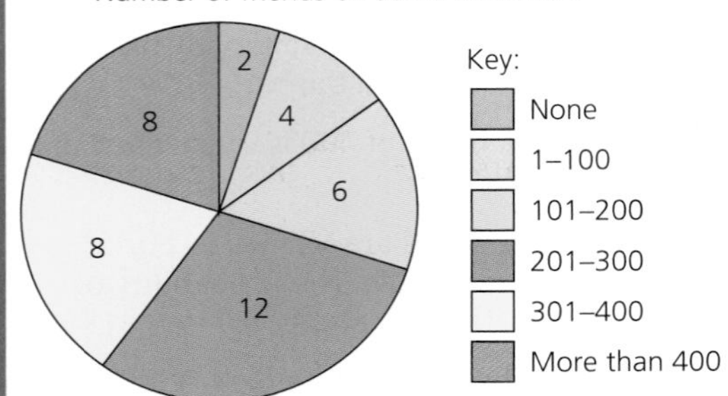

A student is chosen at random. What is the probability that he/she:

a has 101–200 friends on the site
b has friends on the site
c has more than 200 friends on the site?

7 a If I enter a competition and have a 0.00002 probability of winning, how many people entered the competition?
b What assumption do you have to make in order to answer part a?

8 A large bag contains coloured discs. The discs are either completely red (R), completely yellow (Y) or half red and half yellow. The Venn diagram below shows the probability of picking each type of disc.

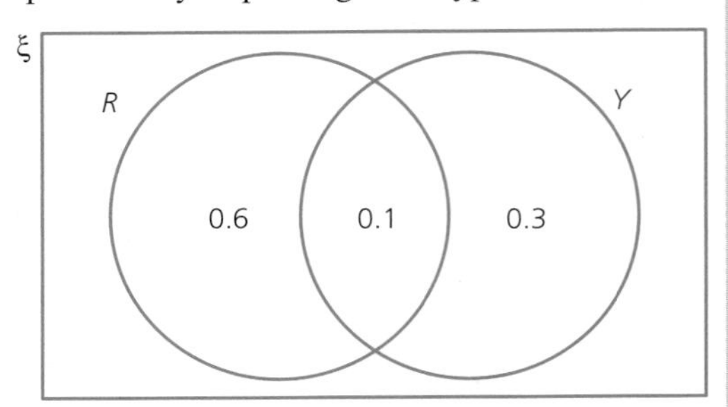

If there are 120 discs coloured completely yellow, calculate:

a the number of discs coloured completely red
b the total number of discs in the bag.

9 A cricket team has a 0.25 chance of losing a game. Calculate, using a tree diagram if necessary, the probability of the team achieving:

a two consecutive wins
b three consecutive wins
c ten consecutive wins.

TOPIC 8

Mathematical investigations and ICT

Probability drop

A game involves dropping a red marble down a chute. On hitting a triangle divider, the marble can bounce either left or right. On completing the drop, the marble lands in one of the trays along the bottom. The trays are numbered from left to right. Different sizes of game exist; the four smallest versions are shown below:

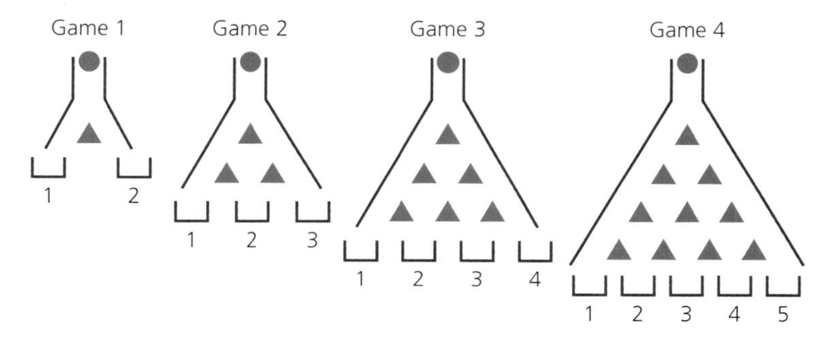

To land in tray 2 in the second game above, the marble can travel in one of two ways. These are: Left – Right or Right – Left. This can be abbreviated to LR or RL.

1 State the different routes the marble can take to land in each of the trays in the third game.
2 State the different routes the marble can take to land in each of the trays in the fourth game.
3 State, giving reasons, the probability of a marble landing in tray 1 in the fourth game.
4 State, giving reasons, the probability of a marble landing in each of the other trays in the fourth game.
5 Investigate the probability of the marble landing in each of the different trays in larger games.
6 Using your findings from your investigation, predict the probability of a marble landing in tray 7 in the tenth game (11 trays at the bottom).

The following question is beyond the scope of the course but is an interesting extension.

7 Investigate the links between this game and the sequence of numbers generated in Pascal's triangle.

Dice sum

Two ordinary dice are rolled and their scores added together. Below is an incomplete table showing the possible outcomes:

		Dice 1					
		1	**2**	**3**	**4**	**5**	**6**
Dice 2	**1**	2			5		
	2						
	3				7		
	4				8		
	5				9	10	11
	6						12

1 Copy and complete the table to show all possible outcomes.
2 How many possible outcomes are there?
3 What is the most likely total when two dice are rolled?
4 What is the probability of getting a total score of 4?
5 What is the probability of getting the most likely total?
6 How many times more likely is a total score of 5 compared with a total score of 2?

Now consider rolling two four-sided dice, each numbered 1–4. Their scores are also added together.

7 Draw a table to show all the possible outcomes when the two four-sided dice are rolled and their scores added together.
8 How many possible outcomes are there?
9 What is the most likely total?
10 What is the probability of getting the most likely total?
11 Investigate the number of possible outcomes, the most likely total and its probability when two identical dice are rolled together and their scores added, i.e. consider 8-sided dice, 10-sided dice, etc.
12 Consider two m-sided dice rolled together and their scores added.
  a What is the total number of outcomes, in terms of m?
  b What is the most likely total, in terms of m?
  c What, in terms of m, is the probability of the most likely total?
13 Consider an m-sided and n-sided dice rolled together, where $m > n$.
  a In terms of m and n, deduce the total number of outcomes.
  b In terms of m and/or n, deduce the most likely total(s).
  c In terms of m and/or n, deduce the probability of getting a specific total out of the most likely total(s).

ICT activity

For this activity, you will be testing the fairness of a spinner that you have constructed.

1 Using card, a pair of compasses and a ruler, construct a regular hexagon.
2 Divide your regular hexagon into six equal parts.
3 Colour the six parts using three different colours, as shown below:

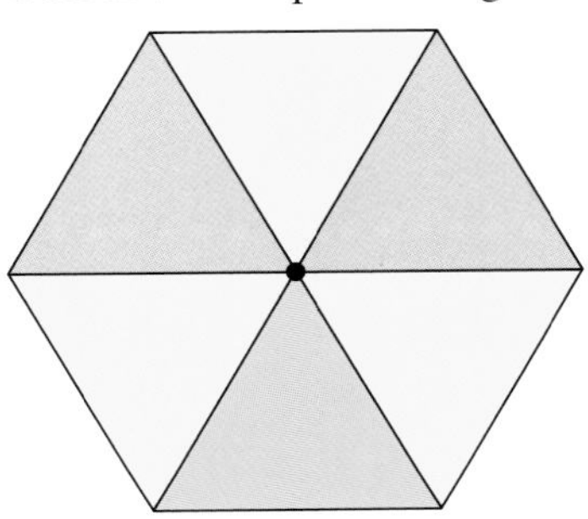

4 Calculate the theoretical probability of each colour. Record these probabilities as percentages.
5 Carefully insert a small pencil through the centre of the hexagon to form a spinner.
6 Spin the spinner 60 times, recording your results in a spreadsheet.
7 Using the spreadsheet, produce a percentage pie chart of your results.
8 Compare the actual probabilities with the theoretical ones calculated in Q.4. What conclusions can you make about the fairness of your spinner?

TOPIC 9

Statistics

Contents

Course

C9.1
Collect, classify and tabulate statistical data.

C9.2
Read, interpret and draw simple inferences from tables and statistical diagrams.

Compare sets of data using tables, graphs and statistical measures.

Appreciate restrictions on drawing conclusions from given data.

C9.3
Construct and interpret bar charts, pie charts, pictograms, stem and leaf diagrams, simple frequency distributions, histograms with equal intervals and scatter diagrams.

C9.4
Calculate the mean, median, mode and range for individual and discrete data and distinguish between the purposes for which they are used.

C9.5
Extended curriculum only.

C9.6
Extended curriculum only.

C9.7
Understand what is meant by positive, negative and zero correlation with reference to a scatter diagram.

C9.8
Draw, interpret and use lines of best fit by eye.

The development of statistics

The earliest writing on statistics was found in a 9th-century book entitled *Manuscript on Deciphering Cryptographic Messages*, written by Arab philosopher al-Kindi (801–873), who lived in Baghdad. In his book, he gave a detailed description of how to use statistics to unlock coded messages.

The *Nuova Cronica*, a 14th-century history of Florence by Italian banker Giovanni Villani, includes much statistical information on population, commerce, trade and education.

Early statistics served the needs of states – *state -istics*. By the early 19th century, statistics included the collection and analysis of data in general. Today, statistics are widely used in government, business, and natural and social sciences. The use of modern computers has enabled large-scale statistical computation and has also made possible new methods that are impractical to perform manually.

29 Mean, median, mode and range

Average

'**Average**' is a word which, in general use, is taken to mean somewhere in the middle. For example, a woman may describe herself as being of average height. A student may think he or she is of average ability in maths. Mathematics is more exact and uses three principal methods to measure average.

Key points

The **mode** is the value occurring the most often.

The **median** is the middle value when all the data is arranged in order of size.

The **mean** is found by adding together all the values of the data and then dividing that total by the number of data values.

Spread

It is often useful to know how spread out the data is. It is possible for two sets of data to have the same mean and median but very different spreads.

The simplest measure of spread is the **range**. The range is simply the difference between the largest and smallest values in the data.

Worked examples

1 a Find the mean, median and mode of the data given:
1, 0, 2, 4, 1, 2, 1, 1, 2, 5, 5, 0, 1, 2, 3

$$\text{Mean} = \frac{1+0+2+4+1+2+1+1+2+5+5+0+1+2+3}{15}$$

$$= \frac{30}{15}$$

$$= 2$$

Arranging all the data in order and then picking out the middle number gives the median:

0, 0, 1, 1, 1, 1, 1, (2), 2, 2, 2, 3, 4, 5, 5

The mode is the number which appeared most often.
Therefore the mode is 1.

b Calculate the range of the data.
Largest value = 5
Smallest value = 0
Therefore the range $= 5 - 0$
$= 5$

2 a The frequency chart shows the score out of 10 achieved by a class in a maths test:

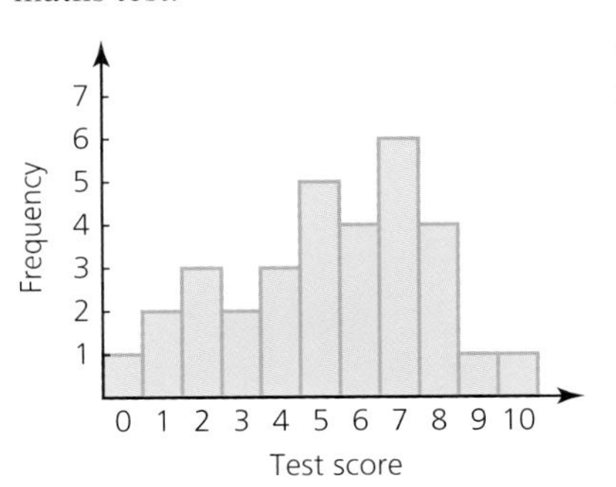

Calculate the mean, median and mode for this data.

Transferring the results to a frequency table gives:

Test score	0	1	2	3	4	5	6	7	8	9	10	Total
Frequency	1	2	3	2	3	5	4	6	4	1	1	32
Frequency × score	0	2	6	6	12	25	24	42	32	9	10	168

From the total column, we can see the number of students taking the test is 32 and the total number of marks obtained by all the students is 168.

Therefore, the mean score $= \frac{168}{32} = 5.25$

Arranging all the scores in order gives:

0, 1, 1, 2, 2, 2, 3, 3, 4, 4, 4, 5, 5, 5, 5, (5, 6), 6, 6, 6, 7, 7, 7, 7, 7, 7, 8, 8, 8, 8, 9, 10

Because there is an even number of students there isn't one middle number. There is a middle pair.

$$\text{Median} = \frac{(5+6)}{2}$$
$$= 5.5$$

The mode is 7 as it is the score which occurs most often.

b Calculate the range of the data.
Largest value = 10
Smallest value = 0
Therefore the range = 10 − 0
= 10

Exercise 29.1

1 Calculate the mean and range of each set of numbers:

a 6 7 8 10 11 12 13
b 4 4 6 6 6 7 8 10
c 36 38 40 43 47 50 55
d 7 6 8 9 5 4 10 11 12
e 12 24 36 48 60
f 17.5 16.3 18.6 19.1 24.5 27.8

Exercise 29.1 (cont)

2 Find the median and range of each set of numbers:
 a 3 4 5 6 7
 b 7 8 8 9 10 12 15
 c 8 8 8 9 9 10 10 10 10
 d 6 4 7 3 8 9 9 4 5
 e 2 4 6 8
 f 7 8 8 9 10 11 12 14
 g 3.2 7.5 8.4 9.3 5.4 4.1 5.2 6.3
 h 18 32 63 16 97 46 83

3 Find the mode and range of each set of numbers:
 a 6 7 8 8 9 10 11
 b 3 4 4 5 5 6 6 6 7 8 8
 c 3 5 3 4 6 3 3 5 4 6 8
 d 4 3 4 5 3 4 5 4
 e 60 65 70 75 80 75
 f 8 7 6 5 8 7 6 5 8

Exercise 29.2

In Q.1–5, find the mean, median, mode and range for each set of data.

1 A hockey team plays 15 matches. The number of goals scored in each match was:
 1, 0, 2, 4, 0, 1, 1, 1, 2, 5, 3, 0, 1, 2, 2

2 The total score when two dice were thrown 20 times was:
 7, 4, 5, 7, 3, 2, 8, 6, 8, 7, 6, 5, 11, 9, 7, 3, 8, 7, 6, 5

3 The ages of girls in a group are:
 14 years 3 months, 14 years 5 months,
 13 years 11 months, 14 years 3 months,
 14 years 7 months, 14 years 3 months,
 14 years 1 month

4 The number of students present in class over a three-week period was:
 28, 24, 25, 28, 23, 28, 27, 26, 27, 25, 28, 28, 28, 26, 25

5 An athlete keeps a record in seconds of her training times for the 100 m race:
 14.0, 14.3, 14.1, 14.3, 14.2, 14.0, 13.9, 13.8, 13.9, 13.8, 13.8, 13.7, 13.8, 13.8, 13.8

6 The mean mass of 11 players in a football team is 80.3 kg. The mean mass of the team plus a substitute is 81.2 kg. Calculate the mass of the substitute.

7 After eight matches, a basketball player had scored a mean of 27 points. After three more matches his mean was 29. Calculate the total number of points he scored in the last three games.

Exercise 29.3

1 An ordinary dice was rolled 60 times. The results are shown in the table:

Score	1	2	3	4	5	6
Frequency	12	11	8	12	7	10

Calculate the mean, median, mode and range of the scores.

2 Two dice were thrown 100 times. Each time, their combined score was recorded and the results put into the table:

Score	2	3	4	5	6	7	8	9	10	11	12
Frequency	5	6	7	9	14	16	13	11	9	7	3

Calculate the mean score.

3 Sixty flowering bushes were planted. At their flowering peak, the number of flowers per bush was counted and recorded. The results are shown in the table:

Flowers per bush	0	1	2	3	4	5	6	7	8
Frequency	0	0	0	6	4	6	10	16	18

a Calculate the mean, median, mode and range of the number of flowers per bush.
b Which of the mean, median and mode would be most useful when advertising the bush to potential buyers?

Student assessment 1

1 Find the mean, median, mode and range of each of the following sets of numbers:
a 63 72 72 84 86
b 6 6 6 12 18 24
c 5 10 5 15 5 20 5 25 15 10

2 The mean mass of 15 players in a rugby team is 85 kg. The mean mass of the team plus a substitute is 83.5 kg. Calculate the mass of the substitute.

3 An ordinary dice was rolled 50 times. The results are shown in the table:

Score	1	2	3	4	5	6
Frequency	8	11	5	9	7	10

Calculate the mean, median and mode of the scores.

4 The bar chart shows the marks out of 10 for an English test taken by a class of students.

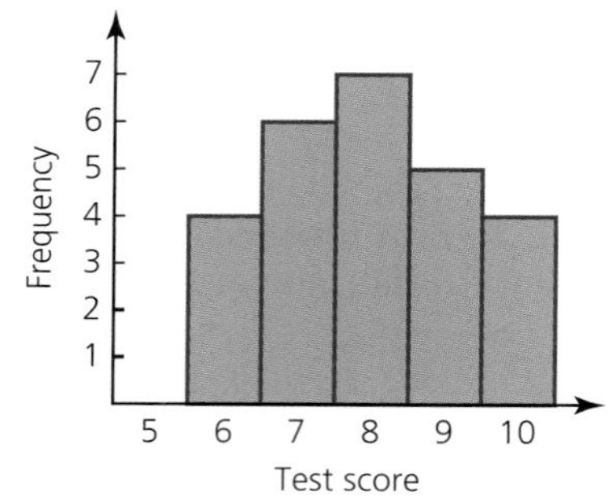

a Calculate the number of students who took the test.

b Calculate for the class:

i the mean test result

ii the median test result

ii the modal test result.

5 A javelin thrower keeps a record of her best throws over 10 competitions. These are shown in the table:

Competition	1	2	3	4	5	6	7	8	9	10
Distance (m)	77	75	78	86	92	93	93	93	92	89

Find the mean, median, mode and range of her throws.

30 Collecting, displaying and interpreting data

Tally charts and frequency tables

The figures in the list below are the numbers of chocolate buttons in each of 20 packets of buttons:

35 36 38 37 35 36 38 36 37 35

36 36 38 36 35 38 37 38 36 38

The figures can be shown on a tally chart:

Number	Tally	Frequency
35	||||	4
36	~~||||~~ ||	7
37	|||	3
38	~~||||~~ |	6

When the tallies are added up to get the frequency, the chart is usually called a **frequency table**. The information can then be displayed in a variety of ways.

Pictograms

● = 4 packets, ◕ = 3 packets, ◑ = 2 packets, ◔ = 1 packet

Buttons per packet	Frequency
35	●
36	● ◕
37	◕
38	● ◑

Bar charts

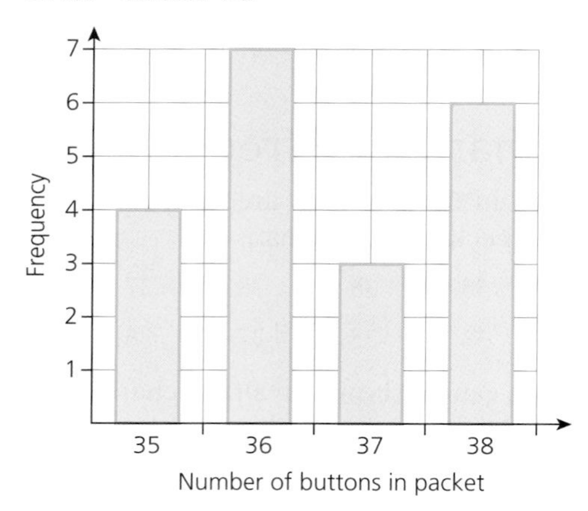

The height of each bar represents the frequency. Therefore, the width of each bar must be the same. To avoid producing a misleading graph, the frequency axis should always start at zero.

Stem and leaf diagrams

Discrete data is data that has a specific, fixed value. A stem and leaf diagram can be used to display discrete data in a clear and organised way. It has an advantage over bar charts as the original data can easily be recovered from the diagram.

The ages of people on a coach transferring them from an airport to a ski resort are as follows:

22	24	25	31	33	23	24	26	37	42
40	36	33	24	25	18	20	27	25	33
28	33	35	39	40	48	27	25	24	29

Displaying the data on a stem and leaf diagram produces the following graph:

Stem	Leaf
1	8
2	0 2 3 4 4 4 4 5 5 5 5 6 7 7 8 9
3	1 3 3 3 3 5 6 7 9
4	0 0 2 8

Key
2 | 5 means 25

A key is important so that the numbers can be interpreted correctly.

In this form the data can be analysed quite quickly.

- The youngest person is 18.
- The oldest is 48.
- The modal ages are 24, 25 and 33.

As the data is arranged in order, the median age can also be calculated quickly. The middle people out of 30 will be the 15th and 16th people. In this case the 15th person is 27 years old and the 16th person is 28 years old. Therefore, the median age is 27.5.

Back-to-back stem and leaf diagrams

Stem and leaf diagrams are often used as an easy way to compare two sets of data. The leaves are usually put 'back-to-back' on either side of the stem.

Continuing from the example given above, consider a second coach from the airport taking people to a golfing holiday. The ages of these people are shown below:

43	46	52	61	65	38	36	28	37	45
69	72	63	55	46	34	35	37	43	48
54	53	47	36	58	63	70	55	63	64

The two sets of data displayed on a back-to-back stem and leaf diagram are shown below:

Golf		Skiing
	1	8
8	2	0 2 3 4 4 4 4 5 5 5 5 6 7 7 8 9
8 7 7 6 6 5 4	3	1 3 3 3 3 5 6 7 9
8 7 6 6 5 3 3	4	0 0 2 8
8 5 5 4 3 2	5	
9 5 4 3 3 3 1	6	
2 0	7	

Key: 5 |3| 6 means 35 to the left and 36 to the right

From the back-to-back diagram it is easier to compare the two sets of data. This data shows that the people on the bus going to the golf resort tend to be older than those on the bus going to the ski resort.

Grouped frequency tables

If there is a big range in the data it is easier to group the data in a **grouped frequency table**.

The groups are arranged so that no score can appear in two groups.

The scores for the first round of a golf competition are:

75	71	75	82	96	83	75	76	82	103	85	79	77	83	85
72	88	104	76	77	79	83	84	86	88	102	95	96	99	102

This data can be grouped as shown:

Score	Frequency
71–75	5
76–80	6
81–85	8
86–90	3
91–95	1
96–100	3
101–105	4
Total	30

Note: it is not possible to score 70.5 or 77.3 at golf. The scores are said to be **discrete**. If the data is **continuous**, for example when measuring time, the intervals can be shown as 0–, 10–, 20–, 30– and so on.

Pie charts

Data can be displayed on a **pie chart** – a circle divided into sectors. The size of the sector is in direct proportion to the frequency of the data.

Worked examples

1 In a survey, 240 English children were asked to vote for their favourite holiday destination. The results are shown on the pie chart.

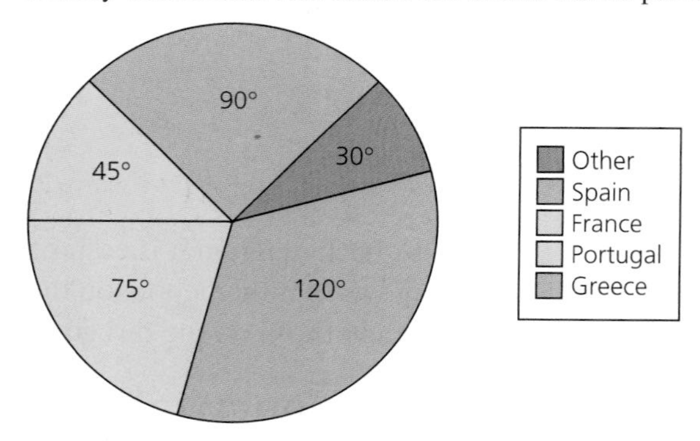

Calculate the actual number of votes for each destination.

The total 240 votes are represented by 360°.

It follows that if 360° represents 240 votes:

There were $240 \times \frac{120}{360}$ votes for Spain

so, 80 votes for Spain.

There were $240 \times \frac{75}{360}$ votes for France

so, 50 votes for France.

There were $240 \times \frac{45}{360}$ votes for Portugal

so, 30 votes for Portugal.

There were $240 \times \frac{90}{360}$ votes for Greece

so, 60 votes for Greece.

Other destinations received $240 \times \frac{30}{360}$ votes

so, 20 votes for other destinations.

Note: it is worth checking your result by adding them:

$80 + 50 + 30 + 60 + 20 = 240$ total votes

2 The table shows the percentage of votes cast for various political parties in an election. If a total of 5 million votes were cast, how many votes were cast for each party?

Party	**Percentage of vote**
Social Democrats	45%
Liberal Democrats	36%
Green Party	15%
Others	4%

The Social Democrats received $\frac{45}{100} \times 5$ million votes

so, 2.25 million votes.

The Liberal Democrats received $\frac{36}{100} \times 5$ million votes

so, 1.8 million votes.

The Green Party received $\frac{15}{100} \times 5$ million votes

so, 750 000 votes.

Other parties received $\frac{4}{100} \times 5$ million votes

so, 200 000 votes.

Check total:

$2.25 + 1.8 + 0.75 + 0.2 = 5$ (million votes)

3 The table shows the results of a survey among 72 students to find their favourite sport. Display this data on a pie chart.

Sport	**Frequency**
Football	35
Tennis	14
Volleyball	10
Hockey	6
Basketball	5
Other	2

72 students are represented by 360°, so 1 student is represented by $\frac{360}{72}$ degrees. Therefore the size of each sector can be calculated as shown:

Football	$35 \times \frac{360}{72}$ degrees	i.e. 175°
Tennis	$14 \times \frac{360}{72}$ degrees	i.e. 70°
Volleyball	$10 \times \frac{360}{72}$ degrees	i.e. 50°
Hockey	$6 \times \frac{360}{72}$ degrees	i.e. 30°
Basketball	$5 \times \frac{360}{72}$ degrees	i.e. 25°
Other sports	$2 \times \frac{360}{72}$ degrees	i.e. 10°

Check total:

$$175 + 70 + 50 + 30 + 25 + 10 = 360$$

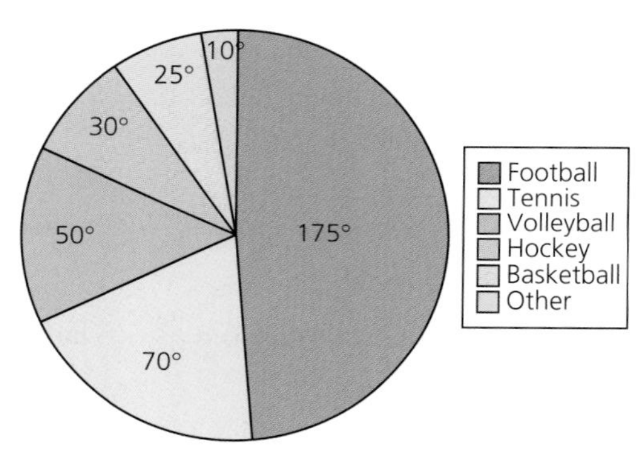

Exercise 30.1

1 The pie charts below show how a girl and her brother spent one day. Calculate how many hours they spent on each activity. The diagrams are to scale.

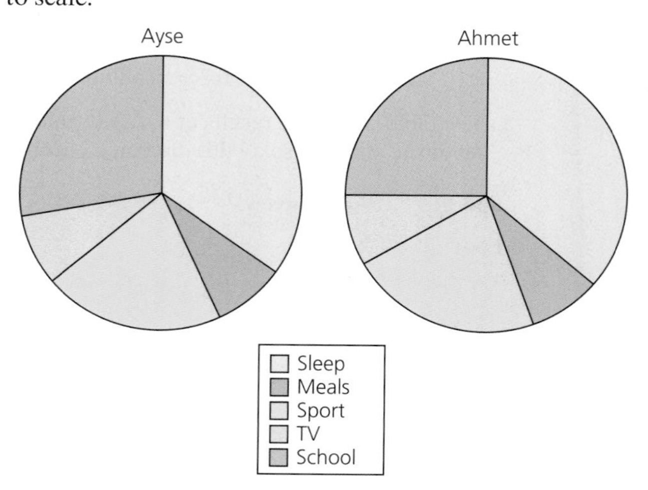

2 A survey was carried out among a class of 40 students. The question asked was, 'How would you spend a gift of $15?'. The results were:

Choice	Frequency
Music	14
Books	6
Clothes	18
Cinema	2

Illustrate these results on a pie chart.

3 A student works during the holidays. He earns a total of $2400. He estimates that the money earned has been used as follows: clothes $\frac{1}{3}$, transport $\frac{1}{5}$, entertainment $\frac{1}{4}$. He has saved the rest.

Calculate how much he has spent on each category, and illustrate this information on a pie chart.

4 A research project looking at the careers of men and women in Spain produced the following results:

Career	Male percentage	Female percentage
Clerical	22	38
Professional	16	8
Skilled craft	24	16
Non-skilled craft	12	24
Social	8	10
Managerial	18	4

a Illustrate this information on two pie charts, and make two statements that could be supported by the data.

b If there are eight million women in employment in Spain, calculate the number in either professional or managerial employment.

5 A village has two sports clubs. The ages of people in each club are:

Ages in Club 1									
38	8	16	15	18	8	59	12	14	55
14	15	24	67	71	21	23	27	12	48
31	14	70	15	32	9	44	11	46	62

Ages in Club 2									
42	62	10	62	74	18	77	35	38	66
43	71	68	64	66	66	22	48	50	57
60	59	44	57	12	–	–	–	–	–

Exercise 30.1 (cont)

a Draw a back-to-back stem and leaf diagram for the ages of the members of each club.
b For each club calculate:
  i the age range of its members
  ii the median age.
c One of the clubs is the Golf club, the other is the Athletics club. Which club is likely to be which? Give a reason for your answer.

6 The heights of boys and girls in a class are plotted as a comparative bar chart (below).

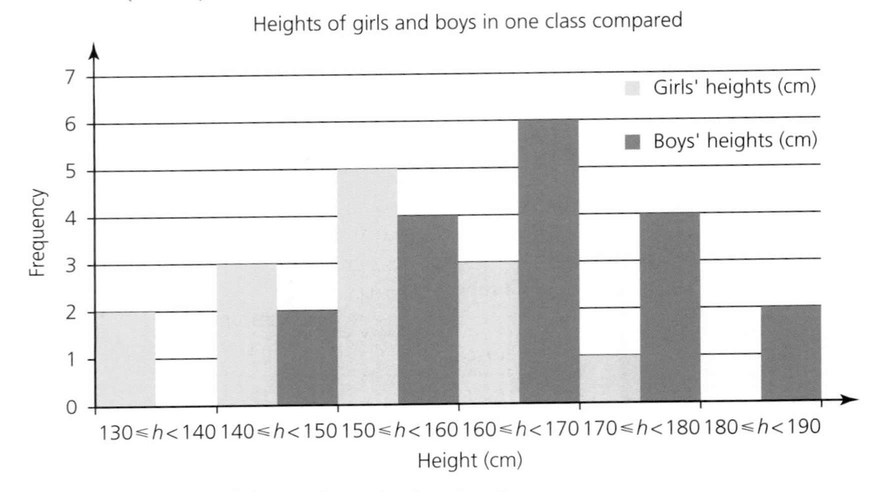

a How many girls are there in the class?
b How many more boys than girls are there in the height range $160 \leqslant h < 170$?
c Describe the differences in heights between boys and girls in the class.
d Construct a comparative bar chart for the heights of boys and girls in your own class.

Scatter diagrams

Scatter diagrams are particularly useful because they can show us if there is a **correlation** (relationship) between two sets of data. The two values of data collected represent the coordinates of each point plotted. How the points lie when plotted indicates the type of relationship between the two sets of data.

Worked example

The heights and masses of 20 children under the age of five were recorded. The heights were recorded in centimetres and the masses in kilograms. The data is shown in a table:

Height	32	34	45	46	52
Mass	5.8	3.8	9.0	4.2	10.1
Height	59	63	64	71	73
Mass	6.2	9.9	16.0	15.8	9.9
Height	86	87	95	96	96
Mass	11.1	16.4	20.9	16.2	14.0
Height	101	108	109	117	121
Mass	19.5	15.9	12.0	19.4	14.3

a Plot a scatter diagram of the above data.

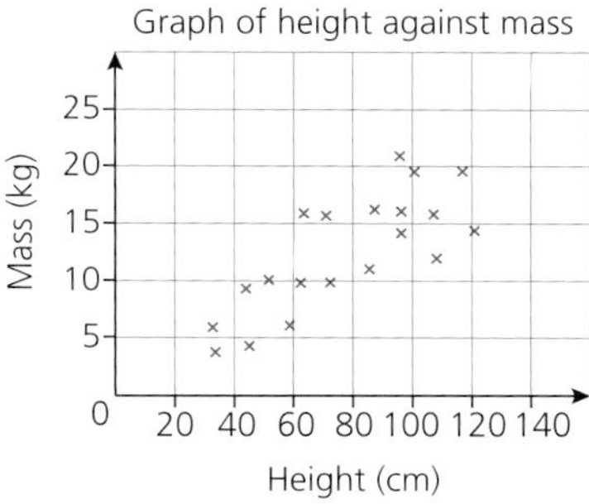

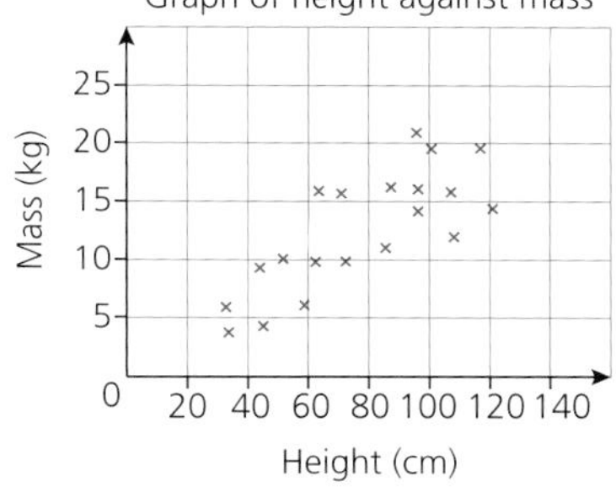

b Comment on any relationship you see.

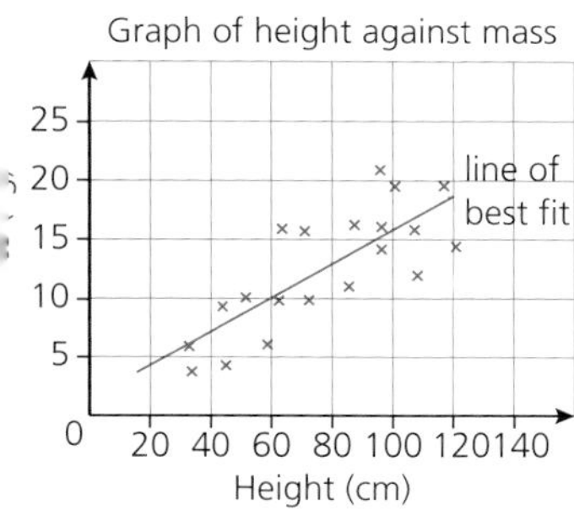

The points tend to lie in a diagonal direction from bottom left to top right. This suggests that as height increases then, in general, mass increases too. Therefore there is a **positive correlation** between height and mass.

c If another child was measured as having a height of 80 cm, approximately what mass would you expect him or her to be?

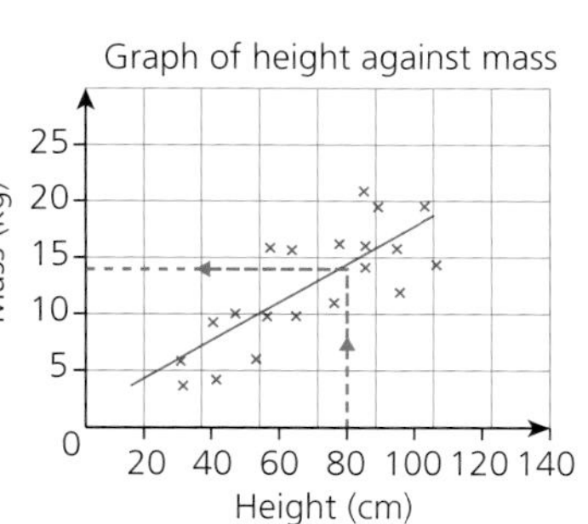

We assume that this child will follow the trend set by the other 20 children. To deduce an approximate value for the mass, we draw a **line of best fit**. This is done by eye and is a solid straight line which passes through the points as closely as possible, as shown.

The line of best fit can now be used to give an approximate solution to the question. If a child has a height of 80 cm, you would expect his or her mass to be in the region of 14 kg.

d Someone decides to extend the line of best fit in both directions because they want to make predictions for heights and masses beyond those of the data collected. The graph is shown below.

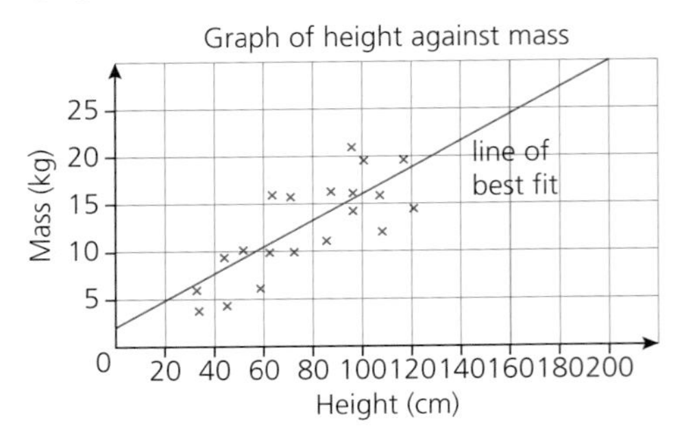

Explain why this should not be done.

The line of best fit should only be extended beyond the given data range with care. In this case it does not make sense to extend it because it implies at one end that a child with no height (which is impossible) would still have a mass of approximately 2 kg. At the other end it implies that the linear relationship between height and mass continues ever upwards, which of course it doesn't.

Types of correlation

There are several types of correlation, depending on the arrangement of the points plotted on the scatter diagram.

A **strong positive correlation** between the variables x and y.

The points lie very close to the line of best fit.

As x increases, so does y.

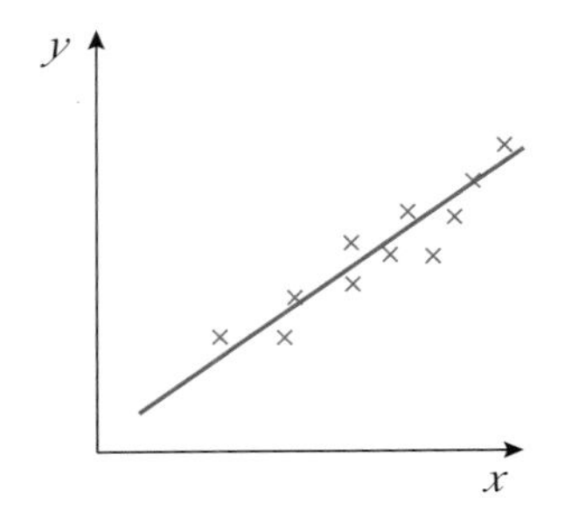

A **weak positive correlation**. Although there is direction to the way the points are lying, they are not tightly packed around the line of best fit.

As x increases, y tends to increase too.

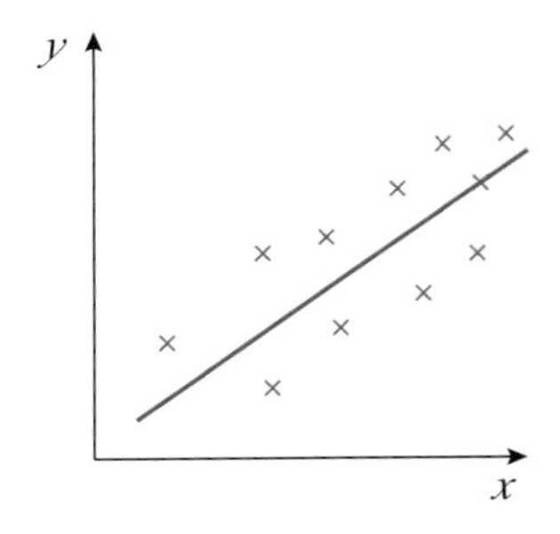

A **strong negative correlation**. The points lie close around the line of best fit. As x increases, y decreases.

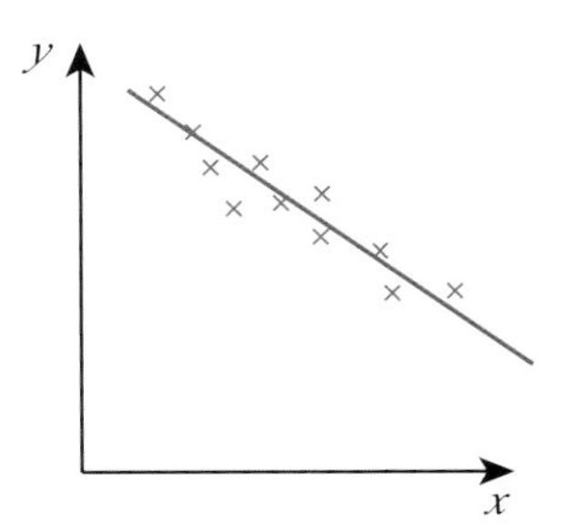

A **weak negative correlation**. The points are not tightly packed around the line of best fit. As x increases, y tends to decrease.

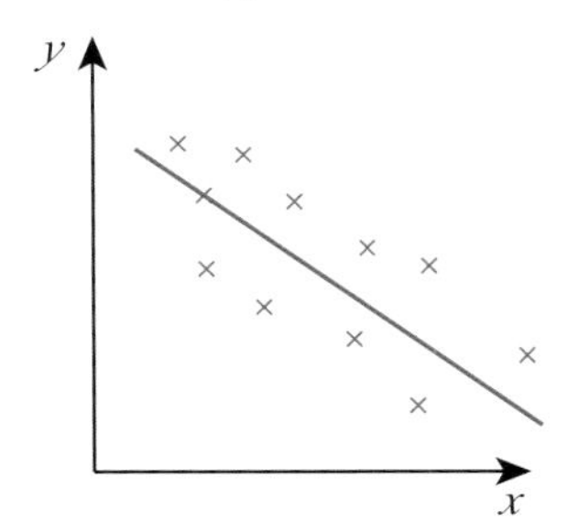

No correlation. As there is no pattern to the way in which the points are lying, there is no correlation between the variables x and y. As a result there can be no line of best fit.

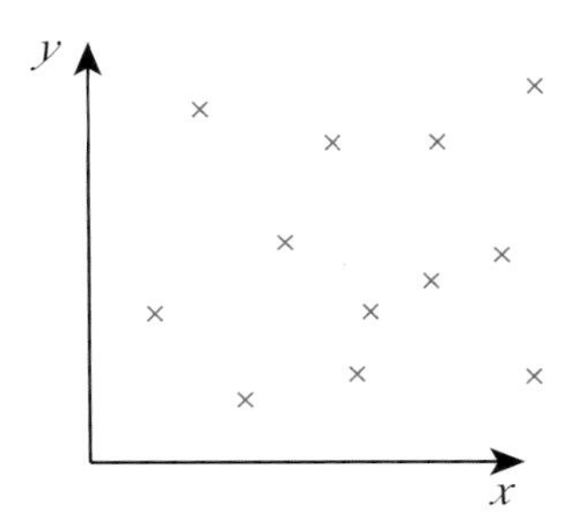

Exercise 30.2

1 State what type of correlation you might expect, if any, if the following data was collected and plotted on a scatter diagram. Give reasons for your answers.
   - **a** A student's score in a maths exam and their score in a science exam.
   - **b** A student's hair colour and the distance they have to travel to school.
   - **c** The outdoor temperature and the number of cold drinks sold by a shop.
   - **d** The age of a motorcycle and its second-hand selling price.
   - **e** The number of people living in a house and the number of rooms the house has.
   - **f** The number of goals your opponents score and the number of times you win.
   - **g** A child's height and the child's age.
   - **h** A car's engine size and its fuel consumption.

Exercise 30.2 (cont)

2 A website gives average monthly readings for the number of hours of sunshine and the amount of rainfall in millimetres for several cities in Europe. The table is a summary for July:

Place	Hours of sunshine	Rainfall (mm)
Athens	12	6
Belgrade	10	61
Copenhagen	8	71
Dubrovnik	12	26
Edinburgh	5	83
Frankfurt	7	70
Geneva	10	64
Helsinki	9	68
Innsbruck	7	134
Krakow	7	111
Lisbon	12	3
Marseilles	11	11
Naples	10	19
Oslo	7	82
Plovdiv	11	37
Reykjavik	6	50
Sofia	10	68
Tallinn	10	68
Valletta	12	0
York	6	62
Zurich	8	136

a Plot a scatter diagram of the number of hours of sunshine against the amount of rainfall. Use a spreadsheet if possible.

b What type of correlation, if any, is there between the two variables? Comment on whether this is what you would expect.

3 The United Nations keeps an up-to-date database of statistical information on its member countries. The table shows some of the information available:

Country	Life expectancy at birth (years, 2005–2010)		Adult illiteracy rate (%, 2009)	Infant mortality rate (per 1000 births, 2005–2010)
	Female	Male		
Australia	84	79	1	5
Barbados	80	74	0.3	10
Brazil	76	69	10	24
Chad	50	47	68.2	130
China	75	71	6.7	23
Colombia	77	69	7.2	19
Congo	55	53	18.9	79
Cuba	81	77	0.2	5
Egypt	72	68	33	35
France	85	78	1	4
Germany	82	77	1	4
India	65	62	34	55
Israel	83	79	2.9	5
Japan	86	79	1	3
Kenya	55	54	26.4	64
Mexico	79	74	7.2	17
Nepal	67	66	43.5	42
Portugal	82	75	5.1	4
Russian Federation	73	60	0.5	12
Saudi Arabia	75	71	15	19
South Africa	53	50	12	49
United Kingdom	82	77	1	5
United States of America	81	77	1	6

a By plotting a scatter diagram, decide if there is a correlation between the adult illiteracy rate and the infant mortality rate.

b Are your findings in part a) what you expected? Explain your answer.

c Without plotting a graph, decide if you think there is likely to be a correlation between male and female life expectancy at birth. Explain your reasons.

d Plot a scatter diagram to test if your predictions for part c) were correct.

Exercise 30.2 (cont)

4 A gardener plants 10 tomato plants. He wants to see if there is a relationship between the number of tomatoes the plant produces and its height in centimetres.
The results are presented in the scatter diagram. The line of best fit is also drawn.

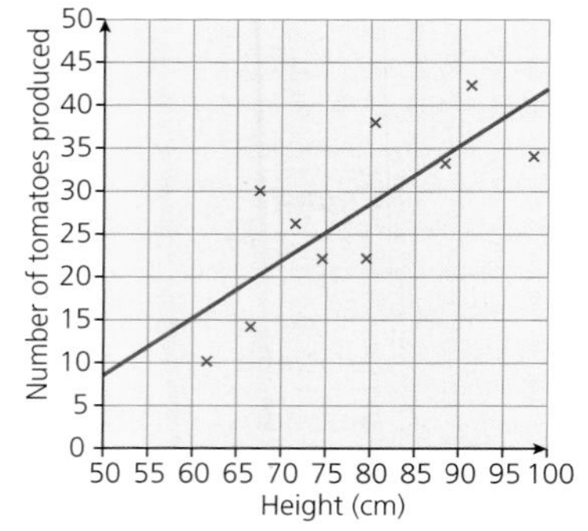

a Describe the correlation (if any) between the height of a plant and the number of tomatoes it produced.

b The gardener has another plant grown in the same conditions as the others. If the height is 85 cm, estimate from the graph the number of tomatoes he can expect it to produce.

c Another plant only produces 15 tomatoes. Estimate its height from the graph.

d The gardener thinks he will be able to make more predictions if the height axis starts at 0 cm rather than 50 cm and if the line of best fit is then extended. By re-plotting the data on a new scatter graph and extending the line of best fit, explain whether the gardener's idea is correct.

5 The table shows the 15 countries that won the most medals at the 2016 Rio Olympics. In addition, statistics relating to the population, wealth, percentage of people with higher education and percentage who are overweight for each country are also given.

Country	Olympic medals			Population (million)	Average wealth per person ($000's)	% with a higher education qualification	% adult population that is overweight	
	Gold	Silver	Bronze				Male	Female
U.S.A.	46	37	38	322	345	45	73	63
U.K.	27	23	17	65	289	44	68	59
China	26	18	26	1376	23	10	39	33
Russia	19	18	19	143	10	54	60	55
Germany	17	10	15	81	185	28	64	49
Japan	12	8	21	127	231	50	29	19
France	10	18	14	664	244	34	67	52
S. Korea	9	3	9	50	160	45	38	30
Italy	8	12	8	60	202	18	66	53
Australia	8	11	10	24	376	43	70	58
Holland	8	7	4	17	184	35	63	49
Hungary	8	3	4	10	34	24	67	49
Brazil	7	6	6	208	18	14	55	53
Spain	7	4	6	46	116	35	67	55
Kenya	6	6	1	46	2	11	17	34

A student wants to see if there is a correlation between the number of medals a country won and the percentage of overweight people in that country. He plots the number of gold medals against the mean percentage of overweight people; the resulting scatter graph and line of best fit is:

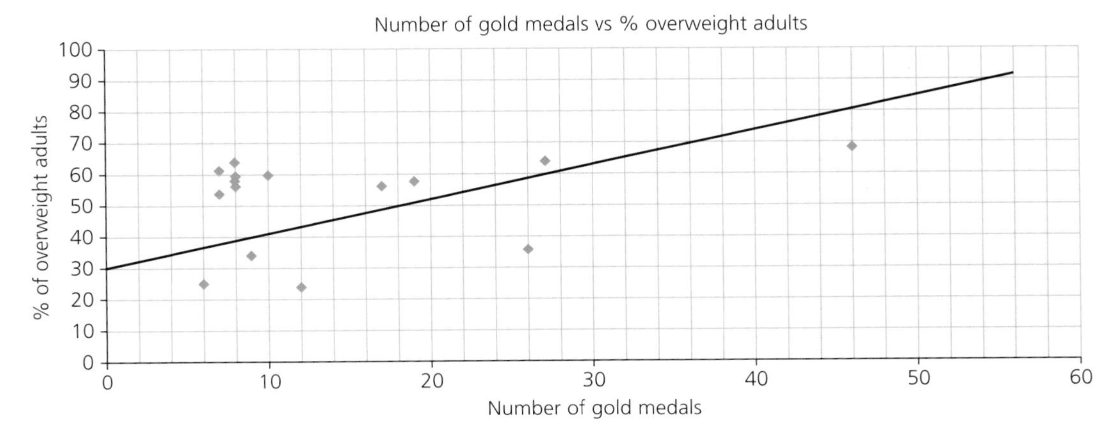

a Describe the type of correlation implied by the graph.

b The student states that the graph shows that the more overweight you are the more likely you are to win a gold medal. Give two reasons why this conclusion may not be accurate.

c Analyse the correlation between two other sets of data presented in the table and comment on whether the results are expected or not. Justify your answer.

Histograms

A **histogram** displays the frequency of either continuous or grouped discrete data in the form of bars. There are several important features of a histogram which distinguish it from a bar chart.

- The bars are joined together.
- The bars can be of varying width.
- The frequency of the data is represented by the area of the bar and not the height (though in the case of bars of equal width, the area is directly proportional to the height of the bar and so the height is usually used as the measure of frequency).

Worked example

The table shows the marks out of 100 in a maths test for a class of 32 students. Draw a histogram representing this data.

Test marks	Frequency
1–10	0
11–20	0
21–30	1
31–40	2
41–50	5
51–60	8
61–70	7
71–80	6
81–90	2
91–100	1

All the class intervals are the same. As a result the bars of the histogram will all be of equal width, and the frequency can be plotted on the vertical axis. The resulting histogram is:

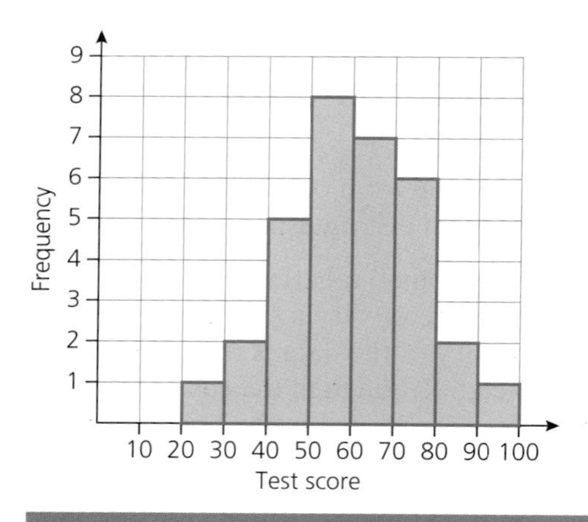

Exercise 30.3

1 The table shows the distances travelled to school by a class of 30 students. Represent this information on a histogram.

Distance (km)	Frequency
$0 \leqslant d < 1$	8
$1 \leqslant d < 2$	5
$2 \leqslant d < 3$	6
$3 \leqslant d < 4$	3
$4 \leqslant d < 5$	4
$5 \leqslant d < 6$	2
$6 \leqslant d < 7$	1
$7 \leqslant d < 8$	1

2 The heights of students in a class were measured. The results are shown in the table. Draw a histogram to represent this data.

Height (cm)	Frequency
145–	1
150–	2
155–	4
160–	7
165–	6
170–	3
175–	2
180–185	1

Note

Both questions in Exercise 30.3 deal with continuous data. In these questions equal class intervals are represented in different ways. However, they mean the same thing. In question 2, 145– means the students whose heights fall in the range $145 \leqslant h < 150$.

Surveys

A survey requires data to be collected, organised, analysed and presented.

A survey may be carried out for interest's sake, for example to find out how many cars pass your school in an hour. A survey could be carried out to help future planning – information about traffic flow could lead to the building of new roads, or the placing of traffic lights or a pedestrian crossing.

Exercise 30.4

1 Below are some statements, some of which you may have heard or read before.
Conduct a survey to collect data which will support or disprove one of the statements. Where possible, use pie charts to illustrate your results.
- **a** Magazines are full of adverts.
- **b** If you go to a football match you are lucky to see more than one goal scored.
- **c** Every other car on the road is white.
- **d** Girls are not interested in sport.
- **e** Children today do nothing but watch TV.
- **f** Newspapers have more sport than news in them.
- **g** Most girls want to be nurses, teachers or secretaries.
- **h** Nobody walks to school any more.
- **i** Nearly everybody has a computer at home.
- **j** Most of what is on TV comes from America.

2 Below are some instructions relating to a washing machine in English, French, German, Dutch and Italian. Analyse the data and write a report. You may wish to comment upon:
- **i** the length of words in each language
- **ii** the frequency of letters of the alphabet in different languages.

Exercise 30.4 (cont)

ENGLISH

ATTENTION

Do not interrupt drying during the programme.

This machine incorporates a temperature safety thermostat which will cut out the heating element in the event of a water blockage or power failure. In the event of this happening, reset the programme before selecting a further drying time.

For further instructions, consult the user manual.

FRENCH

ATTENTION
N'interrompez pas le séchage en cours de programme.

Une panne d'électricité ou un manque d'eau momentanés peuvent annuler le programme de séchage en cours. Dans ces cas arrêtez l'appareil, affichez de nouveau le programme et après remettez l'appareil en marche.

Pour d'ultérieures informations, rapportez-vous à la notice d'utilisation.

GERMAN

ACHTUNG
Die Trocknung soll nicht nach Anlaufen des Programms unterbrochen werden.

Ein kurzer Stromausfall bzw. Wassermangel kann das laufende Trocknungsprogramm annullieren. In diesem Falle Gerät ausschalten, Programm wieder einstellen und Gerät wieder einschalten.

Für nähere Angaben beziehen Sie sich auf die Bedienungsanleitung.

DUTCH

BELANGRIJK
Het droogprogramma niet onderbreken wanneer de machine in bedrijf is.

Door een korte stroom-of watertoevoeronderbreking kan het droogprogramma geannuleerd worden. Schakel in dit geval de machine uit, maak opnieuw uw programmakeuze en stel onmiddellijk weer in werking.

Verdere inlichtingen vindt u in de gebruiksaanwijzing.

ITALIAN

ATTENZIONE
Non interrompere l'asciugatura quando il programma è avviato.

La macchina è munita di un dispositivo di sicurezza che può annullare il programma di asciugaturea in corso quando si verifica una temporanea mancanza di acqua o di tensione. In questi casi si dovrà spegnere la macchina, reimpostare il programma e poi riavviare la macchina.

Per ulteriori indicazioni, leggere il libretto istruzioni.

Student assessment 1

1 The areas of four countries are shown in the table. Illustrate this data as a bar chart.

Country	Nigeria	Republic of the Congo	South Sudan	Kenya
Area in 10 000 km^2	90	35	70	57

2 The table gives the average time taken for 30 pupils in a class to get to school each morning, and the distance they live from the school.

Distance (km)	2	10	18	15	3	4	6	2	25	23	3	5	7	8	2
Time (min)	5	17	32	38	8	14	15	7	31	37	5	18	13	15	8
Distance (km)	19	15	11	9	2	3	4	3	14	14	4	12	12	7	1
Time (min)	27	40	23	30	10	10	8	9	15	23	9	20	27	18	4

a Plot a scatter diagram of distance travelled against time taken.
b Describe the correlation between the two variables.
c Explain why some pupils who live further away may get to school more quickly than some of those who live nearer.
d Draw a line of best fit on your scatter diagram.
e A new pupil joins the class. Use your line of best fit to estimate how far away from school she might live if she takes, on average, 19 minutes to get to school each morning.

3 A class of 27 students was asked to draw a line 8 cm long with a straight edge rather than with a ruler. The lines were measured and their lengths to the nearest millimetre were recorded:

8.8	6.2	8.3	8.1	8.2	5.9	6.2	10.0	9.7
8.1	5.4	6.8	7.3	7.7	8.9	10.4	5.9	8.3
6.1	7.2	8.3	9.4	6.5	5.8	8.8	8.1	7.3

a Present this data using a stem and leaf diagram.
b Calculate the median line length.
c Calculate the mean line length.
d Using your answers to part b) and c), comment on whether students in the class underestimate or overestimate the length of the line.

TOPIC 9

Mathematical investigations and ICT

Reading age

Depending on their target audience, newspapers, magazines and books have different levels of readability. Some are easy to read and others more difficult.

1 Decide on some factors that you think would affect the readability of a text.
2 Write down the names of two newspapers which you think would have different reading ages. Give reasons for your answer.

There are established formulae for calculating the reading age of different texts.

One of these is the Gunning Fog Index. It calculates the reading age as follows:

Reading age $= \frac{2}{5}\left(\frac{A}{n} + \frac{100L}{A}\right)$ where

A = number of words

n = number of sentences

L = number of words with 3 or more syllables

3 Choose one article from each of the two newspapers you chose in Q.2. Use the Gunning Fog Index to calculate the reading ages for the articles. Do the results support your predictions?
4 Write down some factors which you think may affect the reliability of your results.

ICT activity

In this activity, you will be using the graphing facilities of a spreadsheet to compare the activities you do on a school day with the activities you do on a day at the weekend.

1 Prepare a 24-hour timeline for a weekday similar to the one shown below:

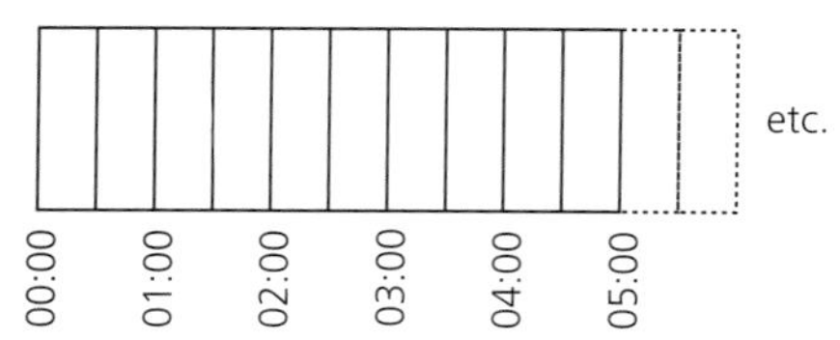

2 By using different colours for different activities, shade in the timeline to show what you did and when on a specific weekday, e.g. sleeping, eating, school, watching TV.
3 Add up the time spent on each activity and enter the results in a spreadsheet like the one below:

	A	B	C
1	Activity	Time spent (hrs)	
2	Sleeping		
3	Eating		
4	School		
5	TV		

4 Use the spreadsheet to produce a fully labelled pie chart of this data.
5 Repeat steps 1–4 for a day at the weekend.
6 Comment on any differences and similarities between the pie charts for the two days.

Index

Q

R

S

T

U

V

X

Y

Z